Statistics for Biology and Health

More information about this series at http://www.springer.com/series/2848

Il Do Ha · Jong-Hyeon Jeong
Youngjo Lee

Statistical Modelling of Survival Data with Random Effects

H-Likelihood Approach

Springer

Il Do Ha
Department of Statistics
Pukyong National University
Busan
Korea (Republic of)

Youngjo Lee
Department of Statistics
Seoul National University
Seoul
Korea (Republic of)

Jong-Hyeon Jeong
Department of Biostatistics
University of Pittsburgh
Pittsburgh, PA
USA

ISSN 1431-8776 ISSN 2197-5671 (electronic)
Statistics for Biology and Health
ISBN 978-981-10-6555-2 ISBN 978-981-10-6557-6 (eBook)
https://doi.org/10.1007/978-981-10-6557-6

Library of Congress Control Number: 2017956741

Printed on acid-free paper

This Springer imprint is published by Springer Nature
The registered company is Springer Nature Singapore Pte Ltd.
The registered company address is: 152 Beach Road, #21-01/04 Gateway East, Singapore 189721, Singapore

Preface

Survival or time-to-event data arise in various research areas such as medicine, epidemiology, genetics, engineering, econometrics, and sociology. Survival data have unique features including incomplete observation such as censoring and/or truncation. Use of semi-parametric models and potential correlation among time-to-events from the same cluster can make the statistical inference further complicated.

Broad classes of multivariate models using random effects have been developed. For inferences about unobserved random variables, the hierarchical (or h-)likelihood has been proposed by Lee and Nelder (1996). This book presents recent works on h-likelihood for the analysis of survival data. The h-likelihood method has been used to make inferences on the random effects models, especially for the frailty model for time-to-event data, where the frailties are treated as unobserved yet realized in the data. The h-likelihood allows an extension to the frailty models under competing risks as well as to the models for joint outcomes, e.g., longitudinal and event time outcomes. The h-likelihood method estimates the population parameters and the random effects simultaneously, with the random effects being updated from the observed data. This book covers the state-of-the-art h-likelihood methods, which include interval estimation of the individual frailty and variable selection of the covariates in the general class for the frailty models with or without competing risks. A beauty of the h-likelihood is that once the statistical model is specified parametrically or nonparametrically, the required inference procedures can be made.

A systematic presentation of the h-likelihood procedures and identification of future directions in survival analysis would be meaningful contributions to the field. Although most of the examples in this book came from biomedical sciences, the methodology is also applicable to engineering, econometrics, and other fields, whenever event times are collected and used for statistical inference.

The targeted audience includes researchers in medicine, graduate students, and Ph.D. (bio)statisticians, interested in working with clustered survival data with or without competing risks. Knowledge of survival analysis at an introductory graduate level is the minimum prerequisite to read this book. To be reader-friendly, the technical details including derivations and proofs are given in the Appendix of each chapter. Real data examples are furnished with R codes to provide readers with useful hands-on tools such as `frailtyHL` in Comprehensive R Archive Network (CRAN). The majority of data sets used in the book are available at URL `http://cran.r-project.org/package=failtyHL` for the R package `frailtyHL` (Ha et al. 2018).

We are grateful to an anonymous reviewer, Prof. Richard Sylvester, Prof. Gilbert MacKenzie, Dr. Maengseok Noh, Mr. Hyunseong Park, Ms. Eunyoung Park, and Mr. Ji Hoon Kwon for their numerous useful comments and suggestions.

Busan, Korea — Il Do Ha
Pittsburgh, USA — Jong-Hyeon Jeong
Seoul, Korea — Youngjo Lee
November 2017

Contents

Abbreviations

AFT	Accelerated failure time
AI	Akaike information
AIC	Akaike information criterion
AR	Autoregressive
BIC	Bayesian information criterion
BN	Bivariate normal
cAIC	Conditional AIC
CHEMO	Chemotherapy
CIF	Cumulative incidence function
CP	Coverage probability
DFI	Disease-free interval
EA	Estimative approach
EB	Empirical Bayesian
EM	Expectation and maximization
GHQ	Gauss–Hermite quadrature
GLM	Generalized linear model
GLMM	Generalized linear mixed model
HGLM	Hierarchical generalized linear model
HL	h-likelihood
h-likelihood	Hierarchical likelihood
HR	Hazard ratio
ILS	Iterative least squares
IPCW	Inverse probability of censoring weighting
IWLS	Iterative weighted least squares
JM	Joint model
K-M	Kaplan–Meier
LASSO	Least absolute shrinkage and selection operator
LMM	Linear mixed model
LQA	Local quadratic approximation
LRT	Likelihood ratio test

LTRC	Left truncated and right censored
mAIC	Marginal AIC
MAR	Missing at random
MHL	Maximum h-likelihood
MHLE	Maximum h-likelihood estimator
MLE	Maximum likelihood estimator
MPPHLE	Maximum penalized partial h-likelihood estimator
N-A	Nelson–Aalen
NPMLE	Nonparametric MLE
pAIC	Partial marginal AIC
PH	Proportional hazards
PMLE	Partial maximum likelihood estimator
PMMLE	Partial maximum marginal likelihood estimator
PPL	Penalized partial likelihood
PQL	Penalized quasi-likelihood
PREMLE	Partial restricted MLE
rAIC	Restricted AIC
REML	Restricted maximum likelihood
REMLE	Restricted MLE
SCAD	Smoothly clipped absolute deviation
sCr	Serum creatinine
SM	Separate model

Chapter 1
Introduction

1.1 Goals

The likelihood, introduced by Fisher (1922), plays an important role in statistical inference about fixed unknowns, namely parameters. The beauty of the likelihood is that once the statistical model is specified parametrically or nonparametrically, the associated inference procedures for the parameters of interest are straightforward. Statistical models have been enriched and actively extended in the literature by allowing random unknowns such as frailties in addition to fixed unknowns. We review recent work on extension of the hierarchical likelihood (h-likelihood) of Lee and Nelder (1996) to time-to-event (survival) data. The h-likelihood overcomes various challenges due to incomplete observations caused by censoring, truncation, and competing events, and presents further extension of existing work, such as complicated structured frailty and joint models.

Survival (time-to-event) data arise in various areas such as medicine, epidemiology, genetics, engineering, econometrics, and sociology, among others. The Cox (1972) proportional hazards (PH) model and the accelerated failure time (AFT) model have been popular for the analysis of survival data and they have been recently extended to multivariate models by incorporating random effects (frailties) to explain dependency and/or heterogeneity among correlated (or multivariate) survival outcomes in the population.

This book presents the h-likelihood approach to statistical inference on correlated survival data. This approach avoids computational difficulties due to intractable integrations that are needed to calculate the marginal likelihood. Moreover, the h-likelihood inference allows for subject-specific inferences on random effects. To be reader-friendly, the technical details including derivations and proofs are given in Appendix of each chapter. Real data examples are furnished with software programs in R to provide readers with useful hands-on tools such as `frailtyHL` in CRAN.

A systematic review of the h-likelihood methods is important for an identification of future direction of the field. This book will also present state-of-the-art statistical methods that were recently developed in likelihood theory and application. Interval

I.D. Ha et al., *Statistical Modelling of Survival Data with Random Effects*,
Statistics for Biology and Health, https://doi.org/10.1007/978-981-10-6557-6_1

estimation of the individual frailty and variable selection of covariates in the general class models with frailties have been of special interest. The interval estimation of frailty could be useful for investigating heterogeneity in treatment effects across centers from multicenter clinical trials and variable selection is useful for models with large number of covariates.

This book is organized as follows. In this chapter, we present several motivating examples of correlated survival data which will be used to fit survival models with random effects such as frailty models, competing-risks frailty models, and mixed-effect models in the later chapters. In Chap. 2, we review the basic methodologies in survival analysis for modeling and analyzing univariate survival data, which include Kaplan–Meier estimator of the survival function, Nelson–Aalen estimator of the cumulative hazard function, the Cox's PH model, and the AFT model. In Chap. 3, we outline the h-likelihood methods for random-effect models in general. Extended likelihood inferences on statistical models with random effects are reviewed from the perspectives of frequentist and Bayesian approaches. In Chap. 4, we present inferences on simple frailty models with one frailty term, together with illustration of `frailtyHL` R-package. We discuss the h-likelihood procedures under right censoring and left truncation. In Chap. 5, we present extensions to multicomponent frailty models, allowing correlation among frailties. We show that h-likelihood methods developed for the single frailty model can be straightforwardly extended to multicomponent models. In Chap. 6, we present inferences on competing-risks frailty models via both cause-specific hazards and subdistribution hazards. In Chap. 7, we present the variable selection procedures using the penalized h-likelihood under the frailty models with and without competing risks. In Chap. 8, we present AFT models with random effects for correlated survival data with real applications. In Chap. 9, we present joint models for time-to-event and repeated measures data. In the last chapter, we present miscellaneous topics, including competing-risks models for multistate data including missing causes of failure, and further extensions. Finally in Appendix A, we summarize specific matrix and vector components in the fundamental formulas of the estimating equations for fixed and random effects used in previous chapters.

1.2 Motivating Examples

Univariate and multivariate survival data usually consist of a single event time and a series of multiple (or recurrent) event times from each individual, respectively. Figure 1.1 displays specific examples of a single event and recurrent events. That is, a single event is a transition from one state (alive) to another state (dead) for a subject, whereas recurrent events are transitions from baseline state (disease-free) to the first state (first recurrence) and the second state (second recurrence) for a subject. Thus, single event times are independent because each (presumably independent) individual experiences only one event, while recurrent event times from the same individual may be correlated. Other types of events include clustered events from

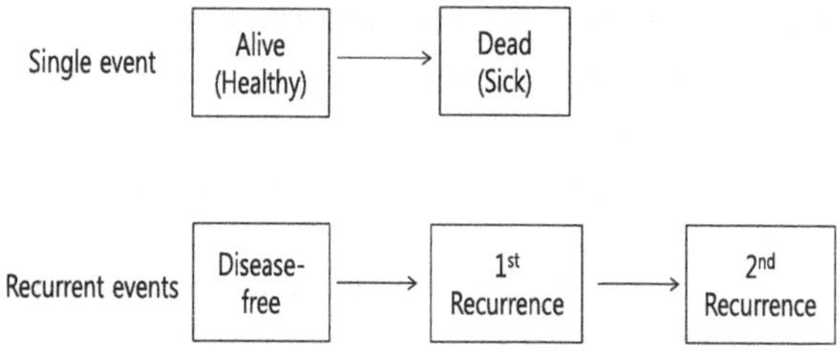

Fig. 1.1 Single event and recurrent events

multicenter clinical trials and competing-risks events as we shall investigate later. Below, we illustrate some practical examples of multivariate (or correlated) survival data which will be used throughout this book.

1.2.1 Kidney Infection Data

The data set consists of times to the first and second recurrences of kidney infection in 38 patients using a portable dialysis machine (McGilchrist and Aisbett 1991). Infections can occur at the location of insertion of the catheter. The catheter is later removed if infection occurs and can be removed for other reasons, which is treated as censoring; about 23.7% of the data were censored. Here, each event time is time to infection since insertion of the catheter. The covariates of interest are Age, Sex (1 = female, 0 = male), and three indicator variables for glumerulonephritis (GN), acute tubular nephropathy (AN) and polycystic kidney disease (PKD) which are different types of kidney disease. The infection times from the same patient, as shown in the case of recurrent events in Fig. 1.1, are likely to be related due to the shared patient effect.

1.2.2 Litter-Matched Rat Data

The data set presented by Mantel et al. (1977) is from a tumorigenesis study of 50 litters of female rats. For each litter, one rat was selected to receive the study drug and the other two rats were treated with placebo. Here, each litter is treated as a cluster. Event time is time to development of tumor, measured in weeks. Death before occurrence of tumor was, for simplicity, treated as a right-censored observation, even if it is clearly a competing event; forty rats developed a tumor, leading to about 73% censoring. Event times for rats within a litter may be correlated due to the shared genetic or environmental effects.

1.2.3 Chronic Granulomatous Disease (CGD) Nested Recurrent Data

The CGD data set (Fleming and Harrington 1991) is from a placebo-controlled randomized trial of gamma interferon in chronic granulomatous disease. The trial aimed to investigate the effectiveness of gamma interferon (γ-IFN) in reducing the rate of serious infections in CGD patients. In total, 128 patients from 13 hospitals were followed for about 1 year. The number of patients accrued per hospital ranged from 4 to 26. Among 63 patients in the treatment group, 14 patients experienced at least one infection and a total of 20 infections were recorded. In the placebo group, 30 out of 65 patients experienced at least one infection, with a total of 56 infections being recorded.

Time to event in this example is the gap time (inter-arrival time) between recurrent infection times. Censoring occurred at the last follow-up for all patients, except one, who experienced a serious infection on the date he left the study. In this study, roughly 63% of the individuals were censored. In Chap. 5, we model the gap times, with the fixed covariates $x_{ijk} = (x_{ijk1}, \ldots, x_{ijk10})^T$, where x_{ijk1} is a treatment indicator (0 = placebo, 1 = γ-IFN), x_{ijk2} pattern of inheritance (0 = autosomal recessive, 1 = X-linked), x_{ijk3} age (in years), x_{ijk4} height (in cm), x_{ijk5} weight (in kg), x_{ijk6} use of corticosteroids at time of study entry (0 = no, 1 = yes), x_{ijk7} use of prophylactic antibiotics at time of study entry (0 = no, 1 = yes), x_{ijk8} sex (0 = male, 1 = female), x_{ijk9} hospital region (0 = U.S., 1 = Europe), and x_{ijk10} a longitudinal variable, representing the accumulated time from the first infection (in years). The rationale for creating such a time-dependent covariate is to investigate how the risk of subsequent infection depends on time from the first infection. A positive coefficient would imply an increasing risk of subsequent infection with elapsed time.

1.2.4 Bladder Cancer Multicenter Data

This data set came from 410 patients with stages Ta and T1 bladder cancer from 21 centers that participated in the EORTC trial 30791 (Sylvester et al. 2006). Time to event is the duration of the disease-free interval (DFI), which is defined as time from randomization to the date of the first recurrence. Patients who did not experience recurrence at the end of the follow-up period were censored at their last date of follow-up; 204 patients (49.8%) were censored. Two covariates of interest are: CHEMO (0 = No, 1 = Yes) and TUSTAT (0 = Primary, 1 = Recurrent), where CHEMO is the treatment indicator representing chemotherapy and TUSTAT is an indicator representing prior recurrent rate. Patients with missing values for TUSTAT were excluded. The numbers of patients enrolled per center varied from 3 to 78, with the mean of 19.5 and the median of 15. Event times (DFI) are expected to be correlated among patients from the same center.

1.2.5 Lung Cancer Multicenter Data

We will also examine the data from the EST 1582 multicenter lung cancer trial (Ettinger et al. 1990). This trial enrolled 579 patients from 31 distinct institutions (centers). The number of patients enrolled per institution ranged from 1 to 56, with the mean of 18.7 and the median of 17. The subjects were randomized to one of two treatment arms, standard chemotherapy (CAV) or an alternating regimen (CAV-HEM). The primary endpoint was the time (in years) from randomization to death. The study had a high mortality rate with the censoring rate of only 1.7%. The median survival time and maximum follow-up were 0.86 years and 8.45 years, respectively. Five dichotomous covariates considered are treatment ($x_{ij1} = 0$ for CAV and 1 for CAV-HEM), presence or absence of bone metastases (x_{ij2}), presence or absence of liver metastases (x_{ij3}), whether the subject was ambulatory or confined to bed or chair (x_{ij4}), and whether there was a weight loss prior to entry (x_{ij5}).

1.2.6 Breast Cancer Competing-Risks Data

We examine a breast cancer dataset from a multicenter clinical trial conducted by the National Surgical Adjuvant Breast and Bowel Project (NSABP; Fisher et al. 1989, 1996), which was one of the National Cancer Institute (NCI) cooperative groups. Total 2,817 eligible patients from 167 distinct centers were followed up for about 20 years since randomization. The number of patients per center varied from 1 to 241, with the mean of 16.9 and the median of 8. The patients were randomized to one of two treatment arms, tamoxifen (1413 patients) or placebo (1404 patients). The average age of patients was 55 and the average tumor size was about 2 centimeters. The aim of the analysis was to investigate the effect of a hormonal treatment (tamoxifen) on local or regional recurrence. Two event types were considered; the first type was local or regional recurrence (Type 1) and the second type was a new primary cancer, distance recurrence or death (Type 2). Only the event that occurs first was of interest in this analysis, so that the repeated event times were not considered. There were 314 Type 1 events (11.15%), 1303 Type 2 events (46.25%), and 1200 patients (42.60%) were censored at the last follow-up. Here, covariates of interest are treatment (tamoxifen = 1, placebo = 0), age, and tumor size.

1.2.5 Lung Cancer Multicenter Data

[illegible]

[illegible]

[illegible]

Chapter 2
Classical Survival Analysis

Let T be time-to-event (failure time), which is a nonnegative random variable. In medicine, a typical example is time from the onset of a condition or an initiation of treatment to death. In studies of reliability of products (or components), time to failure of light bulbs, for example, is often of interest. Rather than using such specific terms, economists refer to *durations* between events (e.g., duration of unemployment). The distribution of failure time is usually non-normal and skewed.

Survival data are typically incomplete because they are subject to *censoring* and/or *truncation*, either from left or right. Survival data can be either univariate or multivariate as shown in Chap. 1. Below, we further elaborate on the features of survival data.

(1) Censoring

A true event time is said to be *right censored* if the event did not occur at the time when the analysis is performed, so it is only known that the true event time is larger than the end of the observation period. Similarly, left censoring occurs when an event is only known to have occurred before an observation begins. Right censoring is commonly encountered in survival data, but left censoring is relatively rare. In particular, right censoring that occurs when the observation period of a study ends is often referred to as *administrative censoring*.

There exist three types of right censoring. Under *Type I censoring*, the duration of censoring time is fixed as the same for all subjects. Under *Type II censoring*, a study continues until the prespecified number of failures (e.g., testing of equipment life) is reached, implying that the censoring times are random. *Random censoring* usually occurs due to staggered entry of patients into the study in clinical trials, where survival time of each patient is measured from study entry time (Cox and Oakes 1984). Figure 2.1 illustrates rearrangement of survival times from calendar time to entry time under random censoring due to staggered entry and loss to follow up or the end of study.

I.D. Ha et al., *Statistical Modelling of Survival Data with Random Effects*,
Statistics for Biology and Health, https://doi.org/10.1007/978-981-10-6557-6_2

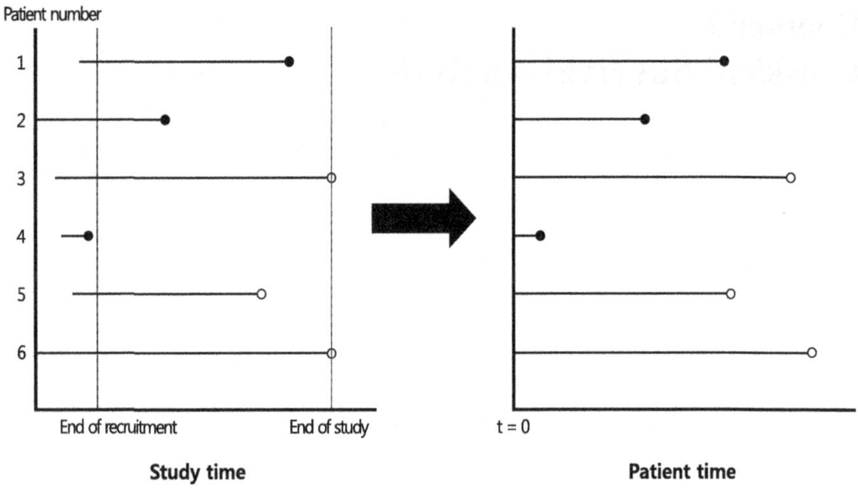

Fig. 2.1 Example of random censoring in six patients; •, death; ○, censoring

Additionally, *interval censoring* occurs when the event time is known to have occurred only within an interval, and *doubly censoring* refers to the case where both left censoring and right censoring occur.

(2) Truncation

Truncation often induces an exclusion of certain subjects from analysis, which might introduce sampling bias into statistical inference.

Left truncation occurs when subjects enter a study at a particular age (not necessarily the origin for the event of interest) and are followed from this delayed entry time until the event occurs or until the subject is censored. Therefore, under left truncation, units that have already experienced the event of interest (e.g., death) before a study begins are excluded (Keiding 1992). This phenomenon is also called "stock sampling with follow up" in econometrics since only those in the "alive state" at a given time are sampled (Lancaster 1990).

Right truncation occurs when only individuals who have experienced the event of interest are included in the study (Klein and Moeschberger 2003). Right truncation can occur in retrospective studies, for example, when studying the incubation period for AIDS in patients who have already developed the disease. In this book, we will mainly focus on random right censoring including *left truncated and right censored* (LTRC). Here, LTRC data occur when individuals enter a study at a particular time point with constraints and are followed from this entry time until the individual is censored or experiences an event.

2.1 Hazard and Survival Function

We first present the basic definitions of survival and hazard function and their relationships, which are the fundamental quantities for parametric and nonparametric inference on survival data.

Assume that failure time T is a nonnegative continuous random variable with a density function $f(t)$ and a corresponding distribution function $F(t) = P(T \leq t)$. The survival function of T, the probability of an individual surviving beyond time t or not experiencing a failure up to time t, is defined by

$$S(t) = P(T > t) = \int_t^{\infty} f(x)dx.$$

For a distribution of lifetimes of an industrial item, $S(t)$ is referred to as the reliability function of T (Crowder et al. 1991). From the definition of $F(t)$, we have that

$$S(t) = 1 - \text{P(an individual fails before or at } t) = 1 - F(t).$$

Notice that $S(t)$ is a monotonically decreasing continuous function with

$$S(0) = 1 \text{ and } S(\infty) = \lim_{t\to\infty} S(t) = 0.$$

The hazard function is defined by

$$\begin{aligned}
\lambda(t) &= \lim_{\Delta t \to 0} \frac{\mathrm{P}(t \leq T < t + \Delta t | T \geq t)}{\Delta t} \\
&= \lim_{\Delta t \to 0} \frac{\mathrm{P}(t \leq T < t + \Delta t)/\Delta t}{\mathrm{P}(T \geq t)} \\
&= \frac{f(t)}{S(t)}
\end{aligned}$$

which is the instantaneous failure rate at time t, given the individual surviving just prior to t. In particular, $\lambda(t)\Delta t$ is the approximate probability of dying in $[t, t + \Delta t)$, given survival just prior to time t. The hazard function is also referred as the hazard rate, failure rate, the force of mortality, and intensity function. The corresponding cumulative (or integrated) hazard function is defined as

$$\Lambda(t) = \int_0^t \lambda(x)dx.$$

From the definition $\lambda(t) = f(t)/S(t)$, we have the following relationships:

$$\lambda(t) = -\frac{d}{dt} \log S(t)$$

since $f(t) = -dS(t)/dt$, and

$$S(t) = \exp\{-\Lambda(t)\}$$

since

$$\Lambda(t) = \int_0^t \lambda(x)dx = \int_0^t \{f(x)/S(x)\}dx = -\log S(t).$$

Note that $\Lambda(\infty) = \lim_{t\to\infty}\Lambda(t) = \infty$ and

$$f(t) = \lambda(t)S(t) = \lambda(t)\exp\{-\Lambda(t)\}.$$

Thus, the hazard function characterizes the probability density function of survival time.

2.1.1 *Parametric Distributions for Survival Times*

The distribution of survival time is often positively skewed. The exponential and Weibull distributions are popular choices for modeling survival data.

Exponential distribution: The exponential distribution is featured with a constant hazard over time:

$$\lambda(t) = \lambda \quad t \geq 0,$$

where $\lambda > 0$, implying that

$$\Lambda(t) = \lambda t \text{ and } S(t) = \exp\{-\Lambda(t)\} = \exp(-\lambda t).$$

Thus, the density is given by

$$f(t) = \lambda(t)\exp\{-\Lambda(t)\} = \lambda\exp(-\lambda t) \quad t \geq 0.$$

Weibull distribution: The Weibull distribution enjoys various hazard shapes characterized by a parameter ϕ:

$$\lambda(t) = \lambda\phi t^{\phi-1} \quad t \geq 0,$$

where $\lambda > 0$ is a scale parameter and $\phi > 0$ is a shape parameter. The Weibull distribution is fairly flexible because its hazard function $\lambda(t)$ is monotone increasing if $\phi > 1$, monotone decreasing if $\phi < 1$, and constant if $\phi = 1$, giving the exponential distribution as a special case. Since

$$\Lambda(t) = \lambda t^{\phi} \text{ and } S(t) = \exp(-\lambda t^{\phi}),$$

Table 2.1 Useful parametric distributions for survival analysis

Distribution	Hazard rate $\lambda(t)$	Survival function $S(t)$	Density function $f(t)$
Exponential ($\lambda > 0$)	λ	$\exp(-\lambda t)$	$\lambda \exp(-\lambda t)$
Weibull ($\lambda, \phi > 0$)	$\lambda\phi t^{\phi-1}$	$\exp(-\lambda t^{\phi})$	$\lambda\phi t^{\phi-1}\exp(-\lambda t^{\phi})$
Log-normal ($\sigma > 0, \mu \in R$)	$f(t)/S(t)$	$1 - \Phi\{(\ln t - \mu)/\sigma\}$	$\varphi\{(\ln t - \mu)/\sigma\}(\sigma t)^{-1}$
Log-logistic ($\lambda > 0, \phi > 0$)	$(\lambda\phi t^{\phi-1})/(1+\lambda t^{\phi})$	$1/(1+\lambda t^{\phi})$	$(\lambda\phi t^{\phi-1})/(1+\lambda t^{\phi})^2$
Gamma ($\lambda, \phi > 0$)	$f(t)/S(t)$	$1 - I(\lambda t, \phi)$	$\{\lambda^{\phi}/\Gamma(\phi)\}t^{\phi-1}\exp(-\lambda t)$
Gompertz ($\lambda, \phi > 0$)	$\lambda e^{\phi t}$	$\exp\{\frac{\lambda}{\phi}(1-e^{\phi t})\}$	$\lambda e^{\phi t}\exp\{\frac{\lambda}{\phi}(1-e^{\phi t})\}$

$\Phi(\cdot)$ [$\varphi(\cdot)$], c.d.f [p.d.f.] of N(0,1); $I(x,\phi) = \frac{1}{\Gamma(\phi)}\int_0^x u^{\phi-1}e^{-u}du$, incomplete gamma function

we have

$$f(t) = \lambda\phi t^{\phi-1}\exp(-\lambda t^{\phi}) \quad t \geq 0.$$

Note that

$$\log\{-\log S(t)\} = \log\lambda + \phi\log t,$$

which is used for checking the Weibull model.

Table 2.1 summarizes useful parametric distributions including exponential, Weibull, log-normal, log-logistic, gamma, and Gompertz. These parametric distributions have been implemented in the `survreg()` function in the R package **survival** as we see in Sect. 2.4.

Percentile of Distribution

In many applications, the percentile of a failure time distribution is of interest, e.g., the median survival time. The $100p$th percentile (or the pth quantile) of the distribution of T is the value t_p satisfying

$$P(T \leq t_p) = p \in (0, 1),$$

which is equivalent to $S(t_p) = 1 - p$. That is, $t_p = F^{-1}(p)$ indicates the time point to which the $100p\%$ of population will fail; in particular, the median survival time $t_{0.5}$ is the median of distribution of T. For example, $t_p = -\log(1-p)/\lambda$ for an exponential distribution and $t_p = \{-\log(1-p)/\lambda\}^{1/\phi}$ for a Weibull distribution.

2.1.2 Nonparametric Estimation of Basic Quantities

In survival analysis, parametric methods based on distributions in Table 2.1 have been well developed and would provide efficient results when the parametric assumptions are satisfied in the data. In practice, however, when the underlying distributional assumption is not testable as in the designing stage of a study or the parametric assumptions are not satisfied in the observed data, nonparametric methods are preferable.

Let T_i $(i = 1, \ldots, n)$ be the potential failure time and C_i be the corresponding potential censoring time for the ith individual. Then, the observable random variables are

$$Y_i = \min(T_i, C_i) \text{ and } \delta_i = I(T_i \leq C_i),$$

where $I(\cdot)$ is the indicator function. The following are the two usual assumptions under noninformative censoring:

Assumption 1: T_i's and C_i's are independent, and pairs (T_i, C_i)'s are also independent $(i = 1, \ldots, n)$.

Assumption 2: C_i's are noninformative of T_i's.

Here, the noninformativeness implies that the censoring distribution does not depend on the parameters of interest from the failure time distribution (Klein and Moeschberger 2003). Under the noninformative censoring, we have the two well-known nonparametric estimators in survival analysis; Kaplan and Meier (1958) estimator for the survival function and Nelson (1969, 1972)–Aalen (1978) estimator for the cumulative hazard function. Note that independence is a probabilistic property, while noninformativeness depends on the relationship between parameters in the model.

Let y_i be the observed value of Y_i. Suppose that there are D $(D \leq n)$ distinct observed event times $y_{(1)} < y_{(2)} < \cdots < y_{(D)}$ among y_i's. Let $d_{(k)}$ be the number of events at $y_{(k)}$ $(k = 1, \ldots, D)$. Let $n_{(k)}$ be the number of individuals who are at risk at $y_{(k)}$, that is, the number of individuals who are alive and uncensored just prior to $y_{(k)}$. The Kaplan–Meier (K–M) estimator of $S(t)$, is defined by

$$\widehat{S}_{K-M}(t) = \prod_{k:y_{(k)} \leq t} \left\{1 - \frac{d_{(k)}}{n_{(k)}}\right\},$$

which is also called the product-limit estimator. The K–M estimator is a step function with jumps at the observed event times and reduces to the empirical survival function estimator under no censoring. The variance of the K–M estimator is usually estimated using Greenwood's formula:

$$\widehat{\text{var}}(\widehat{S}_{K-M}(t)) = \widehat{S}^2_{K\text{-}M}(t) \sum_{k:y_{(k)} \leq t} \frac{d_{(k)}}{n_{(k)}\{n_{(k)} - d_{(k)}\}}.$$

Using the estimated survival function such as $\widehat{S}_{K-M}(t)$, t_p is estimated by the smallest observed survival time such that $S(t_i) \leq 1 - p$. That is,

$$\hat{t}_p = \min\{t_i | \hat{S}(t_i) \le 1 - p\}.$$

The estimation procedure of t_p is implemented in the `quantile()` function in **survival** R package. In addition, the mean survival time $\mu = E(T)$ can be easily estimated by using $\widehat{S}_{K-M}(t)$:

$$\widehat{\mu} = \int_0^\infty \widehat{S}_{K-M}(u)du,$$

which is equal to the area under the estimated survival function.

These nonparametric estimators are illustrated in detail with four examples below.

Example 2.1 Gehan (1965) presented data from a clinical trial comparing drug 6-mercaptopurine (6-MP group) versus placebo (control group) in 42 acute leukemia patients and the treatment allocation for the two groups was randomized by a matching pair. Here, the survival outcome is time to remission (in weeks) as summarized below ("+" denotes censoring by the end of study):

6-MP group: 6, 6, 6, 6^+, 7, 9^+, 10, 10^+, 11^+,13, 16, 17^+, 19^+, 20^+,22, 23, 25^+, 32^+, 32^+, 34^+, 35^+

Placebo group: 1, 1, 2, 2, 3, 4, 4, 5, 5, 8, 8, 8, 8, 11, 11, 12, 12, 15, 17, 22, 23

For simplicity, we consider only 6-MP group. The detailed steps to calculate the K–M estimates and its variances are presented in Table 2.2. From the bottom part of Table 2.2, we can see the K–M estimator is a step function.

The R codes and output for the K–M and quantile estimates using the 6-MP group are as follows.

```
> library(survival)
> data(gehan, package="MASS")
> head(gehan)
  pair time cens   treat
1    1    1    1 control
2    1   10    1    6-MP
3    2   22    1 control
4    2    7    1    6-MP
5    3    3    1 control
6    3   32    0    6-MP
> attach(gehan)
> Six_MP<-subset(gehan,treat=="6-MP") #6-MP group only
> fit1<-survfit(Surv(time,cens)~1,data=Six_MP)
> summary(fit1)
Call: survfit(formula = Surv(time, cens) ~ 1,data = Six_MP)
```

Table 2.2 Construction of the K–M and its SE for the 6-MP group

$y_{(k)}$	$n_{(k)}$	$d_{(k)}$	$\hat{S}(y_{(k)})$	Var
6	21	3	$1-(3/21)=0.857$	0.0058
7	17	1	$0.857\{1-(1/17)\}=0.807$	0.0076
10	15	1	$0.807\{1-(1/15)\}=0.753$	0.0093
13	12	1	$0.753\{1-(1/12)\}=0.690$	0.0114
16	11	1	$0.690\{1-(1/11)\}=0.627$	0.0130
22	7	1	$0.627\{1-(1/7)\}=0.538$	0.0164
23	6	1	$0.538\{1-(1/6)\}=0.448$	0.0181
Time of study (t)			$\hat{S}(t)$	SE
$0 \le t < 6$			1.000	0.000
$6 \le t < 7$			0.857	0.076
$7 \le t < 10$			0.807	0.087
$10 \le t < 13$			0.753	0.096
$13 \le t < 16$			0.690	0.107
$16 \le t < 22$			0.628	0.114
$22 \le t < 23$			0.538	0.128
$23 \le t < 35$			0.448	0.135

```
   time n.risk n.event survival std.err lower  CI upper  CI
      6     21       3    0.857  0.0764        0.720        1.000
      7     17       1    0.807  0.0869        0.653        0.996
     10     15       1    0.753  0.0963        0.586        0.968
     13     12       1    0.690  0.1068        0.510        0.935
     16     11       1    0.627  0.1141        0.439        0.896
     22      7       1    0.538  0.1282        0.337        0.858
     23      6       1    0.448  0.1346        0.249        0.807
> plot(fit1, xlab="time", ylab="Survival function")
> quantile(fit1) # quantile including median
> print(fit1, print.rmean=T) # mean including median
Call: survfit(formula = Surv(time, cens) ~ 1, data = Six_MP)
       n   events   *rmean *se(rmean)   median  0.95LCL  0.95UCL
   21.00   9.00       23.29     2.83      23.00    16.00       NA
    * restricted mean with upper limit =  35
```

Figure 2.2 shows the K–M estimates for the 6-MP group, with their 95% confidence intervals, which visualizes the K–M estimator as a step function with jumps at the observed events (deaths).

On the other hand, the Nelson–Aalen (N–A) estimator of the cumulative hazard function $\Lambda(t)$ is defined by

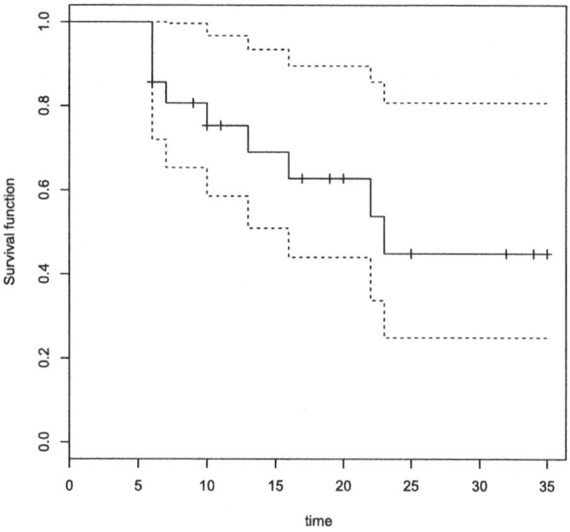

Fig. 2.2 K–M survival function estimates and their 95% confidence intervals for 6-MP group only in Gehan data

Table 2.3 Calculation of the N–A and its SE for 6-MP group

Time t	$\widehat{\Lambda}(t)$	SE
$0 \leq t < 6$	0	0
$6 \leq t < 7$	$3/21 = 0.143$	0.083
$7 \leq t < 10$	$0.143 + (1/17) = 0.202$	0.102
$10 \leq t < 13$	$0.202 + (1/15) = 0.269$	0.121
$13 \leq t < 16$	$0.269 + (1/12) = 0.352$	0.147
$16 \leq t < 22$	$0.352 + (1/11) = 0.443$	0.173
$22 \leq t < 23$	$0.443 + (1/7) = 0.586$	0.224
$23 \leq t < 35$	$0.586 + (1/6) = 0.753$	0.280

$$\widehat{\Lambda}_{N\text{-}A}(t) = \sum_{k:y_{(k)} \leq t} \frac{d_{(k)}}{n_{(k)}}, \tag{2.1}$$

which is identical to the Breslow (1972) estimator of the baseline cumulative hazard function from Cox's (1972) proportional hazards model (see Sect. 2.3 for more details) without covariates. The corresponding variance estimator (Aalen 1978) is given by

$$\widehat{\text{var}}(\widehat{\Lambda}_{N\text{-}A}(t)) = \sum_{k:y_{(k)} \leq t} \frac{d_{(k)}}{n_{(k)}^2}.$$

The detailed steps to calculate the N–A estimator and its variance are presented in Table 2.3.

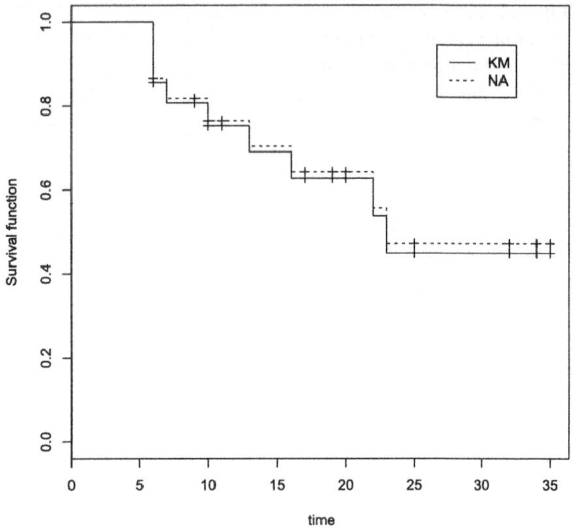

Fig. 2.3 Comparison of K–M and N–A estimates of the survival function for 6-MP group

From $\Lambda(t) = -\log S(t)$, the K–M estimator of $S(t)$ can also be used to estimate $\Lambda(t)$:

$$\begin{aligned}\widehat{\Lambda}_{K-M}(t) &= -\log \widehat{S}_{K-M}(t)\\ &= -\sum_{k:y_{(k)}\leq t} \log\left\{1 - \frac{d_{(k)}}{n_{(k)}}\right\}\\ &\simeq \widehat{\Lambda}_{N\text{-}A}(t).\end{aligned}$$

Since $-\log(1-x) \approx x$ for small x by Taylor expansion, the two estimators, $\widehat{\Lambda}_{K-M}(t)$ and $\widehat{\Lambda}_{N\text{-}A}(t)$, converge to the true cumulative hazard function when the increments $d_{(k)}/n_{(k)}$ are small, that is, when there are many individuals still at risk. Note here that $\widehat{\Lambda}_{K-M}(t) \geq \widehat{\Lambda}_{N\text{-}A}(t)$ in the finite samples since $-\log(1-x) \geq x$. In fact, the two estimators are asymptotically equivalent because the individual increments get arbitrarily smaller as $n \to \infty$ (Breslow and Crowley 1974). Similarly, we have that

$$\widehat{S}_{N\text{-}A}(t) = \prod_{k:y_{(k)}\leq t} \left\{\exp\left(-\frac{d_{(k)}}{n_{(k)}}\right)\right\} \simeq \widehat{S}_{K-M}(t).$$

since $\exp(-x) \approx 1 - x$ for small x. Note that $\widehat{S}_{N\text{-}A}(t) \geq \widehat{S}_{K-M}(t)$. Figure 2.3 shows a comparison of the K–M and N–A estimates of the survival function for 6-MP group, indicating their asymptotic equivalence even in the small sample.

The K–M and N–A estimators possess desirable large sample properties (consistency and asymptotic normality) under Assumptions 1 and 2 (Fleming and Harrington 1991; Andersen et al. 1993). Both estimators are also used as a graphic tool for a model checking. For example, a plot of $\widehat{\Lambda}_{K-M}(t) = -\log \widehat{S}_{K-M}(t)$ versus t will be approximately linear if the exponential distribution with a constant hazard rate, i.e., $-\log S(t) = \lambda t$, fits the data well.

Example 2.2 We show how to compute the N–A estimator for 6-MP group from the Gehan data presented in Example 2.1. The R codes and outputs are as follows:

```
>  fit2<-survfit(coxph(Surv(time,cens)~1, ties="breslow",data=Six_MP))
> summary(fit2)
Call: survfit(formula = coxph(Surv(time, cens) ~ 1, ties ="breslow",
    data = Six_MP))

  time n.risk n.event survival std.err lower 95% CI upper 95% CI
     6     21       3    0.867  0.0715        0.737        1.000
     7     17       1    0.817  0.0828        0.670        0.997
    10     15       1    0.765  0.0927        0.603        0.970
    13     12       1    0.704  0.1035        0.527        0.939
    16     11       1    0.642  0.1111        0.458        0.902
    22      7       1    0.557  0.1249        0.359        0.864
    23      6       1    0.471  0.1317        0.273        0.815
> plot(fit2) # survival plot using the N--A method
>
> ### Comparison of KM and NA estimators for survival function ###
> fit1_KM <-survfit(Surv(time,cens)~1,conf.type="none",data=Six_MP)
> fit2_NA <-survfit(coxph(Surv(time, cens)~1, ties="breslow",
+   data=Six_MP),conf.type="none")
> plot(fit1_KM,xlab="time", ylab="Survival function", lty=1)
> lines(fit2_NA, lty=2)
> legend(locator(1),c("KM","NA"),lty=1:2)
>
> ### N-A cumulative-hazard estimator ###
> Z.NA= -log(fit2$surv)
> Z.NA
[1] 0.1428571 0.2016807 0.2016807 0.2683473 0.2683473 0.3516807
[7] 0.4425898 0.4425898 0.4425898 0.4425898 0.5854469 0.7521136
[13] 0.7521136 0.7521136 0.7521136 0.7521136
```

Example 2.3 (*Proportional case*) We illustrate existing procedures to test equality of failure time distributions, together with a graphical comparison by the K–M estimates. For a k-sample test of equality of survival functions, we can use the log-rank test (Mantel–Haenszel test), Gehan test (generalized Wilcoxon test), or Tarone–Ware test (weighted log-rank test). The log-rank test is popular, and optimal when the hazard functions are proportional between comparison groups, or the hazard ratio is constant (proportional hazards (PH) assumption). Gehan test or Tarone–Ware test could be more efficient for the non-PH data.

Figure 2.4 presents the K–M estimates for 6-MP and control groups, respectively, in Gehan data. This plot suggests the 6-MP patients have overall higher survival probabilities than ones in control group. It is thus clear that the 6-MP group tends to have longer remission times. The corresponding p-values from the log-rank, Gehan and Tarone–Ware tests are all close to zero.

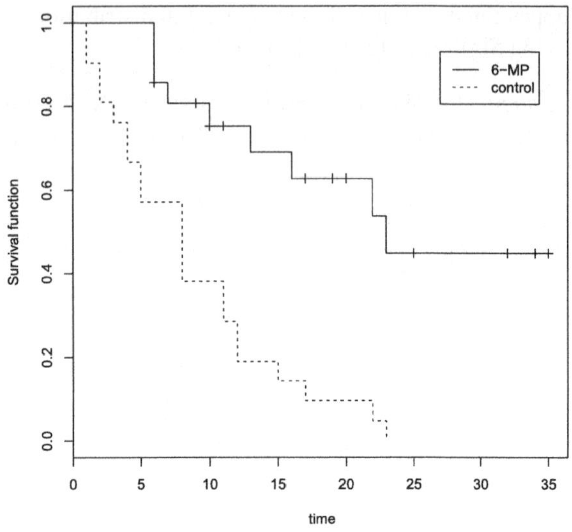

Fig. 2.4 K–M survival function estimates for two groups (6-MP vs. control) in Gehan data

Below is the R codes for the K–M plots and the above-mentioned three tests for equality of the two failure distributions ($k = 2$). The `survdiff()` function implements the Harrington and Fleming family (1982), with the weights $\hat{S}(t)^{\rho}$, where $\hat{S}(t)$ is the K–M estimate of the pooled survival from both groups. This gives the log-rank test if $\rho = 0$, Gehan test if $\rho = 1$ and Tarone–Ware test if $\rho = 0.5$.

```
> fit3<-survfit(Surv(time,cens)~treat,data=gehan)
> summary(fit3)
> plot(fit3, lty=1:2, xlab="time", ylab="Survival function")
> legend(locator(1),c("6-MP","control"),lty=1:2)
> survdiff(formula=Surv(time,cens)~treat,data=gehan) #log-rank test
> survdiff(formula=Surv(time,cens)~treat,data=gehan,rho=1) #Gehan test
> survdiff(formula=Surv(time,cens)~treat,data=gehan,rho=0.5)#Tarone-Ware test
```

Example 2.4 (*Crossing case*) In this example, we compare the three tests again when the two estimated survival curves cross over, indicating a non-proportionality. Consider a data set (in days) from the Gastrointestinal Tumor Study Group (1982), which compared a chemotherapy alone with a combined chemo- and radiation therapy, to treat locally unresectable gastric cancer (Stablein and Koutrouvelis 1985). Each treatment arm had 45 patients, with two patients from the chemotherapy group and six from the combination group censored.

The K–M plot in Fig. 2.5 shows that the two K–M estimates cross over around 2.7 years. The p-values from the log-rank, Gehan and Tarone–Ware tests are 0.635, 0.0465, and 0.168, respectively. We thus see that the Gehan test detects the difference between the two groups most efficiently under this particular circumstances. The following is the R codes for testing the equality of the two survival distributions.

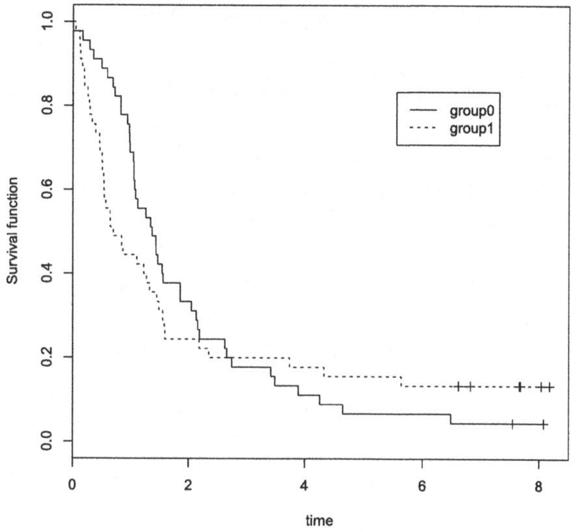

Fig. 2.5 K–M survival function estimates (group0: chemotherapy, group1: combined) in the gastric data

```
> library(YPmodel)
> data(gastric)
> head(gastric)
           V1 V2 V3
1 0.002739726  1  0
2 0.046575342  1  1
3 0.115068493  1  1
4 0.120547945  1  1
5 0.131506849  1  1
6 0.164383562  1  1
> time=gastric$V1 # survival time (unit: year)
> cens=gastric$V2 # censoring indicator
> group=gastric$V3 # group("0",chemotherapy;"1", combined)
>
> fit4<-survfit(Surv(time,cens)~group,data=gastric)
> plot(fit4, lty=1:2, xlab="time", ylab="Survival function")
> legend(locator(1),c("group0","group1"),lty=1:2)
> formula=Surv(time,cens)~group
> survdiff(formula,data=gastric) # log-rank test
> survdiff(formula,data=gastric,rho=1) # Gehan test
> survdiff(formula,data=gastric,rho=0.5)# Tarone-Ware test
```

For the detecting crossing hazards, one could look for a test of interaction between group membership and time (i.e., time-by-covariate interaction) or an alternative modeling approach (Collett, Sect. 4.4, 2015; Burke and MacKenzie 2017).

2.2 Basic Likelihood Inference

In this section, we show a likelihood construction under random right censoring. Let $f_\theta(\cdot)$, $S_\theta(\cdot)$, $\lambda_\theta(\cdot)$, and $\Lambda_\theta(\cdot)$ be density, survival, hazard, and cumulative hazard functions of failure time T with an unknown parameter $\theta \in \Omega$, respectively. Here, Ω is the parameter space. The observable random variables from n individuals consist of the pairs (Y_i, δ_i) $(i = 1, \cdots, n)$, where

$$Y_i = \min(T_i, C_i) \text{ and } \delta_i = I(T_i \le C_i).$$

Let $\mathrm{P}_\theta(y_i, \delta_i)$ be the probability distribution of the pair (y_i, δ_i) of the ith observation. Under the Assumptions in Sect. 2.1.2, the likelihood, denoted by $L_i(\theta; y_i, \delta_i)$, for θ based on the ith observation is given as

$$L_i(\theta; y_i, \delta_i) \equiv \mathrm{P}_\theta(y_i, \delta_i) \propto f_\theta(y_i)^{\delta_i} S_\theta(y_i)^{1-\delta_i}. \tag{2.2}$$

The derivation of (2.2) is as follows. Let $g(\cdot)$ and $G(\cdot)$ be the density function and cumulative distribution function of the censoring time, respectively. From Assumption 1, we have

$$L_i(\theta; y_i, \delta_i = 1) = \mathrm{P}_\theta(Y_i = y_i, \delta_i = 1) = \mathrm{P}_\theta(T_i = y_i, T_i \le C_i) = f_\theta(y_i)[1 - G(y_i)]$$

and

$$L_i(\theta; y_i, \delta_i = 0) = \mathrm{P}_\theta(Y_i = y_i, \delta_i = 0) = \mathrm{P}_\theta(C_i = y_i, T_i > C_i) = S_\theta(y_i)g(y_i).$$

Because $g(\cdot)$ and $G(\cdot)$ do not involve any information about the failure time distribution (and therefore θ), by Assumption 2, we have $L_i(\theta; y_i, \delta_i = 1) \propto f_\theta(y_i)$ and $L_i(\theta; y_i, \delta_i = 0) \propto S_\theta(y_i)$. Thus, the likelihood function $L_i(\theta; y_i)$ for a subject i would include $f_\theta(y_i)$ contributed by the observed event time, i.e., $\delta_i = 1$, or $S_\theta(y_i) = \mathrm{P}_\theta(T_i > y_i)$ contributed by the observed censoring time, i.e., $\delta_i = 0$.

Therefore, the total likelihood function for n independent observations is given by

$$L(\theta) = \prod_i L_i(\theta; y_i, \delta_i) = \prod_i \left[\lambda_\theta(y_i)^{\delta_i} \exp\{-\Lambda_\theta(y_i)\}\right],$$

with the log-likelihood of

$$\ell(\theta) = \log L(\theta) = \sum_i \{\delta_i \log \lambda_\theta(y_i) - \Lambda_\theta(y_i)\}. \tag{2.3}$$

The maximum likelihood estimator (MLE) $\hat{\theta}$ of θ is defined by the value θ maximizing the log-likelihood (2.3), i.e.,

$$\hat{\theta} = \arg\max_{\theta \in \Omega} \ell(\theta),$$

where arg max denotes the argument of the maximum, or equivalently

$$\ell(\hat{\theta}) \geq \ell(\theta) \text{ for all } \theta \in \Omega.$$

In practice, however, it is usually not possible to obtain an explicit form solution for the MLE, especially when the model involves many parameters and its density or estimating equation (i.e., $\partial\ell(\theta)/\partial\theta = 0$) is highly nonlinear. In such situations, the MLE can be numerically obtained by using nonlinear optimization algorithms such as Newton–Raphson method or the `optim()` R function.

Under some regular conditions (Cox and Hinkley, Sect. 9.1, 1974), the MLE $\hat{\theta}$ has the following useful properties:

(1) Consistency: $\hat{\theta}$ is consistent to θ, i.e., for small $\epsilon > 0$,

$$P(|\hat{\theta} - \theta| \geq \epsilon) \to 0 \text{ as } n \to \infty.$$

(2) Invariance: If $g(\theta)$ be a function of θ, not necessary one-to-one or differential, then $g(\hat{\theta})$ is the MLE of $g(\theta)$.

(3) Asymptotic normality: $\hat{\theta}$ is asymptotically normally distributed with mean θ and variance $i^{-1}(\theta)$, i.e.,

$$\hat{\theta} \approx N(\theta, i^{-1}(\theta)) \text{ as } n \to \infty,$$

where $i(\theta) = E(-\partial^2\ell(\theta)/\partial\theta^2)$ is an expected (Fisher) information and its inverse provides an asymptotic variance of $\hat{\theta}$. However, the observed information $I(\theta) = -\partial^2\ell(\theta)/\partial\theta^2$ is usually used in survival analysis because the computation of expectation in $i(\theta)$ is difficult under random censoring and $i(\theta) \approx I(\theta)$ asymptotically.

Note that the three properties above are still applied even if θ is a vector of parameters.

Example 2.5 Let us consider an exponential distribution with a constant hazard; $\lambda(t) = \lambda, t \geq 0$. The log-likelihood based on the observed data (y_i, δ_i) $(i = 1, \ldots, n)$ is given by

$$\ell(\lambda) = \sum_i \{\delta_i \log\lambda - \lambda y_i\} = r\log\lambda - \lambda\sum_i y_i,$$

where $r = \sum_i \delta_i$ is the observed number of events. From $\partial\ell(\lambda)/\partial\lambda = 0$, the MLE of λ is given by $\widehat{\lambda} = r/\sum_i y_i$. The observed information is

$$I(\lambda) \equiv -\partial^2 \ell(\lambda)/\partial\lambda^2 = r/\lambda^2.$$

By the asymptotic normality of the MLE, we have that $\widehat{\lambda} \sim N(\lambda, \lambda^2/r)$ asymptotically. □

Likelihoods under various types of censoring and truncation schemes are summarized in Appendix 2.6.1.

2.3 Cox's Proportional Hazards Models

Nonparametric tests such as the log-rank test can be used to test the equality of failure time distributions among different groups, but they do not typically adjust for confounding factors, which can be included as covariates in a regression setting.

Cox (1972) introduced a regression model for the hazard function, which specifies the relationship between the hazard rate and fixed or time-varying covariates. Let $x = (x_1, \ldots, x_p)^T$ be a vector of covariates for an individual and $\lambda(t; x)$ be the hazard function at time t for an individual with covariates x. Under the Cox model, the hazard function for an individual is of the form

$$\lambda(t; x) = \lambda_0(t) \exp(x^T \beta), \tag{2.4}$$

where $\lambda_0(t)$ is an unspecified baseline hazard function at time t under $x = 0$ and $\beta = (\beta_1, \ldots, \beta_p)^T$ is a vector of regression parameters corresponding to covariates x which can be time-independent or time-dependent. Note here that the exponentiated covariate terms act multiplicatively on individual's hazard rate. For the purpose of identifiability, the term $x^T\beta$ does not include the intercept term. The model in (2.4) is called semiparametric because the form of the baseline hazard part is nonparametric, while that of the covariate part is parametric. This is also called a PH model because the ratio (i.e., hazard ratio (HR) or relative risk) of hazard rates for any two individuals with different covariate vectors, x_1 and x_2, is constant over time t. That is, the hazard ratio is given by

$$\mathrm{HR}(t; x_1, x_2) = \frac{\lambda(t; x_1)}{\lambda(t; x_2)} = \exp\{(x_1 - x_2)^T \beta\}, \tag{2.5}$$

which does not vary with t. Thus, the regression parameters β have an attractive interpretation in terms of the log hazard ratio.

In particular, the interpretation of HR is useful in two sample case with one binary covariate x. Denote $x_1 = 1$ for the new drug and $x_2 = 0$ for the placebo. Then, from (2.5), the HR of a patient in new drug group against one in placebo group is given by

$$\mathrm{HR}(t; x_1, x_2) = \frac{\lambda(t; x_1 = 1)}{\lambda(t; x_2 = 0)} = \exp(\beta_1).$$

Thus, the new drug would be associated with lower (higher) hazard rate relative to the placebo if $\beta_1 < 0$ ($\beta_1 > 0$).

Cox's Partial Likelihood

The Cox model can be directly fitted using the likelihood (2.3) if a parametric form (e.g., Weibull) of $\lambda_0(t)$ is specified. However, when the functional form of $\lambda_0(t)$ in (2.4) is completely unknown, the classical likelihood approach is not directly applicable. Under Assumptions 1 and 2 and no ties, Cox (1972, 1975) introduced the partial (log-)likelihood for estimating β in the absence of information of $\lambda_0(t)$, defined by

$$\ell_C(\beta) = \sum_k \left[x_{(k)}^T \beta - \log \left\{ \sum_{i \in R_{(k)}} \exp(x_i^T \beta) \right\} \right], \tag{2.6}$$

where $x_i = (x_{i1}, \ldots, x_{ip})^T$ is a $p \times 1$ vector of covariates for the ith individual, $y_{(k)}$ is the kth ($k = 1, \ldots, D$) smallest distinct event time among the y_i's, $x_{(k)}$ is the covariate vector corresponding to $y_{(k)}$ and

$$R_{(k)} = R(y_{(k)}) = \{i : y_i \geq y_{(k)}\}$$

is the risk set at time $y_{(k)}$, i.e., the set of all individuals who are alive and uncensored just prior to $y_{(k)}$. The partial likelihood (2.6) depends only on the order in which events occur, not on the exact times of occurrence. It can be viewed as a profile likelihood as shown in Appendix 2.6.2.

Cox (1972) showed that the kth term in the partial likelihood is the conditional probability that an individual fails at time $y_{(k)}$ with covariates $x_{(k)}$, given one of the individuals in $R_{(k)}$ fails at this time. That is, it is expressed as

$$\begin{aligned}
&\text{P(individual fails at } y_{(k)} \mid \text{one failure at } y_{(k)}) \\
&= \frac{\text{P(individual fails at } y_{(k)} \mid \text{survival to } y_{(k)})}{\text{P(one failure at } y_{(k)} \mid \text{survival to } y_{(k)})} \\
&= \frac{\lambda(t; x_{(k)})}{\sum_{i \in R_{(k)}} \lambda(t; x_i)} \\
&= \frac{\exp(x_{(k)}^T \beta)}{\sum_{i \in R_{(k)}} \exp(x_i^T \beta)}.
\end{aligned}$$

Notice here that $\lambda_0(t)$ cancels out and that the partial likelihood is a function only of β.

The partial log-likelihood $\ell_C(\beta)$ is then obtained by taking logarithm of the product of all these conditional probabilities over the D failures. The regression parameters β in the Cox model can be estimated by maximizing the partial log-likelihood. The partial maximum likelihood estimators (PMLEs) $\hat{\beta}$ of β that maximize $\ell_C(\beta)$

are obtained by solving the score equations

$$\frac{\partial \ell_C(\beta)}{\partial \beta} = 0, \tag{2.7}$$

and their variance estimators are obtained from the inverse of observed information matrix, $-\partial^2 \ell_C / \partial \beta^2$. The PMLEs are often called nonparametric MLEs (NPMLEs). The score equations in (2.7) can be usually solved using the Newton–Raphson method with initial values $\hat{\beta}^{(0)} = 0$. Note that the resulting PMLEs $\hat{\beta}$ are consistent and asymptotically normally distributed (Andersen and Gill 1982; Andersen et al. 1993).

Breslow's Likelihood

Several forms of partial likelihoods have been suggested when there are ties among failure times; see, for example, Breslow (1972, 1974), Peto and Peto (1972), and Efron (1977). In particular, from a joint likelihood of β and λ_0, Breslow proposed the following partial likelihood with ties:

$$\ell_B(\beta) = \sum_k \left[s_{(k)}^T \beta - d_{(k)} \log \left\{ \sum_{i \in R_{(k)}} \exp(x_i^T \beta) \right\} \right], \tag{2.8}$$

where $s_{(k)}^T = \sum_{i \in D_{(k)}} x_i^T$ is the sum of the vectors x_i^T over $D_{(k)} = \{i : \delta_i = 1, y_i = y_{(k)}\}$ which is the set of individuals who fail at $y_{(k)}$, and $d_{(k)} = \sum_{i=1}^n I(y_i = y_{(k)})$ is the number of events at $y_{(k)}$. He also proposed an estimator $\widehat{\Lambda}_{0B}(t)$ of the baseline cumulative hazard function $\Lambda_0(t) = \int_0^t \lambda_0(u)du$, given by

$$\widehat{\Lambda}_{0B}(t) = \sum_{k: y_{(k)} \le t} \left\{ \frac{d_{(k)}}{\sum_{i \in R_{(k)}} \exp(x_i^T \widehat{\beta})} \right\}, \tag{2.9}$$

where $\widehat{\beta}$ is a vector of the estimates that maximize $\ell_B(\beta)$. The derivations of (2.8) and (2.9) are given in Appendix 2.6.2.

When there are no ties (i.e., all $d_{(k)} = 1$), Breslow's likelihood $\ell_B(\beta)$ reduces to Cox's partial likelihood $\ell_C(\beta)$. Suppose that there are no covariates, $\exp(x_i^T \beta) = 1$. Breslow's estimator $\widehat{\Lambda}_{0B}(t)$ reduces to the N–A estimator $\widehat{\Lambda}_{N\text{-}A}(t)$ in (2.1) because $n_{(k)} = \sum_i I(i \in R_{(k)})$.

To illustrate notations in Breslow's likelihood (2.8) and Breslow estimator (2.9), we consider a small data set, with five individuals. Suppose that from the ith individual ($i = 1, \ldots, 5$), survival data (y_i, δ_i) are observed with two covariates $x_i^T = (x_{i1}, x_{i2})$. Table 2.4 shows the data set and the steps to calculate the basic quantities, such as $y_{(k)}$, $R_{(k)}$, $n_{(k)}$, $D_{(k)}$, $d_{(k)}$, and $s_{(k)}$.

For example, at the first distinct event time $y_{(1)} = 2$, the remaining quantities are calculated as follows:

Table 2.4 A small data set and calculation of quantities at $y_{(k)}$

Individual i	y_i	δ_i	x_{i1}	x_{i2}		
1	3	0	x_{11}	x_{12}		
2	5	1	x_{21}	x_{22}		
3	5	1	x_{31}	x_{32}		
4	2	1	x_{41}	x_{42}		
5	6	1	x_{51}	x_{52}		
k	$y_{(k)}$	$R_{(k)}$	$n_{(k)}$	$D_{(k)}$	$d_{(k)}$	$s_{(k)}^T$
1	2	$\{1, 2, 3, 4, 5\}$	5	$\{4\}$	1	(x_{41}, x_{42})
2	5	$\{2, 3, 5\}$	3	$\{2, 3\}$	2	$(x_{21} + x_{31}, x_{22} + x_{32})$
3	6	$\{5\}$	1	$\{5\}$	1	(x_{51}, x_{52})

Note: Since $y_{(k)}$ is the kth smallest distinct event time among y_i's, a censoring time $y_1 = 3$ cannot be an event time $y_{(k)}$

$$
\begin{aligned}
R_{(1)} &= R(y_{(1)}) = \{i : y_i \geq y_{(1)}\} = \{i : y_i \geq 2\} = \{1, 2, 3, 4, 5\} \\
n_{(1)} &= 5 \text{ since } n_{(1)} \text{ is the number of elements in } R_{(1)} \\
D_{(1)} &= \{i : \delta_i = 1, y_i = y_{(1)}\} = \{4\} \\
d_{(1)} &= 1 \text{ since } d_{(1)} \text{ is the number of elements in } D_{(1)} \\
s_{(1)}^T &= \sum_{i \in D_{(1)}} x_i^T = x_4^T = (x_{41}, x_{42}).
\end{aligned}
$$

Fitting Procedures

Under the Cox PH model (2.4), the log-likelihood (2.3) becomes

$$\ell(\beta, \lambda_0) = \sum_i \delta_i \{\log \lambda_0(y_i) + \eta_i\} - \sum_i \{\Lambda_0(y_i) \exp(\eta_i)\},$$

where $\eta_i = x_i^T \beta$. Appendix 2.6.2 shows that the profile likelihood $\ell^*(\beta)$ becomes $\ell_B(\beta)$ in (2.8) since

$$
\begin{aligned}
\ell^*(\beta) &= \ell(\beta, \lambda_0)|_{\Lambda_0 = \widehat{\Lambda}_{0B}(\beta)} \\
&= \sum_i \delta_i \eta_i - \sum_k d_{(k)} \log\{\sum_{i \in R_{(k)}} \exp(\eta_i)\}
\end{aligned}
$$

with the constant term being deleted. Note that $\sum_i \delta_i \eta_i = \sum_k s_{(k)}^T \beta$ in (2.8). Below, we present two methods on how to solve $\partial \ell^* / \partial \beta = 0$, which provide the same estimator for β.

• Newton–Raphson method

The usual Newton–Raphson method requires the following two partial derivatives:

$$S^*(\beta_r) = \frac{\partial \ell^*}{\partial \beta_r} = \sum_i \delta_i x_{ir} - \sum_k d_{(k)} \left\{ \frac{\sum_{i \in R_{(k)}} x_{ir} \exp(\eta_i)}{\sum_{i \in R_{(k)}} \exp(\eta_i)} \right\}, \ (r = 1, \ldots, p)$$

$$\begin{aligned} H^*(\beta_{rs}) = -\frac{\partial^2 \ell^*}{\partial \beta_r \beta_s} = \sum_k d_{(k)} \Bigg[& \frac{\sum_{i \in R_{(k)}} x_{ir} x_{is} \exp(\eta_i)}{\sum_{i \in R_{(k)}} \exp(\eta_i)} \\ & - \frac{\{\sum_{i \in R_{(k)}} x_{ir} \exp(\eta_i)\}\{\sum_{i \in R_{(k)}} x_{is} \exp(\eta_i)\}}{\{\sum_{i \in R_{(k)}} \exp(\eta_i)\}^2} \Bigg], \\ & (r, s = 1, \ldots, p), \end{aligned}$$

and $\widehat{\beta}_r$ are obtained by solving iteratively

$$\widehat{\beta}_r^{(k+1)} = \widehat{\beta}_r^{(k)} + [\{H^*(\beta_{rs})\}^{-1} S^*(\beta_r)]|_{\beta_r = \widehat{\beta}_r^{(k)}}.$$

This Newton–Raphson method can be represented by the iterative weighted least squares (IWLS) equation (Appendix 2.6.3), given by

$$(X^T W^* X)\widehat{\beta} = X^T W^* w,$$

where X is a $n \times p$ model matrix for β whose ith row vector is x_i^T, $W^* = W^*(\beta, \lambda_0)$ is a symmetric matrix in (2.19) in Appendix 2.6, and

$$w = \eta + W^{*-1}(\delta - \mu)$$

is an adjusted dependent variable with $\eta = X\beta$. The IWLS equation is popular in the generalized linear models (GLMs; McCullagh and Nelder 1989). However, the matrix W^* in the above IWLS equation is no longer diagonal, so that W^{*-1} is often difficult to be computed (Ha and Lee 2003). Thus, the IWLS equation of the GLMs cannot be directly used for the Cox model, so that it is desirable to develop an alternative iterative procedure without calculating the inverse of W^*.

• A new iterative least squares (ILS) method

The Newton–Raphson procedure can be implemented via the ILS method below, without involving W^{*-1}, by introducing a new adjusted dependent variable

$$w^* = W^* \eta + (\delta - \mu) \ (= W^* w),$$

where $\mu = \exp(\log \Lambda_0(y) + \eta)$.

Theorem 2.1 *The new ILS equation for β in the Cox PH model is given by*

$$(X^T W^* X)\widehat{\beta} = X^T w^*, \tag{2.10}$$

where $w^* = W^*\eta + (\delta - \mu)$.

The proof is given in Appendix 2.6.3, including the form of W^*. Note that the terms λ_{0k} in both W^* and w^* are replaced by their estimates $\widehat{\lambda}_{0k}$ in (2.18). The ILS equation (2.10) is extended to the general frailty models beginning from Chap. 4.

The variance of $\hat{\beta}$ can be estimated by $(X^T W^* X)^{-1}$. It can be shown that the inverse of the second derivative (i.e., $H^*(\beta) = X^T W^* X$) of the profile log-likelihood $\ell^*(\beta)$ gives the same variance estimate of $\hat{\beta}$ as the relevant submatrix of the inverse of the full information matrix derived from the full log-likelihood $\ell(\beta, \lambda_0)$.

- Fitting algorithm:
  - **Step 1**: Take all zeros as initial values $\hat{\beta}^{(0)}$ of β.
  - **Step 2**: Given $\beta^{(0)}$, the new estimates $\hat{\beta}$ are obtained by solving the score equations $\partial\ell^*/\partial\beta = 0$; that is, they are solved using the ILS method with (2.10).
  - **Step 3**: Repeat Step 2 until the maximum absolute difference between the previous and current estimates for β is less than 10^{-6}.

Example 2.6 Results from application of the two methods to Gehan's data are as follows. The Newton–Raphson method is implemented in `coxph()` function in **survival** R package (Therneau 2010) and the ILS method is in `frailtyHL()` function in **frailtyHL** R package (Ha et al. 2018) described in Chap. 4.

```
> ############## Method 1: Fitting Cox model via coxph() ##############
> library(survival)
> gehan$treat=relevel(gehan$treat,ref="control")
> Method1<-coxph(Surv(time, cens)~factor(treat),ties="breslow",data=gehan)
> summary(Method1)
Call:
coxph(formula = Surv(time, cens) ~ factor(treat), data = gehan,
    ties ="breslow")
  n= 42, number of events= 30

                        coef exp(coef) se(coef)      z Pr(>|z|)
factor(treat)6-MP -1.5092     0.2211    0.4096 -3.685 0.000229 ***
Signif. codes:  0 *** 0.001 ** 0.01 * 0.05 . 0.1   1

                     exp(coef) exp(-coef) lower .95 upper .95
factor(treat)6-MP     0.2211       4.523   0.09907    0.4934

Concordance= 0.69   (se = 0.053 )
Rsquare= 0.304    (max possible= 0.989 )
Likelihood ratio test= 15.21  on 1 df,   p=9.615e-05
Wald test            = 13.58  on 1 df,   p=0.0002288
```

```
Score (logrank) test = 15.93  on 1 df,   p=6.571e-05
>
> ############ Method 2: Fitting Cox model via frailtyHL() ############
> library(frailtyHL)
> Method2<-frailtyHL(Surv(time, cens)~treat+(1|pair),
+ varfixed=TRUE,varinit=0,data=gehan)
iteration :
          4
convergence :
  2.731994e-08
[1]"converged"
[1]"Results from the Cox model"
[1]"Number of data : "
[1] 42
[1]"Number of event : "
[1] 30
[1]"Model for conditional hazard : "
Surv(time, cens) ~ treat + (1 | pair)
[1]"Method : HL(0,1)"
[1]"Estimates from the mean model"
          Estimate Std. Error t-value   p-value
treat6-MP   -1.509     0.4096  -3.685 0.0002288
```

Interpretation: The two methods provide identical results using Breslow's method for ties. The output indicates that the estimated 6-MP drug effect is -1.509 with p-value $= 0.00023$. The estimated hazard ratio for 6-MP group relative to placebo group is $\exp(-1.509) = 0.221$, with a corresponding 95% confidence interval of $\exp(-1.509 \pm 1.96 \times 0.4096) = (0.099, 0.493)$. Thus, we see that the 6-MP group has significantly lower hazard rate as compared to the placebo group. □

Remark 2.1
(i) A method for fitting the Cox PH model using Poisson GLM:

Since the maximum likelihood score equations for β become

$$\frac{\partial \ell^*}{\partial \beta} = \frac{\partial \ell}{\partial \beta}|_{\Lambda_0=\widehat{\Lambda}_0},$$

the MLEs are obtained by solving

$$\frac{\partial \ell}{\partial \beta_r}|_{\Lambda_0=\widehat{\Lambda}_0} = \sum_i (\delta_i - \mu_i)x_{ir}|_{\Lambda_0=\widehat{\Lambda}_0} = 0 \ (r = 1, \ldots, p),$$

where $\mu_i = \Lambda_0(y_i)\exp(x_i^T\beta)$ (Appendix 2.6.3). These are also the estimating equations for a Poisson GLM, with the response δ_i and the offset $\log \widehat{\Lambda}_0(y_i)$. Fitting the PH model with parametric baseline hazard $\Lambda_0(\cdot)$ via standard Poisson GLM is straightforward (Aitkin and Clayton 1980). Furthermore, the Poisson GLM fitting of the Cox PH model with nonparametric baseline hazard can be done by using a pseudo-Poisson response variable $y_{i,k}$ having 0 or 1 in (2.20) (Whitehead 1980): for more detailed description, see Appendix 2.6.4.

(ii) PH assumption:

The log-rank statistic can be derived as the score test under the Cox PH model comparing two groups with a single binary covariate. It is asymptotically equivalent to the likelihood ratio and Wald test statistics from the PH model. However, the PH is a strong assumption that clearly needs to be checked in applications because the ratio of the hazard functions between two individuals in different groups can vary over time as in Fig. 2.5. For this case, a time-dependent covariate term $x(t)$ or a time-dependent coefficient term $\beta(t)$ can be introduced into the model to test the PH assumption; `cox.zph()` in survival R package and `PROC PHREG` in SAS are available. If the PH assumption is violated for a discrete covariate, it may be reasonable to stratify on this covariate (i.e., each stratum of this covariate has a different baseline hazard) and employ the PH model with the other covariates within each stratum. □

2.4 Accelerated Failure Time Models

A linear model can be considered for survival data as an alternative to the Cox PH model (2.4) by modelling a direct relationship between the logarithm of failure time and covariates as follows:

$$\log T = x^T\beta + \epsilon, \tag{2.11}$$

where the term $x^T\beta$ includes an intercept term and ϵ is a random error. Note that the logarithmic transformation is often used because T is positive but other transformations can be also used. The model (2.11) can be written as

$$T = \exp(x^T\beta)T_0,$$

where $T_0 = \exp(\epsilon)$, indicating that the role of x is to accelerate (or decelerate) time to failure, T. Thus, this model is referred to as the accelerated failure time (AFT) model. Time-dependent covariates can be also introduced in the AFT model as in the Cox model (Orbe et al. 2002).

Let $S_0(t)$ denote the survival function of $T_0 = \exp(\epsilon)$. Since $T = \exp(x^T\beta + \epsilon)$, we have that

$$S(t) = \mathrm{P}(T > t) = \mathrm{P}\{T_0 > t\exp(-x^T\beta)\} = S_0\{t\exp(-x^T\beta)\}.$$

Because $\lambda(t) = -d\log S(t)/dt$, the hazard function of T under the AFT model can be expressed as

$$\begin{aligned}\lambda(t; x) &= -\frac{S'(t; x)}{S(t; x)}\\ &= \frac{f_0\{t\exp(-x^T\beta)\}\exp(-x^T\beta)}{S_0\{t\exp(-x^T\beta)\}}\end{aligned}$$

$$= \lambda_0\{t\exp(-x^T\beta)\}\exp(-x^T\beta), \qquad (2.12)$$

where $\lambda_0(\cdot) = f_0(\cdot)/S_0(\cdot)$ is the hazard function of T_0 with the density function $f_0(\cdot)$, and is also a function of t, x and β. It is well known that the model (2.12) gives a non-PH model except when $\lambda_0(\cdot)$ follows a Weibull distribution.

Note that the Weibull is the only family of distributions closed under both PH and AFT models. Specifically, if ϵ in the AFT model (2.11) follows an extreme value distribution with scale parameter σ having the density

$$f(e) = \sigma^{-1}\exp\{(e/\sigma) - \exp(e/\sigma)\},$$

for $-\infty < e < \infty$, then T has the Weibull distribution with shape parameter $\phi = 1/\sigma$ and scale parameter $\exp\{-(x^T\beta)\phi\}$, leading to a PH model

$$\lambda(t; x) = \phi t^{\phi-1}\exp(x^T\beta^*), \qquad (2.13)$$

where $\beta^* = -\phi\beta$; this is also easily derived from (2.12) with $\lambda_0(s) = \phi s^{\phi-1}$.

Example 2.7 Parametric regression models for survival data are usually facilitated by location-scale family distributions with an arbitrary transformation of the time variable; the log transformation leads to the AFT models. The `survreg()` in **survival** package fits the parametric AFT models using "exponential", "weibull", "log-normal", "log-logistic", etc. For the SAS, `PROC LIFEREG` is available. Below is an example of fitting the Weibull AFT regression model to the Gehan data set.

```
> AFT_Wei=survreg(Surv(time,cens) ~ factor(treat),dist='weibull',
+ data=gehan)
> summary(AFT_Wei)

Call:
survreg(formula = Surv(time, cens) ~ factor(treat), data = gehan,
    dist ="weibull")
                    Value Std. Error     z        p
(Intercept)         2.248      0.166 13.55 8.30e-42
factor(treat)6-MP   1.267      0.311  4.08 4.51e-05
Log(scale)         -0.312      0.147 -2.12 3.43e-02

Scale= 0.732

Weibull distribution
Loglik(model)= -106.6   Loglik(intercept only)= -116.4
        Chisq= 19.65 on 1 degrees of freedom, p= 9.3e-06
Number of Newton--Raphson Iterations: 5
n= 42
```

Interpretation: In the AFT model (2.11) with a binary covariate (i.e., Ttreat), the linear predictor η can be expressed as

$$\eta = x^T\beta = \beta_0 + \beta_1 I(\text{Treat} = 6-\text{MP}).$$

Thus, survival time (i.e., remission time) in the treatment group (Treat = 6-MP) is increased by a factor of exp(1.267) = 3.55, as compared to the placebo group (Treat = control). Let β be the fixed effects in AFT model (2.11) and β^* be those in Weibull PH model (2.13). Since $\beta^* = -\phi\beta$, the 6-MP effect β_1^* is estimated by $\hat{\beta}_1^* = -(1/\hat{\sigma})\hat{\beta}_1 = -(1/0.732)(1.267) = -1.731$; this is similar to the estimated 6-MP effect (i.e., -1.509) from the Cox PH model in Example 2.6. □

Remark 2.2 The class of semiparametric linear transformation models (Kalbfleisch and Prentice 2002) for T takes the form

$$g(T) = x^T\beta + \epsilon, \tag{2.14}$$

where ϵ is a random error with a distribution function F and $g(\cdot)$ is an increasing function. In case of a known g but an unknown F, the model (2.14) reduces to the semiparametric AFT model (Buckley and James 1979), usually with $g(T) = \log T$. With an unknown g but a known F, it further specifies two classes of models (Cheng et al. 1995); if F is the standard extreme value distribution with

$$F(x) = 1 - \exp(-e^x), \quad -\infty < x < \infty,$$

the model (2.14) becomes the PH model, and if F is the standard logistic function with

$$F(x) = e^x/(1 + e^x), \quad -\infty < x < \infty,$$

it becomes the proportional odds model (Bennett 1983), under which the hazard ratio converges to unity with time.

Cox pointed out "AFT models are in many ways more appealing" than the PH models "because of their quite direct physical interpretation" (Reid 1994). □

2.5 Discussion

Under some regularity conditions, the K–M and N–A estimators are nonparametric MLEs with consistency and asymptotic normality (Johansen 1983; Fleming and Harrington 1991; Andersen et al. 1993). In particular, the N–A estimator of the cumulative hazard function can be easily extended to various regression models such as Cox's PH models and frailty models.

The PH and AFT models are two important classes of regression models for survival data. The Cox PH model is often used in practice because inference on the parameters of interest is possible without assuming any form of the baseline hazard function. However, this model is based on the PH assumption which does not always hold in the observed survival data. If the PH assumption is violated, the inference procedure could provide inefficient results.

The AFT model has several advantages over the PH model as follows:

(i) The AFT model does not need a PH assumption as in the Cox model;
(ii) It models directly the covariate effects on the survival times, so the interpretation is clearer and easier than in the Cox model;
(iii) The estimated regression parameters in the AFT model are relatively robust against mis-specification of the model assumption, while ones in the Cox model can be biased; for more details, see Orbe et al. (2002), Hutton and Monaghan (2002), and Patel et al. (2006).

Inference on the AFT model is typically based on a parametric setting (Hougaard 1999). However, asymptotically efficient methods using rank-based or least squares approaches for the semiparametric AFT models are available in the literature, though likely not widely used (Zeng and Lin 2007; Chiou et al. 2014; Jin 2016).

Furthermore, as an alternative to the Cox PH model, the additive hazards model, where the hazard differences instead of the hazard ratios are constant over time, has been proposed (Aalen 1980; Cox and Oakes 1984; Huffer and McKeague 1991; Lin and Ying 1994). Further, regression models for time-to-event data such as cure-rate models (Farewell 1982; Kuk and Chen 1992), residual life regression models (Oakes and Dasu 1990; Jeong 2014), and non-PH models with generalized time-dependent logistic function (MacKenzie 1996, 1997; Ha and MacKenzie 2010) have been also developed.

For model checking, various residuals have been developed in the literature: generalized residual (Cox and Snell 1968), martingale residual (Barlow and Prentice 1988; Therneau et al. 1990), deviance residual (McCullagh and Nelder 1989; Therneau et al. 1990), and partial residual (Schoenfeld 1982; Grambsch and Therneau 1994).

2.6 Appendix

2.6.1 Construction of Likelihoods of Various Types

The likelihoods under various types of censoring schemes can be expressed by incorporating the following components (Klein and Moeschberger 2003):

(i) Exact survival time t: $f_\theta(t)$
(ii) Right-censored observations C_r: $S_\theta(C_r)$
(iii) Left-censored observations C_l: $1 - S_\theta(C_l)$
(iv) Left-truncated observations b_L: $f_\theta(t)/S_\theta(b_L)$
(v) Right-truncated observations b_R: $f_\theta(t)/\{1 - S_\theta(b_R)\}$
(vi) Interval-censored observations (L, R): $f_\theta(t)/\{S_\theta(L) - S_\theta(R)\}$.

For example, the likelihood for (i), (ii), (iii), and (vi) based on the n observed data can be constructed by putting together the components

$$L(\theta) = \prod_{i\in D} f_\theta(t_i) \prod_{i\in R} S_\theta(C_{ri}) \prod_{i\in L}(1 - S_\theta(C_{li})) \prod_{i\in I}(S_\theta(L_i) - S_\theta(R_i)),$$

where D is the set of death times, R is the set of right-censored observations, L is the set of left-censored observations, and I is the set of interval-censored observations, respectively. For example, for left-truncated data $f(t_i)$ is replaced by $f(t_i)/S(b_{Li})$ and $S(C_{ri})$ is by $S(C_{ri})/S(b_{Li})$ in equation above. For the LTRC data, we observe the data (t_i, δ_i, b_{Li}) having $t_i \geq b_{Li}$ and censoring indicator δ_i. Then, the corresponding likelihood L_i for the ith observation is given by

$$\begin{aligned} L_i(\theta) &= [f_\theta(t_i)/S_\theta(b_{Li})]^{\delta_i} \cdot [S_\theta(t_i)/S_\theta(b_{Li})]^{1-\delta_i} \\ &= [f_\theta(t_i)^{\delta} S_\theta(t_i)^{1-\delta_i}]/S_\theta(b_{Li}), \end{aligned}$$

where $f_\theta(t_i)^{\delta_i} S_\theta(t_i)^{1-\delta_i}$ is the likelihood under right censoring. □

2.6.2 *Derivations of Breslow's Likelihood and Cumulative Hazard Estimator*

Following a profile likelihood argument by Johansen (1983), we derive the Breslow likelihood (hence, the Cox partial likelihood) (2.8) and the Breslow estimator (2.9).

The functional form of $\lambda_0(t)$ in (2.4) is unknown. Following Breslow (1972), we consider the baseline cumulative hazard function $\Lambda_0(t)$ to be a step function with jumps λ_{0k} at the observed event times $y_{(k)}$,

$$\Lambda_0(t) = \sum_{k:y_{(k)}\leq t} \lambda_{0k}, \tag{2.15}$$

where $\lambda_{0k} = \lambda_0(y_{(k)})$. The ordinary log-likelihood (2.3) for censored data corresponding to the ith individual under the Cox model (2.4) is given by

$$\ell_i = \ell_i(\beta, \lambda_0; y_i, \delta_i, x_i) = \delta_i \log \lambda(y_i; x_i) - \Lambda(y_i; x_i), \tag{2.16}$$

where $\Lambda(y_i) = \Lambda_0(y_i)\exp(\eta_i)$ with $\eta_i = x_i^T\beta$ is the cumulative hazard function corresponding to the hazard $\lambda(y_i) = \lambda_0(y_i)\exp(\eta_i)$. From (2.15) and (2.16), the contributions from all individuals are given by

$$\begin{aligned} \ell(\beta, \lambda_0) &\equiv \sum_i \ell_i \\ &= \sum_i \delta_i\{\log \lambda_0(y_i) + \eta_i\} - \sum_i \{\Lambda_0(y_i)\exp(\eta_i)\} \\ &= \sum_k d_{(k)} \log \lambda_{0k} + \sum_i \delta_i\eta_i - \sum_k \lambda_{0k}\left\{\sum_{i\in R_{(k)}} \exp(\eta_i)\right\}, \end{aligned} \tag{2.17}$$

where $\lambda_0 = (\lambda_{01}, \ldots, \lambda_{0D})^T$ and $R_{(k)} = R(y_{(k)}) = \{i : y_i \geq y_{(k)}\}$. Here, note that

$$\sum_i \{\Lambda_0(y_i)\exp(\eta_i)\} = \sum_k \lambda_{0k}\{\sum_{i \in R_{(k)}} \exp(\eta_i)\},$$

since, from (2.15), $\Lambda_0(y_i)$ can be expressed as

$$\Lambda_0(y_i) = \sum_{k:y_{(k)} \leq y_i} \lambda_{0k} = \sum_k \lambda_{0k} I(y_{(k)} \leq y_i) = \sum_k \lambda_{0k} I(i \in R_{(k)}).$$

Following the argument of Johansen (1983), we have that, given β, the score equations

$$\partial\ell/\partial\lambda_{0k} = (d_{(k)}/\lambda_{0k}) - \sum_{i \in R_{(k)}} \exp(\eta_i) = 0 \ (k = 1, \ldots, D),$$

which leads to the NPMLE of $\Lambda_0(t)$ (i.e., Breslow estimator in (2.9)):

$$\widehat{\Lambda}_{0B}(t) = \sum_{k:y_{(k)} \leq t} \widehat{\lambda}_{0k}, \tag{2.18}$$

with

$$\widehat{\lambda}_{0k} = \widehat{\lambda}_{0k}(\beta) = \frac{d_{(k)}}{\sum_{i \in R_{(k)}} \exp(\eta_i)}.$$

Substituting $\widehat{\lambda}_0 = (\widehat{\lambda}_{01}, \ldots, \widehat{\lambda}_{0D})^T$ into $\ell(\beta, \lambda_0)$ of (2.17) yields the profile likelihood, $\ell^*(\beta)$, only depending on β:

$$\begin{aligned} \ell^*(\beta) &= \ell(\beta, \lambda_0)|_{\lambda_0 = \widehat{\lambda}_0(\beta)} \\ &= \sum_i \delta_i \eta_i - \sum_k d_{(k)} \log\{\sum_{i \in R_{(k)}} \exp(\eta_i)\}, \end{aligned}$$

deleting the constant term of $\sum_k \{d_{(k)} \log(d_{(k)}) - d_{(k)}\}$. Note that $\sum_i \delta_i \eta_i = \sum_k s_{(k)}^T \beta$ in (2.8). The log-likelihood $\ell^*(\beta)$ is the kernel of the Breslow likelihood $\ell_B(\beta)$ in (2.8) and also that of the Cox partial likelihood $\ell_c(\beta)$ in (2.6) without ties (i.e., $d_{(k)} = 1$ all k). □

2.6.3 *Proof of Theorem 2.1*

Let $\widehat{\lambda}_0(\beta)$ be a solution of score equation

$$\frac{\partial \ell}{\partial \lambda_0} = 0.$$

From the profile likelihood

$$\ell^* = \ell(\beta, \lambda_0)|_{\lambda_0=\widehat{\lambda}_0(\beta)},$$

after λ_0 being eliminated, following Ha and Lee (2003), we can derive simple matrix forms of the first and second partial derivatives, given by

$$\begin{aligned}\text{(i) } S^*(\beta) &= \partial\ell^*/\partial\beta = [\partial\ell/\partial\beta + (\partial\ell/\partial\lambda_0)(\partial\lambda_0/\partial\beta)]|_{\lambda_0=\widehat{\lambda}_0(\beta)}\\ &= \partial\ell/\partial\beta|_{\lambda_0=\widehat{\lambda}_0(\beta)} = X^T(\delta-\mu)|_{\lambda_0=\widehat{\lambda}_0(\beta)},\end{aligned}$$

$$\begin{aligned}\text{(ii) } H^*(\beta) = -\partial^2\ell^*/\partial\beta^2 &= [H_1 - H_2]|_{\lambda_0=\widehat{\lambda}_0(\beta)}\\ &= [X^T W^* X]|_{\lambda_0=\widehat{\lambda}_0(\beta)},\end{aligned}$$

where $\mu = \exp(\log\Lambda_0(y) + \eta)$ with $\eta = X\beta$, $H_1 = -\,\partial^2\ell/\partial\beta^2 = X^T W_1 X$ with

$$W_1 = \text{diag}(\mu_i) = W_3 B,$$

$W_3 = \text{diag}\{\exp(x_i^T\beta)\}$ and $B = \text{diag}(\Lambda_0(y_i))$, and

$$\begin{aligned}H_2 &= (-\,\partial^2\ell/\partial\beta\partial\lambda_0)(-\,\partial^2\ell/\partial\lambda_0^2)^{-1}(-\,\partial^2\ell/\partial\lambda_0\partial\beta)\\ &= (X^T W_3 M)C^{-1}(M^T W_3 X) = X^T W_2 X.\end{aligned}$$

Here, M is a $n \times D$ indicator matrix with the (i,k)th element $I(y_i \geq y_{(k)})$, and

$$W_2 = (W_3 M)C^{-1}(W_3 M),$$

with $C = \text{diag}\{d(k)/\lambda_{0k}^2\}$. Here, C^{-1} is immediately computed because C is a diagonal matrix. Then, we have

$$W^* = W_1 - W_2. \tag{2.19}$$

Consider one-step Newton–Raphson formula with

$$\hat{\beta} = \beta + H^*(\beta)^{-1}S^*(\beta) = \beta + [(X^T W^* X)^{-1}X^T(\delta-\mu)]|_{\lambda_0=\widehat{\lambda}_0(\beta)}.$$

By some matrix manipulation, we have

$$\begin{aligned}(X^T W^* X)\hat{\beta}|_{\lambda_0=\widehat{\lambda}_0(\beta)} &= [(X^T W^* X)\beta + X^T(\delta-\mu)]|_{\lambda_0=\widehat{\lambda}_0(\beta)}\\ &= X^T(W^* X\beta + (\delta-\mu))|_{\lambda_0=\widehat{\lambda}_0(\beta)},\end{aligned}$$

which leads to (2.10). Note that the terms λ_{0k} in W^* are evaluated at their estimates $\hat{\lambda}_{0k} = \hat{\lambda}_{0k}(\beta) = d_{(k)}/M_k^T\psi$, where M_k is a kth component vector of $M = (M_1, \ldots, M_D)$ and ψ is a vector of $\exp(\eta_i)$'s. This completes the proof. □

2.6.4 Fitting Cox PH Model via a Poisson GLM

We show the semiparametric Cox PH model can be fitted via a Poisson GLM. Let $y_{i,k}$ be 1 if the ith subject fails at $y_{(k)}$ and 0 otherwise. Following Whitehead (1980), let $y_{i,k}$ be an independent random variable that follows a Poisson (Po) distribution:

$$y_{i,k} \sim \text{Po}(\mu_{i,k}),\ i \in R_{(k)}, \tag{2.20}$$

where $\mu_{i,k} = \exp(w_k + x_i^T\beta) = \exp(x_{i,k}^T\gamma)$ with $w_k = \log\lambda_{0k}$. Here, $x_{i,k} = (e_k^T, x_i^T)^T$, e_k is a vector of 0's and 1's such that $e_k^T w = w_k$, and $\gamma = (w^T, \beta^T)^T$. Note that $e_k = (0, \ldots, 1, \ldots, 0)^T$ and $w = (w_1, \ldots, w_k, \ldots, w_D)^T$. Let y denote a vector of $y_{i,k}$'s. Then, this auxiliary model provides the Poisson log-likelihood for $\gamma = (w^T, \beta^T)^T$ of the form

$$\ell_{Po}(\gamma; y) = \sum_k \sum_{i \in R_{(k)}} \{y_{i,k}\log(\mu_{i,k}) - \mu_{i,k}\}.$$

Since $\mu_{i,k} = \lambda_{0k}\exp(\eta_i)$ and

$$\sum_k \sum_{i \in R_{(k)}} y_{i,k}\log(\mu_{i,k}) = \sum_k d_{(k)}\log\lambda_{0k} + \sum_i \delta_i\eta_i,$$

ℓ_{Po} is equivalent to ℓ in (2.17). In fact, the procedures based on ℓ and ℓ^* (or ℓ_{Po} and ℓ^*_{Po}) are equivalent, but ℓ^* would work better for a large sample because the number of nuisance parameters λ_{0k}'s increases with sample size. Here, $\ell_{Po} = \ell^*_{Po}(\beta) = \ell_{Po}(\beta, w)|_{w=\hat{w}(\beta)}$, where $\hat{w}(\beta)$ is the solution of $\partial\ell_{Po}/\partial w = 0$ for given β. □

Chapter 3
H-Likelihood Approach to Random-Effect Models

In this chapter, we introduce an h-likelihood approach to the general class of statistical models with random effects. Consider a linear mixed model (LMM), for $i = 1, \ldots, q$ and $j = 1, \ldots, n_i$

$$y_{ij} = x_{ij}^T \beta + v_i + e_{ij}, \tag{3.1}$$

where y_{ij} is an observed random variable (response), $x_{ij} = (x_{ij1}, \ldots, x_{ijp})^T$ is a vector of covariates, β is a vector of fixed effects, $v_i \sim N(0, \alpha)$ is an i.i.d. random variable for the random effects, $e_{ij} \sim N(0, \phi)$ is an i.i.d random error or measurement error, and v_i and e_{ij} are independent. Parameters ϕ and α are the variance components. In this model, there are two types of unknowns; the fixed unknowns $\theta = (\beta, \phi, \alpha)^T$ and the random unknowns $v = (v_1, \ldots, v_q)^T$.

The statistical model (3.1) consists of three types of objects, data y, parameters (fixed unknowns) θ and unobservables (random unknowns) v. Then, the purpose of statistical inference would be to draw informative scientific explanations about both unknowns θ and v by using the statistical model, based on the observed data $y = (y_{11}, \ldots, y_{qn_q})^T$.

3.1 Three Paradigms of Statistical Inference

In this section, we review various approaches to statistical inferences using the likelihood or probability, which are two important but fundamentally different concepts. We elaborate on how these concepts are used for statistical inference about the fixed and random unknowns.

First let us consider statistical models and associated statistical inferences, based on the two types of objects, the data (random but observed) y and the unobservables (random but unobserved) v, and two related processes between them.

I.D. Ha et al., *Statistical Modelling of Survival Data with Random Effects*,
Statistics for Biology and Health, https://doi.org/10.1007/978-981-10-6557-6_3

- **Statistical model for data generation**: (i) Generate a set of the random quantities v from a probability density function $f(v)$ and then with v fixed, (ii) generate a set of data y from a probability function $f(y|v)$. The hierarchical statistical model is given by the product of the two probability functions

$$f(v)f(y|v).$$

This scenario shows how data y are generated.
- **Statistical inference**: Given data y, we can make inference about v by using a conditional density (predictive probability),

$$f(v|y).$$

The connection between these two processes is given by

$$f(y)f(v|y) = f(v, y) = f(v)f(y|v).$$

On the left-hand side in the above equation, for statistical inference y is fixed while v varies, whereas on the right-hand side both v and y vary. This inferential procedure effectively shows how to update the distribution of v once data y are observed, so that the information in data y can be utilized for the unknown (unobserved) random variables.

Here the predictive probability $f(v|y)$ is updated by applying the Bayes rule, i.e.,

$$f(v|y) = \frac{f(v, y)}{f(y)},$$

which could be used to infer the unobserved random variables v. This probability function is proper in that

$$\int f(v|y)dv = 1.$$

Even if updating the predictive probability for the unobserved random variable based on the observed data through the Bayes rule is feasible, in many situations, the unknowns of interest to be statistically inferred are fixed parameters, not unobserved random variables.

3.1.1 Bayesian Approach

Suppose that a set of data y is generated from a distribution with the probability density function of $f_\theta(y)$ where θ is the fixed parameters. To use the above probability update for statistical inferences about θ, the Bayesian approach uses

$$\pi(\theta)f(y|\theta) = \pi(\theta)f_\theta(y) = f(y)f(\theta|y),$$

where $\pi(\theta)$ is the prior probability and $f(\theta|y)$ is the posterior probability. In this book, we denote $f_\theta(\cdot)$ as the probability density function with fixed parameters θ; the arguments within the parentheses can be either conditional or unconditional. Thus, $f_\theta(y|u)$ and $f_\theta(u|y)$ would have different functional forms even though we use the same $f_\theta()$ to mean a probability density function with parameters θ.

A disagreement arises when a prior density function $\pi(\theta)$ is assumed for the fixed unknowns θ being treated as random, as it should follow a degenerate distribution assigning probability one to a given value and probability zero to all other values. Another conceptual controversy is whether the prior $\pi(\theta)$ can be updated as $f(\theta|y)$ based on the observed data y using the Bayes rule. Specifically, under the LMM the probability density function of the random effects, $f(v)$, allows for correlations among observed data as will be clear in the later chapters, while in the Bayesian model the prior $\pi(\theta)$ does not belong to the assumed statistical model $f(y|\theta) = f_\theta(y)$. In the machine learning field, $-\log \pi(\theta)$ is the penalty in the penalized likelihood approach and the mode of $f(\theta|y)$ is used to estimate θ, as the maximum a posteriori (MAP) estimator. Thus, the main idea of the Bayesian and penalized likelihood approach is to utilize the posterior probability $f(\theta|y)$ to infer θ even though θ is the fixed unknown parameters. We shall discuss further about the penalized likelihood approach for variable selection in Chap. 7.

3.1.2 *Fisher Likelihood Approach*

A solution to inference on the fixed unknowns θ without assuming a prior probability $\pi(\theta)$ was proposed by Fisher (1922). He developed theory based on the likelihood function, the probability of observing the data at hand expressed as a function of the parameters. Again consider a statistical model including data y and $\boldsymbol{\theta}$, fixed unknowns, and two related processes between them:

- **Statistical model for data generation**: Generate a set of data y from a distribution with a probability density function

$$f_\theta(y),$$

  where θ is the fixed unknown parameters.
- **Statistical inference**: Given data y, make inference about $\boldsymbol{\theta}$ in the above statistical model by using the likelihood

$$L(\boldsymbol{\theta}; y).$$

The connection between these two processes in this case is:

$$L(\boldsymbol{\theta}; y) \equiv f_\theta(y),$$

where L and f are algebraically identical, but on the left-hand side y is fixed while $\boldsymbol{\theta}$ varies and on the right-hand side $\boldsymbol{\theta}$ is fixed while y varies. The likelihood $L(\boldsymbol{\theta}; y)$ is not the probability density function of θ since

$$\int L(\theta; y) d\theta \neq 1.$$

3.1.2.1 Exchange Paradox and Likelihood

Probability and likelihood are very different concepts, but the difference is not well understood. Consider the exchange paradox and its likelihood solution (Pawitan and Lee 2017; Lee et al. 2017b, Chap. 4). Unknown fixed θ dollars are put in one envelop and 2θ dollars in another. You are to pick one envelop at random, open it and decide if you would exchange it with another envelop. So you pick one and see 100 dollars. Then you reason that the amount in the other envelop is 50 or 200 with 50–50 chance. If you exchange it, you would expect to get $(50 + 200)/2 = 125$, which is bigger than your current amount of 100. Since the above reasoning holds for any value of money you see in your envelop, you actually do not need to open the envelop in the first place and you would still want to exchange.

This exchange paradox has been analyzed from the Bayesian perspective (Christensen and Utts 1992; Blachman et al. 1996), suggesting that the above intuitive reasoning should be justified by using a uniform prior on $\log \theta$. This implies that from the Bayesian perspective the subjective intuition of an equal probability for the possible amount in either envelop is not an acceptable state of mind. What is this 50–50 chance then if it is not a probability?

Let the unknown fixed amounts in two envelopes be θ and 2θ, and an amount of $Y = y$ (data) is observed, randomly chosen between those two. Noting that on observing $Y = y$, θ can be either y or $y/2$, the likelihood of a fixed unknown θ is then

$$\begin{aligned} L(\theta = y; y) &= P(Y = y|\theta = y) \\ &= P(Y = \theta|\theta = y) = 1/2, \\ L(\theta = y/2; y) &= P(Y = y|\theta = y/2) \\ &= P(Y = 2\theta|\theta = y/2) = 1/2. \end{aligned}$$

This means that the observed data y cannot tell us any preference between the two possible values. Since these are the likelihood values, not probabilities, one cannot use them to take an expectation. In the paradox, the exchange should occur only when the expected value from the exchange is greater than what is observed y. Since the expected value cannot be taken by using the likelihood values, there is no rational basis to exchange. Thus, the paradox is avoided. In contrast to the Bayesian resolution, the equal preference is an acceptable state of mind in this likelihood solution.

3.1.2.2 Likelihood Principle and Likelihood Ratios

Suppose x is one-to-one transformation of the observed data y. If y is continuous, the likelihood based on x is

$$L(\boldsymbol{\theta}; x) = L(\boldsymbol{\theta}; y) \left| \frac{\partial y}{\partial x} \right| .$$

Obviously, x and y should carry the same information about $\boldsymbol{\theta}$, so to compare $\boldsymbol{\theta}_1$ and $\boldsymbol{\theta}_2$ only the likelihood ratio is relevant since it is invariant with respect to the transformation:

$$\frac{L(\boldsymbol{\theta}_2; y)}{L(\boldsymbol{\theta}_1; y)} = \frac{L(\boldsymbol{\theta}_2; x)}{L(\boldsymbol{\theta}_1; x)}.$$

The fact that the likelihood ratio should be the same under the transformation of the data seems a reasonable requirement because the inference results should remain the same with respect to the transformation.

Birnbaum (1962) proves the likelihood principle that the likelihood function contains all the information about the value of the fixed parameter under the true statistical model $f_{\boldsymbol{\theta}}(y)$. This means that to estimate the true value of the fixed parameter, we only need the likelihood function. Fisher advocates the use of the MLEs to infer the parameters. As described in Sect. 2.2, consistency of the MLE of $\boldsymbol{\theta}$ and its asymptotic optimality are well established. The likelihood principle implies that model checking is always important, not just in the likelihood inference because using the likelihood is beneficial only if the assumed model is correct.

One important property of the MLE is its invariance with respect to transformation of the original parameter. By using the likelihood function from the same statistical model, it would make sense to have the same inference results for the original parameter as well as for the transformed one. A trivial example would be the MLE of the variance and standard deviation of the Gaussian distribution. For example, it should not make a difference whether we infer the dispersion parameter in terms of variance σ^2 or standard deviation σ because

$$\frac{L(\sigma_2^2; y)}{L(\sigma_1^2; y)} = \frac{f_{\sigma_2^2}(y)}{f_{\sigma_1^2}(y)} = \frac{L(\sigma_2; y)}{L(\sigma_1; y)} = \frac{f_{\sigma_2}(y)}{f_{\sigma_1}(y)}.$$

However, this does not hold in the Bayesian framework:

$$\frac{f(\sigma_2|y)}{f(\sigma_1|y)} = \frac{\pi(\sigma_2)f_{\sigma_2}(y)}{\pi(\sigma_1)f_{\sigma_1}(y)} = \frac{\sigma_2\pi(\sigma_2^2)f_{\sigma_2^2}(y)}{\sigma_1\pi(\sigma_1^2)f_{\sigma_1^2}(y)} \neq \frac{\pi(\sigma_2^2)f_{\sigma_2^2}(y)}{\pi(\sigma_1^2)f_{\sigma_1^2}(y)} = \frac{f(\sigma_2^2|y)}{f(\sigma_1^2|y)},$$

because of the Jacobian terms in $\pi(\sigma_2) = 2\pi(\sigma_2^2)\sigma_2$.

The likelihood would be a natural choice to infer the fixed unknowns. In computing the likelihood of a transformed parameter, the Jacobian term does not appear. Hence, fundamentally, the likelihood cannot be treated as a probability density function over the parameter space and it does not obey the probability laws, e.g., it does not have to integrate to one.

Since the Bayesian framework gives

$$f(\theta|y) \propto \pi(\theta)f_\theta(y) = \pi(\theta)L(\theta; y),$$

we may view the likelihood as a Bayesian posterior under a uniform prior $\pi(\theta) \equiv 1$. The likelihood approach allows statistical inference about θ without assuming a prior $\pi(\theta)$.

3.1.3 *Extended Likelihood Approach*

Many statistical models involve both parameters and unobservables, and require inferences on both types of unknowns. Consider the LMM presented in (3.1) where the random components are assumed to follow zero-mean normal distributions: (i) $v_i \sim N(0, \alpha)$, and (ii) $e_{ij} \sim N(0, \phi)$. Classical analysis concentrates on estimation of the parameters $\theta = (\beta, \alpha, \phi)$. It is straightforward to write down the likelihood from the multivariate normal distribution $MVN(X\beta, \alpha ZZ^T + \phi I)$ and to obtain their MLEs. Here, X and Z are the model matrices of β and v, respectively. However, in many recent applications the main interest is often in estimation of the conditional (or subject specific) mean given the random effects v_i

$$E(y_{ij}|v_i) = x_{ij}^T\beta + z_{ij}^T v_i.$$

One might be interested in using the Fisher likelihood in making inference about the fixed parameters without assuming the priors, whereas using the probability updates for inference about the unobservables using the Bayes rule. Thus, it is desirable to develop an extended likelihood approach, which gives the likelihood inference for the fixed parameters, while allowing the use of probability updates for inference about the unobserved random variables.

An extended likelihood framework based on three aforementioned objects can be presented as follows. Let θ be the fixed unknown parameters, v be the unobserved random variables and y be the observed data.

- **Statistical model for data generation**: (i) Generate a set of the random quantities v from a probability density function $f_\theta(v)$ and then (ii) with v fixed, generate a set of data y from a probability density function $f_\theta(y|v)$. The combined statistical model is given by the product of the two probability density functions

$$f_\theta(v)f_\theta(y|v). \tag{3.2}$$

- **Statistical inference**: Given data y, we can (i) make inference about θ by using the marginal (Fisher) likelihood

$$L(\theta; y) \equiv f_\theta(y),$$

and (ii) given θ, make inference about v by using a conditional probability (likelihood) of the form

$$L(\theta, v; v|y) \equiv f_\theta(v|y).$$

The extended likelihood for the unknowns (v, θ) is given by

$$L(\theta, v; v, y) = L(\theta; y)f_\theta(v|y) \equiv f_\theta(y)f_\theta(v|y).$$

The connection between these two processes is given by

$$L(\theta; y)f_\theta(v|y) \equiv L(\theta, v; v, y) \equiv f_\theta(v, y) = f_\theta(v)f_\theta(y|v). \tag{3.3}$$

On the left-hand side y is fixed whereas $(v, \boldsymbol{\theta})$ vary, while on the right-hand side θ is fixed while (v, y) vary. In the extended likelihood framework the object v appear in data generation as random sets, but in statistical inference as the unknowns.

3.1.3.1 Wallet Paradox and Extended Likelihood

Consider the wallet game appeared in Gardner (1982) as follows: Two people, equally rich or equally poor, meet to compare the contents of their wallets. Each is ignorant of the contents of the two wallets. Here is the game: whoever has less money receives the contents of the wallet of the other. One of them can reason: "I have a fixed amount u_1 in my wallet; either I lose that or win an amount $u_2 > u_1$ with 50–50 chance. Therefore the game is favorable to me." The other person can reason in the exactly same way. In fact, by symmetry, the game is fair. Where is the mistake in this reasoning?

The wallet game may be modeled as follows: Let U_1 and U_2 be the random amounts of money in two wallets. Let us assume that they are independent and identically distributed (iid) samples from a continuous positive-valued distribution with finite $E(U_1) = E(U_2)$. Now consider a specific realization of $U_1 = u_1$ and $U_2 = u_2$ yet unobserved to both players. Now let $V = I(U_1 < U_2)$, so V is a Bernoulli random variable with $P(V = 1) = 0.5$. The specific realization of $v = 0$ or 1 is unobserved, so we are in a state of uncertainty. However, this uncertainty cannot be a probability, so it does not allow an expectation step that would lead to the paradox: the expected gain-minus-loss is then $P(v = 1)(u_2 - u_1) = (u_2 - u_1)/2 > 0$. In fact, in this problem it is given by the extended likelihood

$$L(v = 1) \equiv P(V = 1) = 1/2$$

and

$$L(v = 0) \equiv P(V = 0) = 1/2, \tag{3.4}$$

so the specific realizations v of V are equally likely. For laymen, such expressions as "50–50 chance", "equally probable" or "equally likely" have all similar meanings. But technically, as in the exchange paradox, we cannot take expectation using the likelihood values, so we have no rational basis to believe that game is favorable to us. The wallet paradox highlights that once we are dealing with realized yet unobserved values, then the step from the probability of a random event to the extended likelihood of realized value becomes necessary.

For the unknowns (θ, v), where θ is fixed and v is an unobserved realization of a random variable, the extended likelihood given data y is

$$L(\theta, v; v, y) = f_\theta(y|v) f_\theta(v).$$

In the wallet game, there is neither data y nor unknown parameter θ, or equivalently for y we can simply generate an independent toss of a fair coin so that $f_\theta(y|v)$ is constant with respect to u, while $f(v)$ is given by (3.4).

Realized yet unobserved random effects are commonly assumed in major areas of statistical applications using the random-effect models. In clinical trials, patient and hospital frailties (unobserved random variables in the hazard or mean) are often of interest where inferences require the extended likelihoods.

3.1.3.2 Extended Likelihood and the H-Likelihood

Bjørnstad (1996) introduced the extended likelihood principle in that all information regarding fixed unknowns θ and random unknowns v in data y resides in the extended likelihood, provided the assumed model is true. This means that for inference about the true value of the fixed parameter and/or unobservable, we only need the extended likelihood function.

In the absence of parameter θ, the extended likelihood (3.3) gives the probability update of the unobservables v by using the Bayes theorem

$$f(v) \Rightarrow f(v|y),$$

and in the absence of the unobservables, it becomes the Fisher likelihood

$$f_\theta(y) \equiv L(\theta; y).$$

This shows that statistical theories have been well developed for these two extreme cases (in the absence of either object). Thus, the extended likelihood principle justifies not only the use of Fisher likelihood for θ but also the probability update for v by using the Bayes theorem.

Recent interest has been in statistical inference on the statistical models such as random-effect models. This book aims to establish the extended likelihood inference about the models for survival data with all three objects, (y, θ, v) present. There have been many attempts previously to use the extended likelihood for statistical inference for (θ, v), but have faced serious difficulties. To overcome these difficulties, the h-likelihood has been introduced by Lee and Nelder (1996).

Example 3.1 Bayarri et al. (1988) showed difficulties in using the extended likelihood. Suppose that there is a single fixed parameter θ, a single unobservable random quantity u and a single observable quantity y. The unobservable random variable u has an exponential probability density

$$f_\theta(u) = \theta \exp(-\theta u), \text{ for } u > 0, \theta > 0,$$

and given u, the observable outcome y also has an exponential density

$$f_\theta(y|u) = f(y|u) = u \exp(-uy), \; y > 0, u > 0,$$

which is free of θ. Then we can derive the following four likelihoods.
(i) The marginal likelihood:

$$L(\theta; y) = f_\theta(y) = \int_0^\infty f(y|u) f_\theta(u) du = \theta/(\theta + y)^2,$$

which gives the MLE $\hat{\theta} = y$, but this classical Fisher likelihood is uninformative about the unknown value of u.
(ii) The conditional likelihood:

$$L(\theta, u; y|u) = f(y|u) = u \exp(-uy),$$

which is uninformative about θ and loses the relationship between u and θ reflected in $f_\theta(u)$. This likelihood carries information only about u in data y. This gives, if maximized, $\hat{u} = 1/y$.
(iii) The extended likelihood:

$$L(\theta, u; y, u) = f(y|u) f_\theta(u) = u\theta \exp\{-u(\theta + y)\},$$

which yields, if jointly maximized with respect to θ and u, the useless estimators $\hat{\theta} = \infty$ and $\hat{u} = 0$.
(iv) Another conditional likelihood:

$$L(\theta, u; u|y) = f_\theta(u|y) = \{f(y|u) f_\theta(u)\}/f_\theta(y) = u(\theta + y)^2 \exp\{-u(\theta + y)\},$$

carries the combined information concerning u from $f_\theta(u)$ and $f(y|u)$. If θ is known, this could be useful for inference about u. However, if θ is unknown, joint maximization yields again the useless estimators $\hat{\theta} = \infty$ and $\hat{u} = 0$.

This example shows difficulties in using the extended likelihood for statistical inference. It also shows that various likelihoods can be formed from the extended likelihood and different likelihoods carry different information. Importantly, though, it clearly demonstrates that a naive joint inference on (θ, u) from an extended likelihood—potentially violating the classical likelihood principle—can be treacherous. In the next section, we outline how to overcome this difficulty by using the h-likelihood, which gives sensible inferences for both θ and u.

3.2 H-Likelihood

The extended likelihood is not in general invariant with respect to the transformation of the unobservables, because a change in this transformation involves a Jacobian term. Thus, the maximum extended likelihood estimator for the unobservable is not invariant with respect to transformation of the unobservables, which can lead to useless estimators. To avoid this difficulty, Lee and Nelder (1996) have introduced the hierarchical (h-)likelihood, an extended likelihood defined on a special scale of v for inference on both fixed and random unknowns.

We first derive a condition that allows a joint inference about (θ, v) from the extended likelihood $L(\theta, v; y, v)$. Let θ_1 and θ_2 be an arbitrary pair of values of the fixed parameter θ. The evidence about these two parameter values is in the likelihood ratio

$$\frac{L(\theta_2; y)}{L(\theta_1; y)}.$$

Suppose that there exists a scale v, such that the likelihood ratio is preserved as follows:

$$\frac{L(\theta_2; y)}{L(\theta_1; y)} = \frac{L(\boldsymbol{\theta}_2, \hat{v}_{\theta_2}; y, v)}{L(\theta_1, \hat{v}_{\theta_1}; y, v)},$$

where $\hat{v}_{\theta_i}$ is the MLE of v at $\theta = \theta_i$ $(i = 1, 2)$. Lee et al. (2017b) called this extended likelihood $L(\theta, v; y, v)$ the h-likelihood if the scale v of the random effects is canonical (i.e., a v-scale satisfying the equation of the likelihood ratio above). The (log)-h-likelihood is defined by the logarithm of the joint density of y and $v = v(u)$ on a particular scale of v among the extended likelihoods (3.2),

$$h = \ell_1(\theta; y|v) + \ell_2(\theta; v), \tag{3.5}$$

where $\ell_1(\theta; y|v) = \log f_\theta(y|v)$ and $\ell_2(\theta; v) = \log f_\theta(v)$.

Below we illustrate how the canonical scale in constructing the h-likelihood avoids problems in the extended likelihood inference and then present the resulting useful properties.

Example 3.2 Consider Example 3.1 again. We showed that the extended likelihood,

$$L(\theta, u; y, u) = f(y|u)f_\theta(u) = u\theta \exp\{-u(\theta + y)\},$$

provides useless estimators. Suppose that we take the scale $v = \log u$ to form the h-likelihood. Then we have

$$f_\theta(v) = f_\theta(u)|du/dv| = \exp(v)\theta \exp(-e^v\theta)$$

and the extended likelihood is given by

$$L(\theta, v; y, v) = f(y|u)f_\theta(v) = e^{2v}\theta \exp\{-e^v(\theta + y)\},$$

or

$$\log L(\theta, v; y, v) = 2v + \log\theta - e^v(\theta + y),$$

to give

$$\hat{u}_\theta = \exp(\hat{v}_\theta) = E_\theta(u|y) = 2/(\theta + y),$$

where $E_\theta(u|y) = \int u f_\theta(u|y)du$. Then, up to a constant term, the profile likelihood is equal to the marginal log-likelihood, $m = \log L(\theta; y)$:

$$\log L(\theta, \hat{v}_\theta; y, v) = 2\log\{2/(\theta + y)\} + \log\theta - 2 = \log L(\theta; y) + \text{constant},$$

so $v = \log u$ is the canonical scale. Thus, this scale yields the (log-)h-likelihood, defined by

$$h = h(\theta, v) = \log L(\theta, v; y, v) = 2v + \log\theta - e^v(\theta + y).$$

This h-likelihood has many interesting properties as in an ordinary likelihood (Lee et al. 2017b) as shown below.
(i) The joint maximization of h with respect to θ and u gives the MLE of θ. That is, from the joint estimating equations

$$\partial h/\partial\theta = 1/\theta - u = 0 \text{ and } \partial h/\partial u = 2/u - (\theta + y) = 0, \tag{3.6}$$

we obtain $\hat{\theta} = y$, exactly the MLE from $L(\theta; y)$.
(ii) From the h-likelihood we derive the observed information matrix

$$I(\hat{\theta}, \hat{u}) = -\partial^2 h/\partial(\theta, u)^2|_{\theta=\hat{\theta}, u=\hat{u}} = \begin{pmatrix} I_{11} = 1/y^2 & I_{12} = 1 \\ I_{21} = 1 & I_{22} = (y + \hat{\theta})^2/2 = 2y^2 \end{pmatrix}.$$

Denote the inverse of $I(\hat{\theta}, \hat{u})$ to be

$$I^{-1}(\hat{\theta}, \hat{u}) = \begin{pmatrix} I^{11} & I^{12} \\ I^{21} & I^{22} \end{pmatrix}.$$

Then the variance estimator of $\hat{\theta}$ is obtained from I^{11}, given by

$$\widehat{\text{var}}(\hat{\theta}) = I^{11} = 2y^2,$$

exactly the same as the one from the marginal likelihood: $-(\partial^2 m/\partial\theta^2|_{\theta=\hat{\theta}})^{-1} = 2y^2$, where $m = \log L(\theta; y)$ is the marginal log-likelihood.
(iii) From the joint estimating equations $\partial h/\partial(\theta, u) = 0$ in (3.6), we have the random-effect estimator

$$\hat{u} = 2/(y + \hat{\theta}) = 1/y,$$

which also becomes $E_\theta(u|y)|_{\theta=\hat{\theta}}$. Since I^{22} yields an estimator of $\text{var}(\hat{u} - u)$ (Lee and Nelder 1996), we also have

$$\widehat{\text{var}}(\hat{u} - u) = I^{22} = 1/y^2,$$

which is larger than the plug-in estimator

$$\widehat{\text{var}}(u|y) = 2/(y + \theta)^2|_{\theta=\hat{\theta}} = 1/(2y^2) = 1/I_{22}$$

obtained from the variance formula when θ is known. This increase reflects the extra uncertainty caused by estimating θ. □

Example 3.3 Suppose that $Y = (Y_{obs}, Y_{cen})$ consists of n realizations from an exponential distribution with mean θ, where Y_{obs} consists of k observed values and Y_{cen} represents $n - k$ censored observations (true failure times unobserved). Suppose that the incomplete data are created by Type I censoring at some known censoring point c (i.e., $Y_{cen} > c$), so that only values less than or equal to c are recorded. Hence, let $t_i = Y_i - c > 0$ for $i > k$, then

$$\begin{aligned} L(\theta; y) &= \int f_\theta(y_{obs}, Y_{cen}) dY_{cen} \\ &= \Pi_{i=1}^k \theta^{-1} \exp\{-y_i/\theta\} \Pi_{i=k+1}^n \int_c^\infty \theta^{-1} \exp\{-Y_i/\theta\} dY_i \\ &= \theta^{-k} \exp\left\{-\sum_{i=1}^k y_i/\theta\right\} \exp\{-\{(n-k)c\}/\theta\}, \end{aligned}$$

which is equal to the likelihood function of the exponential distribution in Example 2.5 under Type I censoring. This shows that Type I censoring does not need Assumptions 1–2 in Sect. 2.1.2 to allow the marginal likelihood above. The resulting MLE is

$$\hat{\theta} = \bar{y} + \{(n - k)c\}/k,$$

where $\bar{y} = \sum_{i=1}^k y_i/k$ is the mean of observed data. Note that if we form the marginal likelihood based on only observed data the MLE becomes $\hat{\theta} = \bar{y}$, which will be severely biased.

Suppose that we use the extended likelihood on the $v_i = t_i = Y_i - c > 0$ scale:

$$\log L(\theta, v; y, v) = -n \log \theta - k\bar{y}/\theta - (n-k)c/\theta - \sum_{i=k+1}^{n} v_i/\theta,$$

which has the maximum at $\hat{v}_i = 0$ (giving $\hat{Y}_i = c$) for $i > k$, with a wrong MLE

$$\hat{\theta} = \{k\bar{y} + (n-k)c\}/n.$$

In this model, the scale $v_i = \log t_i$ is canonical to form the h-likelihood as follows,

$$h = -n \log \theta - k\bar{y}/\theta - (n-k)c/\theta - \sum_{i=k+1}^{n} (e^{v_i}/\theta - v_i).$$

For $i = k+1, \ldots, n$ we have $\partial h/\partial v_i = -e^{v_i}/\theta + 1$ to give $\hat{t}_i = e^{\hat{v}_i} = \theta > 0$ (giving $\hat{Y}_i = \theta + c$) with a joint maximum

$$\widehat{\theta} = \bar{y} + \{(n-k)c\}/k$$

giving the correct MLE. Note that from the marginal likelihood, $m = \log L(\theta; y)$, the variance estimator for $\widehat{\theta}$ is $-(\partial^2 m/\partial\theta^2|_{\theta=\hat{\theta}})^{-1} = \hat{\theta}^2/k$.

From the h-likelihood, we have

$$I(\hat{\theta}, \hat{v}) = -\partial^2 h/\partial(\theta, v)^2|_{\theta=\hat{\theta}, v=\hat{v}} = \begin{pmatrix} I_{11} = n/\hat{\theta}^2 & I_{12} = -(1/\hat{\theta})1_{n-k}^T \\ I_{21} = -(1/\hat{\theta})1_{n-k} & I_{22} = I_{n-k} \end{pmatrix},$$

where 1_{n-k} is a $(n-k) \times 1$ vector of ones and I_{n-k} is a $(n-k) \times (n-k)$ identity matrix. Thus, from the h-likelihood, the variance estimator for $\widehat{\theta}$ is

$$\widehat{\text{var}}(\hat{\theta}) = I^{11} = (I_{11} - I_{12}I_{22}^{-1}I_{21})^{-1} = \hat{\theta}^2/k,$$

which is exactly the same as that from the marginal likelihood. □

Historically, there have been many attempts to establish an extended likelihood inference in vain, because the joint maximization of an extended likelihood in an arbitrary scale yields nonsensible estimators. These examples show that for an extended likelihood inference, it is important to define the h-likelihood on a particular scale $v(u)$ of the random effects u. However, the canonical scale does not exist in general. In the next section, we show how to make inference about θ and v using the h-likelihood in general when there is no canonical scale.

3.3 Hierarchical Generalized Linear Models

Lee and Nelder (1996) introduced the hierarchical GLMs (HGLMs), which are the GLMs where the linear predictor contains both fixed and random effects as follows:
(i) Conditional on random effects u, the responses y follow a GLM family of distributions, satisfying

$$E(y|u) = \mu \text{ and } \mathrm{var}(y|u) = \phi V(\mu),$$

with a linear predictor of the form

$$\eta = g(\mu) = X\beta + Zv,$$

where $g(\cdot)$ is the GLM link function, X and Z are the model matrices for fixed effects β and random effects v, and $v = v(u)$ with some strictly monotonic function $v(\cdot)$. Here, $V(\cdot)$ is the variance function.
(ii) The random effects u follow some distribution with a parameter α.
The distribution of u follows any conjugate distribution of the GLM family of distributions. Here, $\theta = (\beta, \phi, \alpha)^T$ are the fixed unknown parameters with dispersion parameters (ϕ, α).
Let us consider two simple examples of the HGLMs.

Example 3.4 (*Normal-Normal HGLM*) The normal LMM in (3.1) is an HGLM because
(i) $y|u$ follows a normal distribution with

$$V(\mu) = 1 \quad \text{to give} \quad \mathrm{var}(y|u) = \phi,$$

and the identity link provides the linear predictor

$$\eta = \mu = X\beta + Zv,$$

where $v = u$.
(ii) $u \sim N(0, \alpha)$.
This normal LMM can be written

$$y = X\beta + Zv + e,$$

where $e \sim N(0, \phi I)$, which covers model (3.1). We call this model the normal-normal HGLM, where the first adjective refers to the distribution of the $y|u$ component and the second to the u component. Besides this model, the error component e does not appear in the model as below.

Example 3.5 (*Poisson HGLM*) Suppose that $y|u$ follows a Poisson distribution with mean

$$\mu = E(y|u) = \exp(X\beta)u.$$

With the log link, we have the linear predictor

$$\eta = \log \mu = X\beta + v,$$

where $v = \log u$. Here, $V(\mu) = \mu$ and $\phi = 1$. If the distribution of u is gamma, v has a log-gamma distribution and we call the model the Poisson-gamma HGLM. The generalized linear mixed model (GLMM) assumes a normal distribution for v (the conjugate distribution of normal GLM family), so the distribution of u is log-normal. The corresponding Poisson GLMM could be called the Poisson-log-normal HGLM under the $v = \log u$ transformation. Both Poisson-gamma model and Poisson GLMM belong to the class of HGLMs.

In the HGLMs, the random effects v are combined additively with the fixed effects in the linear predictor η. Such a scale is called a weak canonical scale and it can always be defined as long as we can define the linear predictor for the HGLMs (Lee et al. 2017b). From (3.5) the h-likelihood for the HGLMs with $v = v(u)$ is of the form

$$h = \ell_1(\beta, \phi; y|v) + \ell_2(\alpha; v), \tag{3.7}$$

where $\ell_1(\beta, \phi; y|v) = \log f_{\beta,\phi}(y|v)$ and $\ell_2(\alpha; v) = \log f_\alpha(v)$. We now show how to construct the h-likelihood for the two Poisson HGLMs (i.e., Poisson-log-normal and Poisson-gamma models). Since, for those two models, the first term of the h-likelihood (3.7) is identically given by

$$\ell_1 = \ell_1(\beta; y|v) = y \log \mu - \mu - \log \Gamma(y + 1)$$

with $\mu = \exp(X\beta)u$, we present only the second term ℓ_2 in the following examples.

Example 3.6 (*Normal random effect*) If $v_i \sim N(0, \alpha)$, its density function becomes $f_\alpha(v_i) = (2\pi\alpha)^{-1/2} \exp\{-v_i^2/(2\alpha)\}$. Thus, the log-likelihood for v_i is given by

$$\ell_{2i} = \ell_{2i}(\alpha; v_i) = \log f_\alpha(v_i) = -\log(2\pi\alpha)/2 - v_i^2/(2\alpha).$$

Example 3.7 (*Gamma random effect*) If u_i follows a gamma distribution with mean 1 and variance α, the density function of u_i is given by

$$f_\alpha(u_i) = \{\Gamma(1/\alpha)\alpha^{1/\alpha}\}^{-1} u_i^{1/\alpha - 1} \exp(-u_i/\alpha).$$

Thus, the density of $v_i = \log u_i$ becomes $f(v_i) = f(u_i)|du_i/dv_i|$, where $du_i/dv_i = \exp(v_i)$. Accordingly, the log-likelihood for v_i is given by

$$\ell_{2i} = \ell_{2i}(\alpha; v_i) = \log f_\alpha(v_i) = (v_i - u_i)\alpha^{-1} + c(\alpha),$$

where $c(\alpha) = -\log \Gamma(1/\alpha) - \alpha^{-1} \log \alpha$.

In the normal LMMs and Poisson-gamma model with single random effects, v is canonical, so that the joint maximization of h provides the MLEs for β (Appendix 3.5.1). However, in general, v is weak canonical.

3.3.1 Inferences on the Fixed Unknowns

In the HGLMs, Lee et al. (2017b) proposed to use the modes of various likelihoods, derived from the h-likelihood:
(i) for the random effects v, use the h-likelihood h
(ii) for the fixed effects β, use the marginal (Fisher) likelihood

$$L(\theta; y) = f_\theta(y) = \int f_\theta(v) f_\theta(y|v)dv,$$

by integrating out the nuisance unobservables v.
(iii) for the variance components $\psi = (\alpha, \phi)$, use the conditional likelihood

$$r = L(\psi; y|\tilde{\beta}) = f_\psi(y|\tilde{\beta}) = f_\theta(y)/f_\theta(\tilde{\beta}), \tag{3.8}$$

where $\tilde{\beta}$ are the MLE of β and $\theta = (\beta, \psi)$. Because $\tilde{\beta}$ is asymptotically sufficient for β (Cox and Reid 1987), we may use this conditional likelihood for estimation of ψ.

The integration to obtain the Fisher likelihood is in general intractable. Thus, Lee and Nelder (1996, 2001a) advocated to use the adjusted profile likelihood. Let ℓ be either the log-h-likelihood h or the log likelihood $m = \log L(\theta; y)$, with nuisance parameters ξ. Lee and Nelder (2001a) considered a function $p_\xi(\ell)$, defined by

$$p_\xi(\ell) = \left[\ell - \frac{1}{2}\log\det\{H(\ell; \xi)/(2\pi)\}\right]\Bigg|_{\xi=\widehat{\xi}}, \tag{3.9}$$

where $H(\ell; \xi) = -\partial^2\ell/\partial\xi^2$ and $\widehat{\xi}$ solves $\partial\ell/\partial\xi = 0$. The function $p_\xi(\cdot)$ produces an adjusted profile likelihood, eliminating nuisance parameters ξ, which can be either fixed effects β or random effects $v = (v_1, \dots, v_q)^T$ or both. Note that $p_v(h)$ is the first-order Laplace approximation to

$$m = \log\left\{\int \exp(h)dv\right\},$$

i.e.,

$$m = p_v(h) + O(N^{-1})$$

as $N = \min_{1 \le i \le q} n_i \to \infty$ (Lee and Nelder 2001a): see also Appendix 3.5.1. In the LMMs, $p_\beta(m) = \log L(\psi; y|\tilde{\beta}) \equiv \log f_\psi(y|\tilde{\beta})$ of Smyth (2002). In general, $p_\beta(m)$ is the Cox and Reid (1987) adjusted profile likelihood, approximating $L(\psi; y|\tilde{\beta})$. In the LMMs

$$m \equiv p_v(h) \text{ and } r \equiv p_\beta(m) \equiv p_{\beta,v}(h), \tag{3.10}$$

where r is the restricted likelihood of Patterson and Thompson (1971), which reduces bias, especially in the finite samples (Harville 1977).

• **Inferential procedure of the h-likelihood**

In general, the marginal likelihood is hard to compute. Thus, for estimation of the fixed parameters, Lee et al. (2017b) proposed to use the following likelihoods:

(i) for the random effects v, use the h-likelihood h
(ii) for the fixed effects β, use $p_v(h)$
(iii) for the variance components $\psi = (\alpha, \phi)$, use $p_{\beta,v}(h)$.

In the binary HGLMs, the joint maximization of h over (v, β) gives non-ignorable biases in estimating β, which is reduced most by using $p_v(h)$. Throughout this book, the modes of $p_v(h)$ and $p_{\beta,v}(h)$ are called the MLE and restricted MLE (REMLE), respectively.

Example 3.8 Suppose that $Y = (Y_{obs}, Y_{cen})$ consists of n realizations from a regression with mean $X\beta$ and variance σ^2, where Y_{obs} consists of k observed values and Y_{cen} represents $n - k$ censored values. The censored data are created by Type I censoring at some known censoring point c (i.e., $Y_{cen} > c$), so that only values less than or equal to c are recorded. Then, similar to Example 3.3 we have the marginal likelihood

$$\begin{aligned} L(\theta; y) &= \Pi_{i=1}^k (\sqrt{2\pi}\sigma)^{-1} \exp\{-(y_i - x_i\beta)^2/2\sigma^2\}\Pi_{i=k+1}^n P(x_i\beta + e_i > c) \\ &= \Pi_{i=1}^k (\sqrt{2\pi}\sigma)^{-1} \exp\{-(y_i - x_i\beta)^2/2\sigma^2\}\Pi_{i=k+1}^n \Phi((x_i\beta - c)/\sigma). \end{aligned}$$

However, there is no canonical scale here. Following Example 3.3, we take the log-h-likelihood on the $v_i = \log(Y_i - c)$ scale to have

$$\begin{aligned} h = \log L(\theta, v; y, v) = &- (n/2)\log(2\pi\sigma^2) - \sum_{i=1}^k (y_i - x_i\beta)^2/(2\sigma^2) \\ &- \sum_{i=k+1}^n \{(Y_i - x_i\beta)^2/(2\sigma^2) - \log(Y_i - c)\}. \end{aligned}$$

For $i = k + 1, \ldots, n$, $\partial h/\partial v_i = 0$ gives

$$\tilde{Y}_i = \{x_i\beta + c + \sqrt{(x_i\beta - c)^2 + 4\sigma^2}\}/2 > c.$$

Because v is not a canonical scale, the joint maximization of h and the use of the adjusted profile log-likelihood $p_v(h)$ lead to different estimators. Numerically the

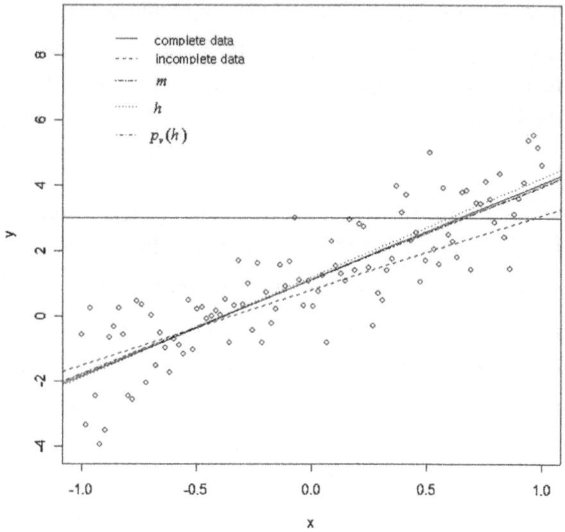

Fig. 3.1 Tobit regression. Complete data (solid line); incomplete data using simple regression (dashed line), using m (two-dashed line), using h (dotted line), using $p_v(h)$ (dot-dashed line)

former is easier to compute, but the latter gives a better approximation to the MLE. To check performance of these joint estimators, we generate a data set from a Tobit model; for $i = 1, \ldots, 100$

$$Y_i = \beta_0 + \beta_1 x_i + e_i$$

where $\beta_0 = 1$, $\beta_1 = 3$, $x_i = -1 + 2i/100$, $e_i \sim N(0, 1)$ and $c = 3$.

Figure 3.1 shows the result from a simulated data set. The use of $p_v(h)$ gives a better approximation to the marginal log-likelihood. In Fig. 3.1 the solid line is the simple regression fit using all the data (this is possible only in simulation not in practice) and the dotted line is that using only the observed data. Figure 3.1 shows that ignoring the censoring mechanism can result in a bias. The marginal MLE, accounting for the censoring mechanism, corrects such bias. The marginal MLEs based upon Gauss–Hermite quadrature (GHQ) and the adjusted profile likelihood $p_v(h)$ are almost identical. The numerical method such as Gauss–Hermite quadrature cannot be used if the dimension of integration is large, but the Laplace approximation has no such limitation. In this example, where a canonical scale does not exist, the use of $p_v()$ gives an estimation for β, essentially without a bias. We see that the joint maximization of the h-likelihood leads to a slightly biased estimation of β, but practically a satisfactory estimation. □

Table 3.1 shows historical evolvement of estimating criteria for the HGLMs. Here, **mord** and **dord** are the orders of approximations to fit the mean parameters (**mord** = 0 or 1) and the dispersion parameters (**dord** = 1 or 2), respectively. For the dispersion parameters, we need a further elaboration to reduce biases. The first-order approximation $p_{\beta,v}(h)$ often gives very accurate approximation. However, it could introduce a non-ignorable bias to the dispersion estimator in case of small cluster

Table 3.1 Estimation criteria for the h-likelihood, HL(mord, dord)

Method	Criterion for v	Criterion for β	Criterion for ψ	Literature
HL(0,1)	h	h	$p_{\beta,v}(h)$	Lee and Nelder (1996)
HL(0,2)	h	h	$s_{\beta,v}(h)$	Lee and Nelder (2001a)
HL(1,1)	h	$p_v(h)$	$p_{\beta,v}(h)$	Yun and Lee (2004)
HL(1,2)	h	$p_v(h)$	$s_{\beta,v}(h)$	Noh and Lee (2007)

sizes under some models. To reduce the bias further in estimating the dispersion parameters, the second-order approximation $s_{\beta,v}(h)$ needs to be used as follows:

$$s_{\beta,v}(h) = p_{\beta,v}(h) - \{F(h)/24\}, \tag{3.11}$$

where

$$F(h) = \mathrm{tr}[-\{3(\partial^4 h/\partial v^4) + 5(\partial^3 h/\partial v^3)H(h,v)^{-1}(\partial^3 h/\partial v^3)\}H(h,v)^{-2}]|_{v=\widehat{v}}.$$

The HL(0,1) for the normal random effects and HL(0,2) for the gamma random effects often perform well when cluster sizes are not very small (e.g., $n_i \geq 3$). As the orders in **mord** and **dord** increase, the bias of estimators is reduced, but the calculation can be computationally intensive due to the extra terms, particularly when the number of random components is greater.

• HL(0,1) method

We present how to implement parameters via the simple HL(0,1) method. Given $\psi = (\phi, \alpha)^T$, the estimates of $\tau = (\beta^T, v^T)^T$ are obtained by solving

$$\partial h/\partial \beta = 0 \text{ and } \partial h/\partial v = 0,$$

which lead to the maximum h-likelihood (MHL) score equations for $\hat{\tau} = (\hat{\beta}^T, \hat{v}^T)^T$:

$$\begin{pmatrix} X^T W X & X^T W Z \\ Z^T W X & Z^T W Z + Q \end{pmatrix} \begin{pmatrix} \widehat{\beta} \\ \widehat{v} \end{pmatrix} = \begin{pmatrix} X^T W w \\ Z^T W w + R \end{pmatrix}, \tag{3.12}$$

where W is the GLM weight matrix with a diagonal element

$$w_{ij} = [\mathrm{var}(y_{ij}|v_i)\{g'(\mu_{ij})\}^2]^{-1}$$

with $\mathrm{var}(y_{ij}|v_i) = \phi V(\mu_{ij})$ and $g'(\mu_{ij}) = \partial g(\mu_{ij})/\partial \mu_{ij}$, $Q = \mathrm{diag}(-\partial^2 \ell_2/\partial v_i^2)$ is a diagonal matrix, $w = \eta + (y - \mu)g'(\mu)$ is the GLM adjusted dependent variable, and

$R = Qv + (\partial \ell_2/\partial v)$; $R = 0$ if $v \sim N(0, \alpha)$. Note that the asymptotic covariance matrix (Lee and Nelder 1996) of $\hat{\tau} - \tau$ is given by

$$\text{var}(\hat{\tau} - \tau) = H^{-1},$$

where $H = H(h, \tau) = -\partial^2 h/\partial \tau^2$ is the square matrix on the left-hand side of (3.12). So, the upper left-hand corner of H^{-1} provides the covariance matrix of $\hat{\beta}$, given by

$$\text{var}(\hat{\beta}) = (X^T \Sigma^{-1} X)^{-1},$$

where $\Sigma = W^{-1} + ZQ^{-1}Z^T$. In the LMM (3.1), the MHL score Eq. (3.12) becomes Henderson's (1975) score equation

$$\begin{pmatrix} X^T X & X^T Z \\ Z^T X & Z^T Z + (\phi/\alpha) I_q \end{pmatrix} \begin{pmatrix} \widehat{\beta} \\ \widehat{v} \end{pmatrix} = \begin{pmatrix} X^T y \\ Z^T y \end{pmatrix}, \tag{3.13}$$

leading to the MLE of β

$$(X^T \Sigma^{-1} X)\hat{\beta} = X^T \Sigma^{-1} y,$$

where $\Sigma = \text{var}(y) = \phi I_n + \alpha Z Z^T$. Here, I_q denotes a $q \times q$ identity matrix.

Let

$$\mathbf{P} = \begin{pmatrix} X & Z \\ \mathbf{0} & I_q \end{pmatrix} \text{ and } \mathbf{V} = \begin{pmatrix} W & \mathbf{0} \\ \mathbf{0} & Q \end{pmatrix}.$$

Then the joint Eq. (3.12) reduce to a simple matrix form

$$(\mathbf{P}^T \mathbf{V} \mathbf{P})\hat{\tau} = \mathbf{P}^T \mathbf{V} \mathbf{y_0}, \tag{3.14}$$

where $\mathbf{y_0} = (w^T, R^T Q^{-1})^T$. Note here that $H = \mathbf{P}^T \mathbf{V} \mathbf{P}$. In fact, the estimating Eq. (3.14) can be viewed as the IWLS equations from an augmented weighted linear model (Lee and Nelder 2006, p. 154):

$$\mathbf{y_0} = \mathbf{P}\tau + \boldsymbol{\epsilon}^*,$$

where an error term $\boldsymbol{\epsilon}^* \sim N(\mathbf{0}, \mathbf{V}^{-1})$.

For estimation of the dispersion parameters $\psi = (\phi, \alpha)^T$, we use the adjusted profile h-likelihood, $p_{\beta,v}(h)$, of ψ after eliminating (β, v), defined by

$$p_{\beta,v}(h) = \left[h - \frac{1}{2} \log \det\{H(h; \tau)/(2\pi)\} \right] \Bigg|_{(\beta,v)=(\widehat{\beta},\widehat{v})}, \tag{3.15}$$

where $H(h; \tau) = -\partial^2 h/\partial \tau^2$ and $\widehat{\tau} = \widehat{\tau}(\psi) = (\widehat{\beta}(\psi), \widehat{v}(\psi))$. The REMLEs of ψ are obtained by solving the estimating equations

$$\partial p_{\beta,v}(h)/\partial\psi = 0. \tag{3.16}$$

Note here that in implementing the above REML equation we allow the $\partial\hat{v}/\partial\psi$ terms (Lee and Nelder 2001a). Note also that the estimated standard errors (SEs) for $\hat{\beta}$ and $\hat{\psi}$ are obtained from the inverses of corresponding Hessian matrices, $H(h,\tau) = -\partial^2 h/\partial\tau^2$ and $-\partial^2 p_{\beta,v}(h)/\partial\psi^2$, respectively.

Particularly, in the normal LMMs (3.1) the estimating Eq. (3.16) provides simple REML estimators for ϕ and α, given by

$$\widehat{\phi} = \frac{\sum_{ij}(y_{ij}-\widehat{\mu}_{ij})^2}{n-(p+q-\gamma_0)} \quad \text{and} \quad \widehat{\alpha} = \frac{\sum_i \widehat{v}_i^2}{q-\gamma_1}, \tag{3.17}$$

where $\widehat{\mu}_{ij} = x_{ij}^T\widehat{\beta} + \widehat{v}_i$, $\gamma_0 = \phi\text{tr}\{\widehat{H_0}^{-1}(\partial\widehat{H_0}/\partial\phi)\}$, $\gamma_1 = -\alpha\text{tr}\{\widehat{H_0}^{-1}(\partial\widehat{H_0}/\partial\alpha)\}$, and $\widehat{H_0} = H_0(h,\tau)|_{\tau=\widehat{\tau}(\psi)}$ with $H_0 = \phi H$. Note that in the GLMM where $v_i \sim N(0,\alpha)$, the REMLE of α has the same form as that of α in (3.17). For the general HGLMs, see Lee and Kim (2016). □

3.3.2 *Inferences on the Unobservables*

Inference on the unobservables is not possible from the Fisher likelihood as they are removed from the likelihood function by integration. In this section, we study how to make inference about them using the h-likelihood. The Bayesian approach has been successful in drawing inferences about the unobservables. As Efron (2013) pointed out, however, the use of Bayes theorem in the absence of prior is an unresolved but important problem. We believe that this can be accomplished via the h-likelihood approach.

Suppose that our interest is only in the unobservables. When θ is known, since $f_\theta(y)$ is a known constant, all the information about v in the extended likelihood is in $f_\theta(v|y)$. Thus, when the true value of θ is known, inference about v can be made using $f_\theta(v|y)$. However, since θ is unknown in practice, we may make inference about v by using $f_{\hat{\theta}}(v|y)$ with $\hat{\theta}$ being the MLE. This is the so-called "estimative approach" (EA) or empirical Bayesian (EB) approach. This approach gives an asymptotically correct inference, but it often results in a poor inferential performance in the finite samples because it cannot account for uncertainty, caused by estimating θ. Such an uncertainty is in $f_\theta(y)$, so that the drawback of the EB approach can be overcome by using the whole h-likelihood.

3.3.2.1 Wald Interval

It is important to investigate the heterogeneity in the outcomes among clusters (e.g., centers) in order to understand and interpret the variability in the data. Such heterogeneity can be accounted for by the random cluster effects. In addition to the estimation of the random effects, a measure of uncertainty for these point estimates is useful and necessary. We introduce the Wald interval for the individual random effects. To investigate and explain the sources of such heterogeneities, interval estimation for individual cluster effects has been used (Carlin and Louis 2000). A standard method is the EB confidence interval, which has been criticized for not maintaining the nominal level and hence fully Bayesian methods have been developed.

Meng (2009, 2011) established Bartlett-like identities for the h-likelihood. That is, the score for the unobservables has zero expectation and the variance of the score is the expected negative Hessian under easily verifiable conditions. Paik et al. (2015) studied the conditions that the asymptotic normality holds for $\hat{v} - v$, for example, when v is the cluster effects. Lee and Nelder (2009), Lee and Ha (2010) and Paik et al. (2015) proposed the Wald intervals for the random effects.

We first show how to construct the h-likelihood interval of the random effects and show its relationship with EB and fully Bayesian credible intervals. Given θ, let $\hat{v}(\theta)$ be a maximum h-likelihood estimator (MHLE) for the random effects solving $\partial h/\partial v = 0$, which gives the EB-mode estimator for v, without computing $f_\theta(v|y)$. In the HGLMs, (v, β) and dispersion parameters are asymptotically orthogonal (Lee and Nelder 1996); therefore, in estimating (v, β), we do not need to consider the information loss caused by estimating the dispersion parameters. The negative Hessian matrix of β and v based on h is given by

$$H(h; \beta, v) \equiv \begin{pmatrix} -\partial^2 h/\partial\beta\partial\beta^T & -\partial^2 h/\partial\beta\partial v^T \\ -\partial^2 h/\partial v\partial\beta^T & -\partial^2 h/\partial v\partial v^T \end{pmatrix} = \begin{pmatrix} H_{11} & H_{12} \\ H_{12}^T & H_{22} \end{pmatrix}. \tag{3.18}$$

For the LMMs, Henderson (1975) showed that the lower right-hand corner of $H(h; \beta, v)^{-1}$ gives an estimate of the unconditional mean squared error (UMSE):

$$\begin{aligned} \text{UMSE} &\equiv E_\theta[\{\hat{v}(\hat{\theta}) - v\}\{\hat{v}(\hat{\theta}) - v\}^T] \\ &= E_\theta[\{\hat{v}(\theta) - v\}\{\hat{v}(\theta) - v\}^T] + E_\theta[\{\hat{v}(\hat{\theta}) - \hat{v}(\theta)\}\{\hat{v}(\hat{\theta}) - \hat{v}(\theta)\}^T], \end{aligned} \tag{3.19}$$

where $\hat{v}(\hat{\theta}) \equiv \hat{v}(\theta)|_{\theta=\hat{\theta}}$ and $\hat{\theta}$ is either the MLE or REMLE of θ. Note that, generally, as $N = \min_{1\le i\le q} n_i \to \infty$, we have (Lee and Nelder 1996; Booth and Hobert 1998)

$$E(v|y) = \hat{v}\{1 + O(N^{-1})\} \text{ and } \text{var}(v|y) = H_{22}^{-1}\{1 + O(N^{-1})\}.$$

The second term in (3.19) is the inflation caused in the UMSE because of the estimation of θ by $\hat{\theta}$. Lee and Nelder (1996) and Lee et al. (2017b) showed that this holds generally in the HGLMs. The above UMSE could be used to construct the h-likelihood confidence intervals for v when the asymptotic normality holds. For example, let A be the lower right-hand corner of $H(h; \hat{\beta}, \hat{v})^{-1}$ corresponding to v, with the kth diagonal element a_{kk}. Then, $A = \{H_{22} - H_{12}(H_{11})^{-1}H_{21}\}^{-1}|_{\beta=\hat{\beta}, v=\hat{v}}$, which provides a generally good estimators for $\text{var}(\hat{v}-v)$. Thus, a $(1-\lambda)$ h-likelihood interval for v_k is

$$\hat{v}_k \pm z_{\lambda/2} \cdot \text{SE}(\hat{v}_k - v_k), \tag{3.20}$$

where $z_{\lambda/2}$ is the standard normal quantile with a probability of $\lambda/2$ in the right tail, and $\mathrm{SE}(\hat{v}_k - v_k)$ is $\sqrt{a_{kk}}$. This confidence interval will work well if $\hat{v} - v$ is approximately normal.

Note that

$$\mathrm{var}(v|y) = E_{\theta|y}\{\mathrm{var}(v|y, \theta)\} + \mathrm{var}_{\theta|y}\{E(v|y, \theta)\}.$$

Carlin and Gelfand (1990) noted that the EB variance estimate only approximates the first term in the above equation, and ignores the second. Kass and Steffey (1989) used a Laplace approximation to show that under the uniform prior $\pi(\theta) = 1$, the estimator for $\mathrm{var}(v|y)$ can be obtained from $H(h; \theta, v)^{-1} = \{-\partial h^2/\partial(\theta, v)^2\}^{-1}$. Thus, the h-likelihood interval improves the EB interval and it can be interpreted either as a Bayesian credible interval (under the uniform prior) or as a frequentist confidence interval (Lee and Kim 2016).

3.3.2.2 Interval Based on the Predictive Distribution

The Wald interval works well when the distribution of $\hat{v} - v$ is approximately normal. If the distribution of $\hat{v} - v$ is skewed, however, it may not work well. In general, the fully *Bayesian credible interval* is proposed based on the (Bayesian) predictive distribution

$$\pi(v|y) = \int f_\theta(v|y)\pi(\theta|y)d\theta, \tag{3.21}$$

where θ is integrated out. For the frequentist interval, Lee and Kim (2016) proposed to use the (frequentist) predictive distribution

$$P(v|y) = \int f_\theta(v|y)c(\theta|y)d\theta, \tag{3.22}$$

where $c(\theta|y)$ is the frequentist confidence density such as the bootstrap distribution (Lee et al. 2017b, Chap. 4). This leads to the bootstrap method to get an estimate of the predictive distribution

$$P^B(v|y) \equiv \frac{1}{B}\sum_{j=1}^{B} f_{\theta_j^*}(v|y),$$

where $\theta_1^*, \ldots, \theta_B^*$ are the bootstrap replicates of $\widehat{\theta}$. For interval estimation, they proposed to use the percentiles. When the distribution of $v|y$ is very skewed, we need to use the bootstrap method.

In this book, we use the Wald interval in the previous section because, in many applications, we found that its performance is satisfactory and it is computationally easy because all necessary quantities are computed to obtain the MHLEs. Better interval estimators can be made by using numerically intensive boostrap method. For the performances of the boostrap method for interval estimation of random effects see Lee and Kim (2016) and of subject-specific function estimators see Cao et al. (2017).

3.3.2.3 Example for the Predictive Distribution

Estimation of the predictive distribution is crucial for inference about the random effects. Suppose that we have the number of epileptic seizures in an individual for five weeks, $y = (3, 2, 5, 0, 4)$. Suppose that these counts are iid from a Poisson distribution with mean θ. Here, $\hat{\theta} = (3+2+5+0+4)/5 = 2.8$ is the MLE of θ, which maximizes the Fisher likelihood $f_\theta(y)$. Inference about θ can be made by using the likelihood. Now we want to find a good predictive distribution for the seizure counts for the next week, v. Then, because $f_\theta(v = i|y) = f_\theta(v = i)$, the plug-in technique gives the predictive distribution of the seizure count v for the next week as

$$P^E(v|y) = f_{\hat{\theta}}(v = i|y) = f_{\hat{\theta}}(v = i) = \exp(-2.8)2.8^i/i!.$$

This gives an asymptotically correct inference because $f_{\hat{\theta}}(v = i|y) \to f_{\theta_0}(v = i|y)$ where θ_0 is the true value of θ. However, its finite sampling property is in doubt. Pearson (1920) pointed out that the limitation of the Fisher likelihood using the plug-in method is its inability to account for the uncertainty in estimating θ. This plug-in technique is the so-called EB method.

He suggested using the fully Bayesian predictive distribution (3.21). To reduce dependence on priors, noninformative priors such as Jeffreys' prior can be considered to form the predictive distribution (Lee and Kim 2016); Jeffreys' prior under Poisson with mean θ is $\pi(\theta) \propto \sqrt{i(\theta)} = \theta^{-1/2}$. The resulting predictive distribution gives a predictive probability with higher probabilities for larger y. Pearson (1920) pointed out that this Bayesian procedure handles the uncertainty caused by estimating θ. However, this Bayesian procedure depends upon the choice of a prior and it would be difficult to justify the specific choice of Jeffreys' prior. Here, the h-likelihood is proportional to

$$f_\theta(3, 2, 5, 0, 4, v) = \exp(-6\theta)\theta^{3+2+5+0+4+v}/(3!2!5!0!4!v!),$$

where $\hat{\theta}(v) = (3+2+5+0+4+v)/6$. Then, the normalized profile likelihood $f_{\hat{\theta}(v)}(3, 2, 5, 0, 4, v)$ gives the predictive distribution of Mathiasen (1979), almost identical to Pearson's but without assuming a prior on θ (Fig. 3.2); for more discussion, see Bjørnstad (1990). This example shows that the standard methods for the likelihood inference can be used for the prediction problem by using the h-likelihood. Lee and Kim (2016) studied various predictive distributions based on the frequentist confidence densities. All of them account for the uncertainty caused by estimating θ. They found that the use of the normalized likelihood as a confidence density in (3.22) gives the best interval, maintaining the stated level.

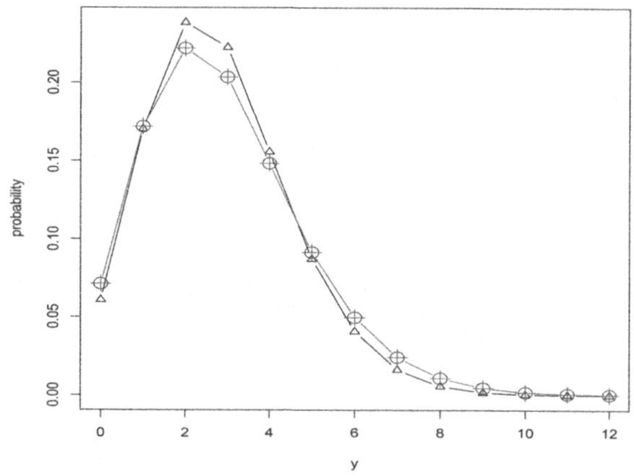

Fig. 3.2 Predictive density of the number of seizure counts: Plug-in method (△), Bayesian method (∘), and h-likelihood method (+)

3.4 A Practical Example: Epilepsy Seizure Count Data

We illustrate the h-likelihood approach using the epilepsy seizure count data from a clinical trial carried out by Leppik et al. (1985) and previously analyzed by Thall and Vail (1990). The data come from a randomized clinical trial conducted among patients suffering from simple or complex partial seizures to receive either the antiepileptic drug progabide or a placebo, as an adjuvant to the standard chemotherapy. Primary outcome of interest (y) is the number of seizures occurring over the previous 2 weeks measured at each of four successive postrandomization clinic visits. Thall and Vail (1990) took a quasi-likelihood approach and focused on comparing various types of overdispersion models. In this analysis, we assume the extra variation is due to individual-specific seizure propensity and conduct a secondary analysis to quantify the seizure propensity. We formally identify patients with high seizure propensity using the inferential procedure described in Sect. 3.3. Specifically, we assume that the inherent seizure propensity exists and is realized (subject was born with it) but cannot be observed. We would like to draw inference about the realized seizure propensity and apply the Wald interval described in the Eq. (3.20).

- **Model considered: Poisson-gamma HGLM**

The data consist of four repeated measures ($n_i = 4$) for $K = 59$ epileptic patients, with covariates Constant, Base (x_1), Trt (x_2; placebo $= 0$, progabide $= 1$), Base.Trt (x_3), Age (x_4) and Visit ($x_5 = -0.3, -0.1, 0.1, 0.3$ for each visit). We assume a Poisson-gamma HGLM as follows:
(i) $y_{ij}|u_i$ ($i = 1, \ldots, 59; j = 1, 2, 3, 4$) follows a Poisson distribution with mean $\mu_{ij} = \exp(\eta_{ij})$, where

$$\eta_{ij} = \beta_0 + \beta_1 x_{1ij} + \beta_2 x_{2ij} + \beta_3 x_{3ij} + \beta_4 x_{4ij} + \beta_5 x_{5ij} + v_i$$

is the linear predictor with $v_i = \log u_i$.

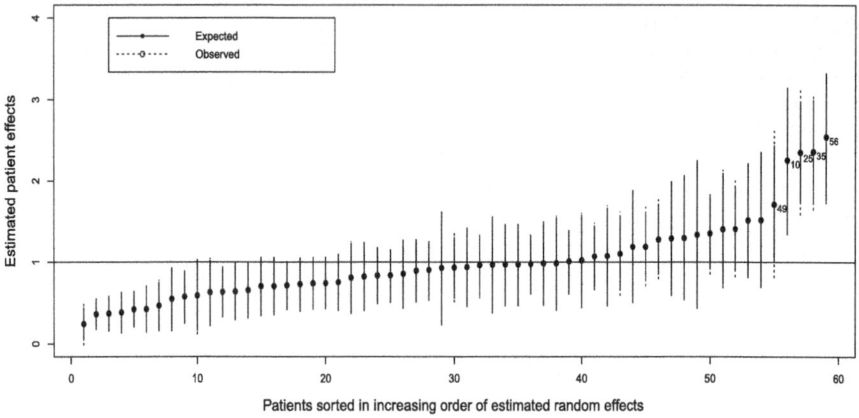

Fig. 3.3 Estimated random effects of 59 patients in the epileptic data and their 95% confidence intervals, under Poisson-gamma model; Expected and Observed mean confidence intervals based on expected and observed information matrices

(ii) The random effect u_i is assumed to arise from a gamma distribution having mean 1 and variance α.

- **Estimation of parameters**

The Poisson-gamma model was fitted using the HL(0,2) method. The estimates of the fixed parameters and their standard errors (SEs) are $\hat{\beta}_0 = -1.32$ (SE = 1.25), $\hat{\beta}_1 = 0.88$ (SE = 0.13), $\hat{\beta}_2 = -0.90$ (SE = 0.40), $\hat{\beta}_3 = 0.35$ (SE = 0.20), $\hat{\beta}_4 = 0.50$ (SE = 0.37), $\hat{\beta}_5 = -0.29$ (SE = 0.10), and $\hat{\alpha} = 0.28$ (SE = 0.06), yielding a significant difference between the two treatment groups.

- **Inference on heterogeneity**

Lee and Ha (2010) have found that in constructing the HL interval, using the u-scale gives a better coverage probability than using the v -scale (i.e., $v = v(u)$) for the HGLMs: see also Paik et al. (2015). Now, we focus on investigating the heterogeneity among patients and construct 95% confidence intervals for the realized values of patient effects $u_i (i = 1, \ldots, 59)$ under the Poisson-gamma model:

$$\{\hat{u}_i - 1.96\text{SE}(\hat{u}_i - u_i), \hat{u}_i + 1.96SE(\hat{u}_i - u_i)\}.$$

Note here that $\hat{u}_i$ is a solution of $\partial h/\partial u_i = 0$ and that $SE(\hat{u}_i - u_i)$ is obtained from the inverse of the observed information $H(h; \beta, u_i) = -\partial^2 h/\partial(\beta, u_i)^2$ in (3.18).

Figure 3.3 gives 95% confidence intervals for the realized but unobserved individual seizure propensity ($K = 59$). The intervals are plotted against the increasing order of estimated random effects. The confidence intervals obtained from both 'observed' and 'expected' versions (Paik et al. 2015) of the variance estimates of $\hat{u}_i - u_i$ show similar trends. Here, the expected version indicates the inverse of $E\{H(h; \beta, u_i)\}$. Figure 3.3 demonstrates substantial variations in seizure propensity among patients.

Especially, for four patients (patient id = 10, 25, 35, and 56), 95% confidence intervals for u_i do not contain 1, suggesting that the seizure propensity is significantly different from the norm. Patient id 49's interval excludes 1 using the variance estimate via the expected information. These patients were identified as outliers via the residual analyses by Thall and Vail (1990), Breslow and Clayton (1993) and by Ma and Jørgensen (2007), but there were no formal inferential procedures. We also identify patients with low propensity significantly different from 1, which the previous analyses could not. Lee and Ha (2010) plotted the h-likelihood surface of u, which is reasonably symmetric, so that a similar conclusion was drawn based on the interval from the predictive distribution. Thus, we recommend to check the symmetry of the h-likelihood surface if the Wald interval is expected to work well.

3.5 Appendix

3.5.1 *Proof of Approximation in Poisson-Gamma HGLM*

Consider the Poisson-gamma model. That is, assume that response variables y_{ij} given the random effect u_i of the ith individual are independent and that they have a Poisson distribution with mean $E(y_{ij}|u_i) = \mu_{ij}u_i$ $(i = 1, \ldots, q; j = 1, \ldots, n_i)$:

$$y_{ij}|u_i \sim \text{Poisson}(\mu_{ij}u_i) \text{ with } \mu_{ij} = \exp(x_{ij}^T\beta),$$

where $x_{ij} = (x_{ij1}, \ldots, x_{ijp})^T$ is a $p \times 1$ covariate vector for the jth observation of the ith individual and $\beta = (\beta_1, \ldots, \beta_p)$ is corresponding regression parameters. Here, u_i's follow a gamma distribution with $E(u_i) = 1$ and $\text{var}(u_i) = \alpha$.

The marginal likelihood, denoted by L_i, of the ith individual is defined by

$$L_i = L_i(\beta, \alpha) = f_{\beta,\alpha}(y_{i1}, \ldots, y_{in_i}) = \int \exp(h_i)dv_i,$$

where $f_{\beta,\alpha}(\cdot)$ is the joint density of $y_{i1}, \ldots, y_{in_i}$, $h_i = \sum_j \ell_{1ij} + \ell_{2i}$ is the contribution of the ith individual to the h-likelihood h in (3.7), and $v_i = \log u_i$. Then, the marginal likelihood for all individuals becomes $L = \prod_i L_i$. From Example 3.7 we have an explicit marginal log-likelihood:

$$\begin{aligned} m = m(\beta, \alpha) = \log L &= \sum_i \log\left\{\int \exp(h_i)dv_i\right\} \\ &= \sum_{ij}\{y_{ij}x_{ij}^T\beta - \log\Gamma(y_{ij}+1)\} + \sum_i\{-(\alpha^{-1} + y_{i+})\log(\alpha^{-1} + \mu_{i+}) \\ &+ \log\Gamma(\alpha^{-1} + y_{i+}) + c(\alpha)\}, \end{aligned} \tag{3.23}$$

where $y_{i+} = \sum_j y_{ij}$, $\mu_{i+} = \sum_j \mu_{ij} = \sum_j \exp(x_{ij}^T\beta)$ and $c(\alpha) = -\log\Gamma(\alpha^{-1}) - \alpha^{-1}\log\alpha$. Here, the h-likelihood is given by

$$h = h(\beta,\alpha) = \sum_{ij}\{y_{ij}x_{ij}^T\beta - \log\Gamma(y_{ij}+1)\} + \sum_i\{(\alpha^{-1}+y_{i+})v_i - (\alpha^{-1}+\mu_{i+})u_i + c(\alpha)\}.$$

From

$$\partial h/\partial v_i = (y_{i+}+\alpha^{-1}) - (\mu_{i+}+\alpha^{-1})u_i = 0,$$

we have

$$\hat{u}_i = \frac{\alpha^{-1}+y_{i+}}{\alpha^{-1}+\mu_{i+}},$$

which also becomes $E(u_i|y_i)$ since the conditional distribution of u_i given the ith observed data y_i is gamma. Note here that the ith component of adjustment term for $p_v(h)$,

$$H(h;\, v_i)|_{u_i=\hat{u}_i} = -\partial^2 h/\partial v_i^2|_{u_i=\hat{u}_i} = (\alpha^{-1}+\mu_{i+})\hat{u}_i = \alpha^{-1}+y_{i+}\,,$$

is free of β but depends upon α. Since $H(h,v)|_{u=\hat{u}} = \mathrm{diag}(\alpha^{-1}+y_{i+})$ is a $q\times q$ diagonal matrix, we have that

$$\begin{aligned} p_v(h) &= \left[h - \frac{1}{2}\log\det\{H(h;\,v)/(2\pi)\}\right]|_{u=\widehat{u}} \\ &= \sum_{ij}\{y_{ij}x_{ij}^T\beta - \log\Gamma(y_{ij}+1)\} + \sum_i\{-(\alpha^{-1}+y_{i+})\log(\alpha^{-1}+\mu_{i+}) \\ &+ (\alpha^{-1}+y_{i+})\log(\alpha^{-1}+y_{i+}) - (\alpha^{-1}+y_{i+}) - \log(\alpha^{-1}+y_{i+})/2 \\ &+ \log(2\pi)/2 + c(\alpha)\}, \end{aligned}$$

which is equivalent to approximating m of (3.23) by the first-order Stirling approximation

$$\log\Gamma(x) \doteq (x-1/2)\log(x) + \log(2\pi)/2 - x$$

for $\Gamma(\alpha^{-1}+y_{i+})$. Thus, the marginal ML estimator for β (maximizing $p_v(h)$) can be obtained by maximization of h. Note that given α, the MHL estimator for β is the same as the ML estimator (Lee and Nelder 1996). Furthermore, a good approximation to the ML estimator for α can be obtained by using $p_v(h)$ if the first-order Stirling approximation works well. It can be further shown that the second-order Laplace approximation, $s_v(h) = p_v(h) - \{F(h)/24\}$, is equivalent to approximating m by the second-order Stirling approximation

$$\log \Gamma(x) \doteq (x - 1/2)\log(x) + \log(2\pi)/2 - x + 1/(12x).$$

Here, under the HGLM with one-random component v, the term, $F(h)$, in (3.11) is given by

$$F(h) = \sum_{i=1}^{q} \left\{ -3\left(\frac{\partial^4 h}{\partial v_i^4}\right) h_{ii}^2 - 5\left(\frac{\partial^3 h}{\partial v_i^3}\right)^2 h_{ii}^3 \right\}\Bigg|_{v=\hat{v}},$$

where h_{ii} is the ith diagonal element of $H(h, v)^{-1}$, Under the Poisson-gamma model it gives a simple form:

$$F(h) = -2\sum_{i=1}^{q}(\alpha^{-1} + y_{i+})^{-1}.$$

Note that in general, $m = s_v(h) + O(N^{-2})$ as $N = \min_{1 \le i \le q} n_i \to \infty$ (Lee and Nelder 2001a). For the REML estimator for α, we use $p_{\beta,v}(h)$ or $s_{\beta,v}(h)$.

Chapter 4
Simple Frailty Models

The concept of frailty was first introduced by Vaupel et al. (1979) to account for the impact of individual heterogeneity in univariate (independent) survival data. In this chapter, we introduce the frailty model, an extension of the Cox PH model, for analyzing correlated survival data. The frailty is modeled by an unobserved random effect acting multiplicatively on the individual hazard rate to describe the individual heterogeneity and the correlation (dependency) among survival data from the same subject or cluster (Clayton 1978; Hougaard 2000; Duchateau and Janssen 2008). Even if heterogeneity and correlation are different concepts, they can both be modeled by frailties.

Various inferential procedures for the frailty model have been proposed in the literature. In this book, we focus on the likelihood-based approaches: We discuss the h-likelihood, penalized partial likelihood, and marginal likelihood procedures and compare them. All of these procedures can be derived from the h-likelihood. Through real data analyses, we demonstrate the h-likelihood procedures, available in the **frailtyHL** (Ha et al. 2018) R package. We also review recent developments in the model-selection procedures and interval estimation of the individual frailties.

4.1 Features of Correlated Survival Data

Survival data, namely time-to-event data, are often collected as a single observed event per individual. This simple form of data is called independent survival data and has been often analyzed using the Cox PH model when there are confounding factors that need to be adjusted for. However, in many biomedical studies, a correlation among times-to-events can be induced by clustering as follows:

- **Recurrent or multiple events**: Each subject can potentially experience more than one event; for example, a patient can have times to recurrent events of the same

I.D. Ha et al., *Statistical Modelling of Survival Data with Random Effects*,
Statistics for Biology and Health, https://doi.org/10.1007/978-981-10-6557-6_4

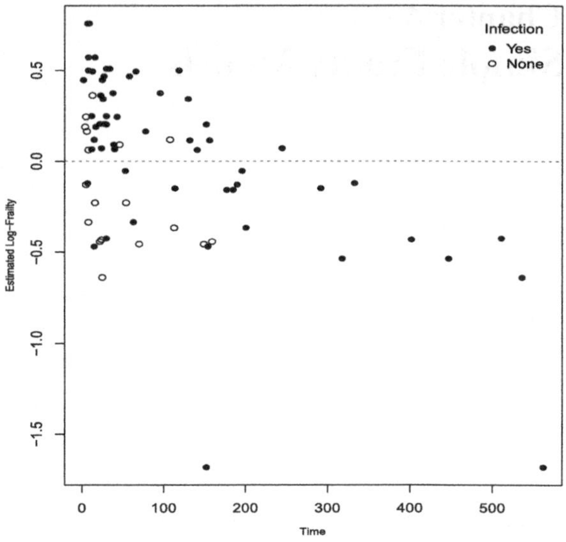

Fig. 4.1 Estimated log-frailty versus observed event time for each patient in kidney infection data; "Yes", infection and "None", censoring

type (e.g., recurrences of tumor or recurrent infections of disease) or multiple events of different types (e.g., death after recurrences).

- **Events by pair or family**: Various clustered time-to-event studies from twin or family study, matched pair study, and study of organ systems (e.g., left and right eyes).
- **Events from multicenter**: In multicenter clinical trials, survival times of patients from the same center may have common medical characteristics and practice patterns.

These event times within a cluster can be correlated due to a common genetic or environment effect on the same cluster. We call such data correlated (i.e., multivariate or clustered) survival data. This correlation can be modeled by introducing a frailty (an unobserved random effect) term into the PH model. The frailty model, an extension of the Cox PH model allowing random effects, has been widely used for modeling dependency within a cluster as well as heterogeneity between clusters. Fitting the Cox PH model ignoring the correlation can lead to underestimation of the fixed covariate effects as we will see later. The frailty model can also be used to describe heterogeneity in the independent survival data. The idea is that individuals who are more frail will die (or experience an event) earlier than ones who are less frail. Or a family with a larger frailty value will experience their events at earlier times than a family with a smaller frailty value. Figure 4.1 displays the meaning of frailty, using the log-normal frailty model in Example 4.1 with kidney infection data. This indicates that the estimated frailties are overall larger (i.e., more frail) for patients who had an event (infection) early than those who had an event later.

4.2 The Model and H-Likelihood

In this section, we first define the univariate frailty model, which includes a frailty term from a univariate distribution, and present the h-likelihood for the model. Then, we present various existing likelihood-based methods, derived from the h-likelihood. The results presented in this section will be extended to the multivariate frailty models in the later chapters, especially under competing risks.

4.2.1 Univariate Frailty Model

Suppose that data consists of observed times to an event subject to censoring, and collected from q subjects (or clusters) with n_i observations per cluster ($i = 1, \ldots, q$; $j = 1, \ldots, n_i$). Let T_{ij} be the potential event time for the jth observation for the ith subject and C_{ij} be the corresponding potential censoring time. Here, $n = \sum_i n_i$ is the total sample size. In the multicenter clinical trials, n_i would be the number of patients in the ith center and n would be the total number of patients coming from all q centers. In the independent and bivariate data, $n_i = 1$ and $n_i = 2$, respectively, for all i. Denote by U_i the unobserved univariate frailty for the ith subject. The two assumptions for the (Fisher) likelihood construction in Sect. 2.1.2 under the Cox PH model are extended to the frailty model as follows.

Assumption 3:
Given $U_i = u_i$, the pairs $\{(T_{ij}, C_{ij}), j = 1, \ldots, n_i\}$ are conditionally independent and both T_{ij} and C_{ij} are also conditionally independent for $j = 1, \ldots, n_i$.

Assumption 4:
Given $U_i = u_i$, $\{C_{ij}, j = 1, \ldots, n_i\}$ are conditionally noninformative of T_{ij}.

• **Univariate (or one-component) frailty model**: Given an unobserved frailty for the ith subject $U_i = u_i$, suppose that the conditional hazard function of T_{ij} takes the form of

$$\lambda_{ij}(t|u_i) = \lambda_0(t)\exp(x_{ij}^T\beta)u_i, \tag{4.1}$$

where $\lambda_0(\cdot)$ is an arbitrary baseline hazard function, $x_{ij} = (x_{ij1}, \ldots, x_{ijp})^T$ is a vector of fixed covariates and $\beta = (\beta_1, \ldots, \beta_p)^T$ is a vector of the corresponding regression parameters. For the purpose of identifiability, the term $x_{ij}^T\beta$ does not include the intercept term as in the Cox PH model (2.4). The random variable U_i is assumed to be independently and identically distributed (iid). Popular distributions for U_i are gamma and log-normal.

In the frailty model (4.1), $E(U_i)$ (or $E(V_i)$) is confounded with the baseline hazard $\lambda_0(t)$ since, for example,

$$\lambda_0(t)u_i = (a\lambda_0(t))(u_i/a) \text{ for all } a > 0.$$

Lee and Nelder (1996) proposed to impose constraints on the random effects such that $E(U_i) = 1$ or $E(V_i) = 0$, rather than putting constraints on the fixed parameters, which is convenient in the multiple random-effect models. Elbers and Ridder (1982) showed that the frailty model (4.1) with a vector of covariates x_{ij} is identifiable if the frailty distribution has the finite mean (i.e., $E(U_i) < \infty$). Thus, it is traditionally assumed that $E(U_i) = 1$ and $\text{var}(U_i) = \alpha$ for the gamma frailty model and that $E(V_i) = 0$ and $\text{var}(V_i) = \alpha$ for the log-normal frailty model, i.e., $V_i = \log U_i \sim N(0, \alpha)$. Note that model (4.1) reduces to the Cox PH model when $u_i = 1$ for all i (i.e., $\alpha = \text{var}(U_i) = 0$).

For the gamma frailty model, the marginal likelihood is explicitly available after integrating out the frailty term, but not for the log-normal frailty model. However, the log-normal frailty model is useful, particularly to incorporate correlated frailties. It is worthy to note that the h-likelihood can easily accommodate other frailty distributions (Duchateau and Janssen 2008; Wienke 2011).

Under the model (4.1), the conditional survival function of T_{ij} given $U_i = u_i$ can be expressed as

$$S(t|u_i) = P(T_{ij} > t|U_i = u_i) = B(t)^{u_i},$$

where $B(t) = \exp\{-\Lambda_0(t)e^{x_{ij}^T\beta}\}$ with baseline cumulative hazard function $\Lambda_0(t) = \int_0^t \lambda_0(k)dk$. This means that the random variable U_i is a frailty which describes the individual unobserved risk. Note that a larger value of u_i implies a smaller survival probability $S(t|u_i)$, indicating a poor survival. In other words, individuals in a group i with $u_i > 1$ [$u_i < 1$] are more frail or at higher risk [less frail or at lower risk], respectively (Fig. 4.1).

T_{ij} and T_{ik}, for $j \neq k$, (i.e., survival times of individuals within cluster i) are dependent unconditionally because of their shared frailty u_i. The multivariate survival function of $T_{i1}, \cdots, T_{in_i}$ can be derived by integrating out the frailty from the conditional survival function, $S(t|u_i) = P(T_{ij} > t|u_i)$: From Assumption 3, we have

$$\begin{aligned} S(t_1, \ldots, t_{n_i}) &= P(T_{i1} > t_1, \ldots, T_{in_i} > t_{n_i}) \\ &= \int \Big\{ \prod_j S(t_j|u_i) \Big\} f(u_i)du_i, \end{aligned}$$

where $f(u_i)$ is the density function of the frailty U_i.

Remark 4.1 In Appendix 4.7.1, we show that under the frailty model (4.1) the marginal hazard function with covariates x, denoted by $\lambda^{\text{M}}(t; x)$, becomes

$$\lambda^{\mathrm{M}}(t; x) = \lambda_0(t)\exp(x^T\beta)E(U|T > t; x),$$

where the conditional expectation $E(U|T > t; x)$ indicates the expected frailty among the survivors at time t. Under the gamma frailty, we have

$$E(U|T > t; x) = \{1 + \alpha\Lambda_0(t)\exp(x^T\beta)\}^{-1},$$

which decreases with time t (see Fig. 4.1). Here $\Lambda_0(t) = \int_0^t \lambda_0(k)dk$ is the baseline cumulative hazard function. This leads to the Burr model (Burr 1942), given by

$$\lambda^{\mathrm{M}}(t; x) = \frac{\lambda_0(t)\exp(x^T\beta)}{1 + \alpha\Lambda_0(t)\exp(x^T\beta)},$$

and the hazard ratio is not constant over time (non-proportional hazards). Furthermore, for any frailty distribution the hazard ratio with a single covariate x is given by

$$\frac{\lambda^{\mathrm{M}}(t; x = 1)}{\lambda^{\mathrm{M}}(t; x = 0)} = \exp(\beta)\frac{E(U|T > t; x = 1)}{E(U|T > t; x = 0)},$$

which will be time dependent except under specific circumstances. Thus, the frailty models generally lead to non-PH models. □

4.2.2 H-Likelihood and Related Likelihoods

The observable random variables are

$$Y_{ij} = \min(T_{ij}, C_{ij}) \text{ and } \delta_{ij} = I(T_{ij} \le C_{ij}).$$

Let y_{ij} be an observed value of Y_{ij}.

• **Definition of the h-likelihood** (Appendix 4.7.2): Under Assumptions 3 and 4, the h-likelihood for the frailty model becomes

$$h = h(\beta, v, \lambda_0, \alpha) = \sum_{ij} \ell_{1ij} + \sum_i \ell_{2i}, \tag{4.2}$$

where

$$\ell_{1ij} = \ell_{1ij}(\beta, \lambda_0; y_{ij}, \delta_{ij}|u_i) = \delta_{ij}\{\log\lambda_0(y_{ij}) + \eta_{ij}\} - \{\Lambda_0(y_{ij})\exp(\eta_{ij})\}$$

is the logarithm of the conditional density function for Y_{ij} and δ_{ij} given $U_i = u_i$, the so-called ordinary log-likelihood for censored survival data given u_i, $\ell_{2i} = \ell_{2i}(\alpha; v_i)$

is the logarithm of the density function for $V_i = \log(U_i)$, $\Lambda_0(t)$ is the baseline cumulative hazard function, and

$$\eta_{ij} = x_{ij}^T\beta + v_i$$

is a linear predictor on the log-hazard with $v_i = \log(u_i)$. As mentioned in Chap. 3, the h-likelihood can be interpreted as the Bayesian posterior under the uniform prior.

• **Profile h-likelihood**: Suppose that the functional form of $\lambda_0(t)$ in (4.1) is unknown. Let $\lambda_{0k} = \lambda_0(y_{(k)})$ be the baseline hazard function at $y_{(k)}$, where $y_{(k)}$ is the kth $(k = 1, \ldots, D)$ smallest distinct event time among y_{ij}'s. Following Breslow (1972), we define the baseline cumulative hazard function $\Lambda_0(t) = \int_0^t \lambda_0(u)du$ to be a step function with jumps λ_{0k} at observed event times $y_{(k)}$:

$$\Lambda_0(t) = \sum_{k:y_{(k)}\le t} \lambda_{0k} = \sum_k \lambda_{0k} I(y_{(k)} \le t). \tag{4.3}$$

Following Appendix 2.6.2, by substituting (4.3) into (4.2) the first term in (4.2) becomes

$$\sum_{ij} \ell_{1ij} = \sum_k d_{(k)} \log \lambda_{0k} + \sum_{ij} \delta_{ij}\eta_{ij} - \sum_k \lambda_{0k}\left\{ \sum_{(i,j)\in R_{(k)}} \exp(\eta_{ij})\right\},$$

where $d_{(k)}$ is the number of deaths at $y_{(k)}$ and

$$R_{(k)} = R(y_{(k)}) = \{(i, j) : y_{ij} \ge y_{(k)}\}$$

is the risk set at $y_{(k)}$. As the number of the terms λ_{0k}'s in $\sum_{ij} \ell_{1ij}$ increases with the number of events, the dimension of the function $\lambda_0(t)$ is potentially high (Zeng and Lin 2007; Ha et al. 2010) when $\lambda_0(t)$ is unknown. In survival analysis, however, λ_{0k}'s are not often of interest. Thus, we may eliminate them using the profile h-likelihood

$$h^* = h|_{\lambda_0=\widehat{\lambda}_0},$$

leading to

$$h^* = h^*(\beta, v, \alpha) = \left\{\sum_k d_{(k)} \log \widehat{\lambda}_{0k} + \sum_{ij} \delta_{ij}\eta_{ij} - \sum_k d_{(k)}\right\} + \sum_i \ell_{2i} \tag{4.4}$$

where

$$\widehat{\lambda}_{0k}(\beta, v) = \frac{d_{(k)}}{\sum_{(i,j)\in R_{(k)}} \exp(\eta_{ij})}$$

are the solutions of the estimating equations, $\partial h/\partial\lambda_{0k} = 0$, for $k = 1, \ldots, D$. Note that from (4.3) we have

$$\widehat{\Lambda}_0(t) = \sum_{k:y_{(k)} \le t} \widehat{\lambda}_{0k}$$

is a nonparametric MHLE, which is an extension of Breslow's (1974) estimator of the baseline cumulative hazard function for the Cox model to the frailty model.

• **Relation to the penalized partial likelihood (PPL)**: Therneau and Grambsch (2000, p. 251) and Ripatti and Palmgren (2000) proposed to construct the h-likelihood (4.2) using the partial log-likelihood (Cox 1972; Breslow 1974) for ℓ_{1ij}. They call the resulting h-likelihood the penalized partial likelihood (PPL), defined by

$$h_p = h_p(\beta, v, \alpha) = \ell_p + \sum_i \ell_{2i}, \tag{4.5}$$

where

$$\ell_p = \sum_{ij} \delta_{ij}\eta_{ij} - \sum_k d_{(k)} \log\left\{ \sum_{(i,j)\in R_{(k)}} \exp(\eta_{ij}) \right\}$$

is Breslow's likelihood given $U_i = u_i$. In the penalized likelihood approach, the likelihood for the model is ℓ_p and $\sum_i \ell_{2i}$ is associated with a penalty term. Thus, the frailty parameter α is a tuning parameter (Therneau and Grambsch 2000, p. 233), but not the model parameter. On the other hand, in the frailty model, α is the model parameter to describe a correlation among event times from the same subject, so that we need to extend the ML and REML estimators in the LMMs to the frailty models. In the LMMs, the frailty parameter α is the variance component, which is an important quantity for statistical inference. Thus, in this book, we call h_p the partial h-likelihood and we pay a proper attention to inference about the frailty parameter.

Note that the profile h-likelihood h^* is proportional to the partial h-likelihood h_p because from (4.4) and (4.5) we have

$$h^* = h_p + \sum_k d_{(k)}\{\log d_{(k)} - 1\}, \tag{4.6}$$

where $d_{(k)}\{\log d_{(k)} - 1\}$ is a constant depending upon only observed data, not depending upon unknown parameters. In this book, we call both h_p and h^* the partial h-likelihood.

The PPL procedure estimates the frailty parameter by modifying the existing variance-component estimation in the LMMs, which is in spirit similar to the penalized quasi-likelihood (PQL) method of Breslow and Clayton (1993). However, the PQL procedure is severely biased with binary data, while the h-likelihood approach does not introduce such severe bias (Lee et al. 2017b, Sect. 6.4). Therneau et al. (2003) improved the PQL procedure, but it is still not satisfactory because it omits some necessary terms as we shall see later. In the h-likelihood approach, the frailty parameters are estimated by using the likelihood method, while in the penalized likelihood approach, they are tuning parameters, but not the model parameters. Ha et al. (2010) discussed this difference between the PPL and h-likelihood procedures and showed the superiority of the h-likelihood procedure over the PPL procedure. The PPL differs from the h-likelihood in estimating the frailty parameters, which affects estimation of both fixed and random effects.

• **Relation to the marginal likelihood**: Following Appendix 3.5.1, the marginal likelihood of the ith individual is

$$L_i = L_i(\beta, \lambda_0, \alpha) = f_{\beta,\lambda_0,\alpha}(y_{i1}^*, \ldots, y_{in_i}^*) = \int \exp(h_i) dv_i,$$

where $f_{\beta,\lambda_0,\alpha}(\cdot)$ is the joint density of $y_{i1}^*, \ldots, y_{in_i}^*$ with $y_{ij}^* = (y_{ij}, \delta_{ij})$ for $j = 1, \ldots, n_i$, $h_i = \sum_j \ell_{1ij} + \ell_{2i}$ is a contribution from the ith individual in (4.2), and $v_i = \log u_i$. Thus, the marginal log-likelihood for all individuals, denoted by m, can be obtained by integrating out the frailties from the h-likelihood:

$$m = m(\beta, \lambda_0, \alpha) = \sum_i \log L_i(\beta, \lambda_0, \alpha) = \sum_i \log\left\{\int \exp(h_i)\, dv_i\right\}. \tag{4.7}$$

This marginal likelihood (i.e., observed-data likelihood) has been often used for inference (e.g., Klein 1992; Nielsen et al. 1992). In the semiparametric frailty models, the number of nuisance parameters increases with sample size. With binary data, it causes severe biases in the MLEs (Andersen 1970). In the finite sample, the resulting MLEs from the frailty models could suffer from a substantial bias caused by the presence of many nuisance parameters λ_0, when cluster size n_i or the censoring rate is small (Ha et al. 2010). However, in the frailty models, when cluster size $n_i \geq 2$ for all i and $q \to \infty$, Parner (1998) and Gamst et al. (2009) have shown the consistency and asymptotic normality of the (marginal) MLEs under the gamma and log-normal frailty models, respectively. When cluster size $n_i = 1$ for all i, Barker and Henderson (2005) showed that the MLEs can be substantially biased in the finite sample. Thus, it is of interest to find a modification of m for the frailty models to eliminate such bias. Ha et al. (2010) proposed an adjusted profile marginal likelihood $p_w(m)$ with $w = \log \lambda_0$,

$$p_\omega(m) = \left[m - \frac{1}{2}\log\det\{D(m; \omega)/(2\pi)\}\right]\Bigg|_{\omega=\widetilde{\omega}},$$

where $\omega = (\omega_1, \ldots, \omega_D)^T$ with $\omega_k = \log \lambda_{0k}$, $D(m; \omega) = -\partial^2 m/\partial\omega^2$ is the adjustment term for eliminating λ_0, and $\widetilde{\omega}$ solves $\partial m/\partial\omega = 0$. They showed the relationship between those two likelihoods under the univariate gamma frailty model is given by, as $N = \min_{1\leq i\leq q} n_i \to \infty$,

$$m_p \approx p_w(m),$$

where

$$m_p = m_p(\beta, \alpha) = \log\left\{\int \exp(h_p)\, dv\right\} \tag{4.8}$$

is the partial marginal likelihood.

For parameter estimation in the frailty models, it is desirable to eliminate both nuisance parameters λ_{0k} (fixed) and frailties v_i (random). In the partial marginal likelihood m_p, λ_{0k}'s are first eliminated from the h-likelihood h by the profiling method and then eliminate v_i via integration $\int \exp(h_p)\, dv$. In the marginal likelihood approach, the nuisance random frailties v_i are first eliminated via integration $\int \exp(h)\, dv$. Then, the next step is to eliminate by profiling out nuisance parameters λ_{0k} from m. When the number of fixed nuisance parameters increase with the sample size, the simple profile likelihood $m^* = m|_{\lambda_0=\hat{\lambda}_0}$ (i.e., using m for estimating fixed parameters) may not work well. Here an adjusted profile likelihood such as $p_w(m)$ comes to rescue (Ha et al. 2010).

The partial marginal likelihood m_p does not involve nuisance parameters λ_{0k}, so that it might be intriguing to attempt to obtain the partial maximum marginal likelihood estimators (PMMLEs). However, m_p is not useful in practice due to intractable integration, not allowing a closed form even under the univariate gamma frailty model. Moreover, m_p involves high-dimensional integration with the dimension being the number of the frailties, so that numerical method such as Gauss–Hermit cannot be used. Even in the gamma frailty models, the EM algorithm is difficult to apply (Gu et al. 2004). Thus, in this book, we use the Laplace approximation $p_v(h_p)$ in (4.9) for m_p: for the details of Laplace approximations see Sect. 3.3.1.

4.3 Inference Procedures Using R

In this section, we review various likelihood procedures for fitting a semiparametric frailty model (4.1) with an arbitrary baseline hazard function. Then, we illustrate the R package **frailtyHL** with two well-known data sets and compare with various alternative likelihood procedures in R.

4.3.1 Review of Estimation Procedures

Table 4.1 shows historical evolvement of the estimating criteria for the log-normal and gamma frailty models.

• **H-likelihood versus PPL procedures**: From the definition of $p_\xi(\ell)$ in (3.9), two adjusted profile h-likelihoods $p_v(h_p)$ and $p_{\beta,v}(h_p)$ are defined as follows:

Table 4.1 Estimation criteria for the h-likelihood (HL(`mord, dord`)), PPL (`coxph, coxme`), and marginal likelihood (ML; `phmm`) for log-normal (LN) and gamma frailty models (FMs)

Method	Criterion		Literature
	β	α	
HL			
HL(0,1)	h_p	$p_{\beta,v}(h_p)$	Ha and Lee (2003)
HL(0,2)	h_p	$s_{\beta,v}(h_p)$	Ha and Lee (2003)
HL(1,1)	$p_v(h_p)$	$p_{\beta,v}(h_p)$	Ha et al. (2012)
HL(1,2)	$p_v(h_p)$	$s_{\beta,v}(h_p)$	Ha et al. (2012)
PPL			
`coxph`	h_p	$p_{\beta,v}(h_p)$	Therneau (2010) for LN FM
`coxph`	h_p	m	Therneau (2010) for gamma FM
`coxme`	h_p	$p_v(h_p)$	Therneau (2011) for LN FM
ML			
`phmm`	m	m	Donohue and Xu (2012) for LN FM

$$p_v(h_p) = \left[h_p - \frac{1}{2} \log \det\{H(h_p; v)/(2\pi)\} \right] \Bigg|_{v=\hat{v}}, \tag{4.9}$$

where $H(h_p; v) = -\partial^2 h_p/\partial v^2$ and $\hat{v}$ solves $\partial h_p/\partial v = 0$, which is the first-order Laplace approximation of m_p in (4.8), and

$$p_{\beta,v}(h_p) = \left[h_p - \frac{1}{2} \log \det\{H(h_p; \beta, v)/(2\pi)\} \right] \Bigg|_{\beta=\hat{\beta}, v=\hat{v}}, \tag{4.10}$$

where $H(h_p; \beta, v) = -\partial^2 h_p/\partial(\beta, v)^2$ and $(\hat{\beta}, \hat{v})$ solves $\partial h_p/\partial(\beta, v) = 0$. This becomes Cox and Reid's (1987) adjusted profile marginal likelihood eliminating fixed effects β by conditioning their asymptotic sufficient statistics $\hat{\beta}$, in addition to eliminating random effects v by the first-order Laplace approximation. For estimation of β, the h-likelihood methods allow for the Laplace approximation $p_v(h_p)$ to m_p, but the PPL procedures always use h_p. For estimation of α, the PPL methods use adjusted profile h-likelihoods $p_v(h_p)$ and $p_{\beta,v}(h_p)$ that give the partial maximum likelihood estimators (PMLEs) and partial restricted maximum likelihood estimators (PREMLEs), respectively. However, the PPL method does not compute all the terms necessary to implement these adjusted profile likelihoods as we shall discuss. In this chapter, the MLEs and REMLEs for the parametric baseline hazard models are extended to the PMLEs and PREMLEs for the nonparametric baseline hazard models, respectively.

Furthermore, the h-likelihood method allows the partial restricted likelihood based on the second-order Laplace approximation $s_{\beta,v}(h_p)$ for the PREMLEs. The corresponding second-order approximation is

$$s_{\beta,v}(h_p) = p_{\beta,v}(h_p) - \{F(h_p)/24\}, \quad (4.11)$$

where

$$F(h_p) = \mathrm{tr}[-\{3(\partial^4 h_p/\partial v^4) + 5(\partial^3 h_p/\partial v^3)H(h_p, v)^{-1}(\partial^3 h_p/\partial v^3)\}H(h_p, v)^{-2}]|_{v=\widehat{v}}.$$

To reduce the computational burden, we use $F(h)$ instead of $F(h_p)$. It is recommended to use the second-order approximation in the gamma frailty model (Ha et al. 2010).

For handling tied event times, the h-likelihood procedures use the Breslow's method, while the PPL procedures allow for the Efron's method.

• **Comparison of the h-likelihood procedures**: The `frailtyHL()` function provides estimators based on various orders of Laplace approximations for the fixed effects and dispersion parameters. As the orders in **mord** and **dord** increase, the biases of the estimators decrease, but the procedures become computationally more intensive due to calculation of the extra terms. Denote `HL(a,b)` for the h-likelihood method using order `"a"` in **mord** and order `"b"` for **dord**. We recommend using HL(1,1) with the log-normal frailty and HL(1,2) with the gamma frailty. However, for the log-normal frailty model, HL(0,1), and for the gamma frailty model, HL(0,2) often perform well if α is not large. Note that the asymptotic variance matrices of $\hat{\tau} = (\hat{\beta}, \hat{v})$ and $\hat{\alpha}$ are directly obtained from the inverses of Hessian matrix $\{-\partial^2 h_p/\partial\tau^2\}^{-1}$ and $\{-\partial^2 p_{\beta,v}(h_p)/\partial\alpha^2\}^{-1}$, respectively; the **frailtyHL** package provides the standard errors (SEs) of $\hat{\alpha}$ as well as $\hat{\beta}$.

• **PPL and ML procedures**: Based on the PPL methods, the `coxph()` and `coxme()` functions, respectively, implement the PREMLE and PMLE of α for the log-normal frailty model, and the `coxph()` function also implements the MLEs, maximizing the marginal likelihood m, for α for the gamma frailty model. For comparison, we present the Breslow's and Efron's methods for handling ties in survival times in the `coxph()` and `coxme()` functions in Example 4.1; Therneau (2010) recommended the Efron's method. For the log-normal frailty model, the MLE maximizing m is available via the `phmm()` function, but care must be taken to ensure that the MCEM algorithm converges (Donohue and Xu 2012). However, the MLE from `phmm()` can be biased in finite sample, particularly for smaller cluster sizes (e.g., when cluster size n_i is 1 or 2) (Ha et al. 2010).

Furthermore, with the log-normal frailty, the `coxph()` function uses the existing codes in the LMMs so that it misses the term $\partial\hat{v}/\partial\alpha$ in (4.41) of Appendix 4.7 when solving the score equation $\partial p_{\beta,v}(h_p)/\partial\alpha = 0$; the resulting PREMLE can lead to an underestimation of the parameters, especially when the cluster size n_i is small or the censoring proportion is high (Ha et al. 2010; Lee et al. 2017b). To overcome this problem, for the gamma frailty model, Therneau and Grambsch (2000) have developed the codes for the MLE for α because the gamma frailty allows an explicit form of the marginal likelihood. For the frailty models, Ha et al. (2010) showed that the h-likelihood method yields the least biased estimators.

Remark 4.2 In the gamma frailty model (4.1), given α the MHLE for β that maximizes h_p is the same as the MLE, which is obtained by maximizing the marginal likelihood m (hence $p_v(h)$). However, the two methods are different in the estimation of α. The proofs of these two statements are given in Appendix 4.7.3. Ha and Lee (2003) and Ha et al. (2011) showed that inference on β is less sensitive against a misspecification of the frailty distribution. □

• **Numerical example**: We investigate the utility of various likelihoods, using independent survival data with $n_i = 1$ for all i as an extreme case. For this purpose, we considered a simulated data set from the gamma frailty model (4.1) with $(q, n_i) = (100, 1)$. That is, we generated data assuming an exponential baseline hazard $\lambda_0(t) = 1$, one covariate following the standard normal distribution with $\beta = 1$, and the variance of the gamma frailty $\alpha = 1$, ensuring existence of the frailty. The corresponding censoring times were generated from an exponential distribution to achieve about 5% censoring.

We fitted the gamma frailty model to the simulated data set. For implementation of the likelihood methods, we used a simple grid search method; in the inner loop, given α, we maximize h_p for (β, v), and in the outer loop, given (β, v), the following eight likelihoods are maximized for α (Ha et al. 2010):

(i) MLEs: m, $p_v(h)$, $s_v(h)$, $p_w(m)$,
(ii) PMLEs: $p_v(h_p)$, $s_v(h_p)$,
(iii) PREMLEs: $p_{\beta,v}(h_p)$, $s_{\beta,v}(h_p)$.

The eight profile likelihoods are plotted against α in Fig. 4.2. Here we find, compared to the true value $\alpha = 1$, that three likelihoods ($p_w(m)$, $s_v(h_p)$ and $s_{\beta,v}(h_p)$) retrieve the true frailty, whereas the remaining likelihoods (m, $p_v(h)$, $s_v(h)$, $p_v(h_p)$ and $p_{\beta,v}(h_p)$) do not. Thus, in the univariate gamma frailty model, further elaborate approximations ($s_v(h_p)$, $s_{\beta,v}(h_p)$ or a modification ($p_w(m)$) provide more accurate estimation results, together with retrieval of the frailty term. These results also confirm the simulation results from Barker and Henderson (2005), Ha (2007) and Ha et al. (2010). With the bivariate survival data, we have found that all eight likelihoods retrieve the true frailty well (not shown). Thus, care is necessary, particularly in the independent data cases.

4.3.2 Fitting Algorithm and Inference

Given α, the joint maximization of h_p for (β, v) (i.e., $\partial h_p / \partial(\beta, v) = 0$) leads to the iterative least squares (ILS) score equations (Appendix 4.7.4):

$$\begin{pmatrix} X^T W^* X & X^T W^* Z \\ Z^T W^* X & Z^T W^* Z + Q \end{pmatrix} \begin{pmatrix} \widehat{\beta} \\ \widehat{v} \end{pmatrix} = \begin{pmatrix} X^T w^* \\ Z^T w^* + R \end{pmatrix}, \quad (4.12)$$

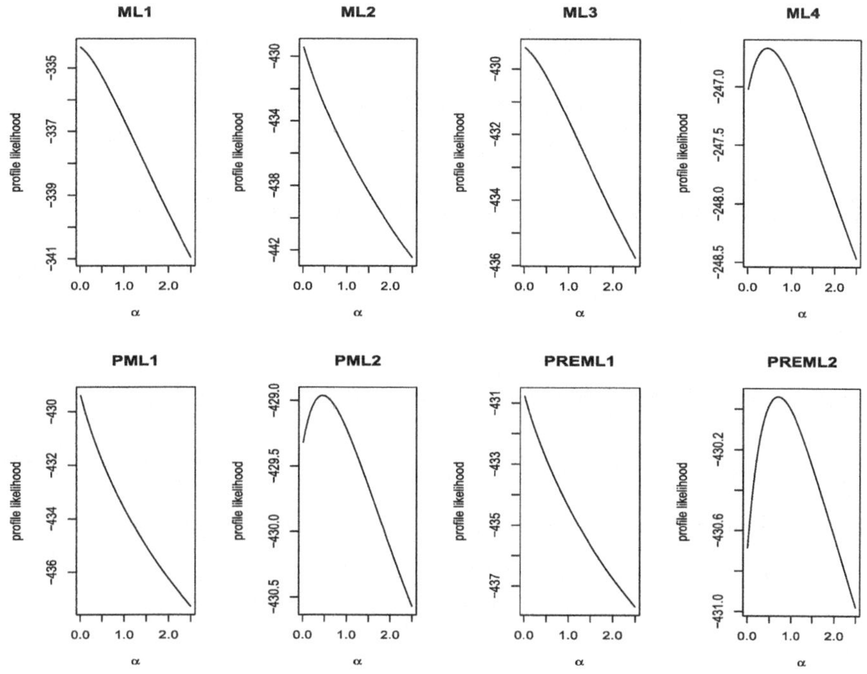

Fig. 4.2 Profile likelihoods for frailty parameter α in a simulated data set; ML1, m; ML2, $p_v(h)$; ML3, $s_v(h)$; ML4, $p_w(m)$ with $w = \log \lambda_0$; PML1, $p_v(h_p)$; PML2, $s_v(h_p)$; PREML1, $p_{\beta,v}(h_p)$; PREML2, $s_{\beta,v}(h_p)$

where X and Z are $n \times p$ and $n \times q$ model matrices for β and v whose ijth row vectors are x_{ij}^T and z_{ij}^T, respectively, and $z_{ij} = (z_{ij1}, \ldots, z_{ijq})^T$ is a $q \times 1$ group indicator vector whose rth element is $\partial\eta_{ij}/\partial v_r$, $W^* = -\partial^2 h_p/\partial\eta^2$ is the symmetric weight matrix given in (4.35) of Appendix 4.7.4, and

$$Q = \text{diag}(-\partial^2 \ell_{2i}/\partial v_i^2)$$

is a $q \times q$ diagonal matrix. Here

$$w^* = W^*\eta + (\delta - \mu)$$

with $\eta = X\beta + Zv$, $\mu = \exp(\log \Lambda_0 + \eta)$ and

$$R = Qv + (\partial\ell_2/\partial v).$$

Note here that $R = 0$ under the log-frailty, if $v \sim N(0, \alpha I_q)$. In particular, under the Cox PH model without the frailty, the joint score Eq. (4.12) reduce to the ILS Eq. (2.10), given by

$$(X^T W^* X)\widehat{\beta} = X^T w^*.$$

Let

$$\mathbf{P} = \begin{pmatrix} X & Z \\ \mathbf{0} & I_q \end{pmatrix} \text{ and } \mathbf{V} = \begin{pmatrix} W^* & \mathbf{0} \\ \mathbf{0} & Q \end{pmatrix}.$$

Then the ILS Eq. (4.12) can be written in a new simple matrix form as in (3.14):

$$(\mathbf{P}^T \mathbf{V} \mathbf{P})\hat{\tau} = \mathbf{P}^T \mathbf{y}_0^*, \tag{4.13}$$

where $\mathbf{y}_0^* = (w^{*T}, R^T)^T$. In particular, if $v \sim N(0, \alpha I_q)$, then $R = \mathbf{0}$ and $\mathbf{y}_0^* = (w^{*T}, \mathbf{0}^T)^T$. Note that $H_p = H(h_p; \beta, v) = -\partial^2 h_p / \partial(\beta, v)^2 = \mathbf{P}^T \mathbf{V} \mathbf{P}$.

In Appendix 4.7.5, we outline how to obtain the PREMLE for α, by solving

$$\partial p_{\beta,v}(h_p)/\partial\alpha = 0. \tag{4.14}$$

For the log-normal frailty model with $v_i \sim N(0, \alpha)$, the PREMLE is given by

$$\hat{\alpha} = \frac{\hat{v}^T \hat{v}}{q - \gamma}, \tag{4.15}$$

where $\gamma = -\alpha \mathrm{tr}\{\widehat{H}_p^{-1}(\partial \widehat{H}_p/\partial\alpha)\}$ and $\widehat{H}_p$ is given in Appendix 4.7.5. This is an extension of (3.17) for the REMLE from the h-likelihood h under HGLMs to that for the PREMLE from the partial h-likelihood h_p under semiparametric frailty models.

- **Fitting algorithm**:
  Suppose that HL(0,1) is used. The fitting algorithm is as follows.

- **Step 1**: Take (0,0,0.1) as the initial guesses of components of (β, v, α).
- **Step 2**: Given $\hat{\alpha}$, the new estimates $(\hat{\beta}, \hat{v})$ are obtained by solving the ILS equations (4.12), i.e., (4.13). Then, given $(\hat{\beta}, \hat{v})$, a new estimate $\hat{\alpha}$ is obtained by solving $\partial p_{\beta,v}(h_p)/\partial\alpha = 0$. In particular, for the log-normal frailty with $v_i \sim N(0, \alpha)$, we have a simple PREMLE, given in (4.15).
- **Step 3**: Repeat Step 2 until the maximum absolute difference between the previous and current estimates for (β, v) and α is less than 10^{-6}.

After the convergence criterion has met, we compute the estimates of the SEs of $\hat{\beta}$, $\hat{v} - v$ and $\hat{\alpha}$. Note that the estimates of $\mathrm{var}(\hat{\tau} - \tau)$ and $\mathrm{var}(\hat{\alpha})$ are obtained from the inverses of $H_p = -\partial^2 h_p/\partial\tau^2 = \mathbf{P}^T \mathbf{V} \mathbf{P}$ and $-\partial^2 p_\tau(h_p)/\partial\alpha^2$, respectively.

- **Test for the frailty parameter**:
  Testing the absence of a frailty effect,

$$H_0 : \alpha = \mathrm{var}(v_i) = 0,$$

is equivalent to test $v_i = 0$ for all i. Care is necessary because such a null hypothesis is on the boundary of the parameter space ($\alpha \geq 0$). Thus, the standard chi-square distribution cannot be applied. The null distribution for the likelihood ratio test (LRT) statistic follows an asymptotic chi-square mixture distribution, i.e., a mixture of χ_0^2 and χ_1^2 with equal weights of 0.5. Here the χ_0^2 distribution gives probability mass 1 to the value 0 (Self and Liang 1987; Stram and Lee 1994; Vu et al. 2001; Vu and Knuiman 2002; Ha and Lee 2005a; Xu et al. 2009). For testing the need for a random component (i.e., a frailty term), we use the LRT statistic, denoted by LR, based on the partial restricted likelihood (Ha et al. 2011, 2016a); it is calculated as

$$\text{LR} = -2[p_\beta(h_p) - p_{\beta,v}(h_p)],$$

where $p_\beta(h_p)$ is the likelihood under $H_0 : \alpha = 0$. At a 5% significance level, the critical value of $\chi^2_{1,0.10} = 2.71$ under the chi-square mixture distribution should be used.

Generally, we denote the mixture of two chi-square distributions with k_1 and k_2 degrees of freedom, with equal weights 0.5, by $\chi^2_{k_1:k_2}$ (Verbeke and Molenberghs 2009). Then, the p-value is calculated as

$$\begin{aligned} p &= P(\chi^2_{k_1:k_2} > \text{LR}) \\ &= \frac{1}{2}P(\chi^2_{k_1} > \text{LR}) + \frac{1}{2}P(\chi^2_{k_2} > \text{LR}). \end{aligned}$$

For example, the p-value for LRT with the $\chi^2_{0:1}$ distribution is given by

$$\begin{aligned} p &= P(\chi^2_{0:1} > \text{LR}) \\ &= \frac{1}{2}P(\chi^2_0 > \text{LR}) + \frac{1}{2}P(\chi^2_1 > \text{LR}) \\ &= \frac{1}{2}P(\chi^2_1 > \text{LR}) \end{aligned}$$

since χ^2_0 is the distribution defined by $P(\chi^2_0 = 0) = 1$.

4.3.3 Implementation Using R

In this section, we outline the **frailtyHL** package (Ha et al. 2018) to fit the semiparametric frailty model (4.1). The main function, `frailtyHL()`, fits the log-normal frailty model as a default as follows:

```
> frailtyHL(Surv(time, status) ~ x + (1|id),
+            RandDist = "Normal",
+            mord = 0, dord = 1,
+            Maxiter = 200, convergence = 10^-6,
+            varfixed = FALSE, varinit= 0.1)
```

Inclusion of the option `RandDist="Gamma"` allows to fit the gamma frailty model. The first argument is a formula object, with the response on the left-hand side of the $\sim$ operator, and the terms for the fixed and random effects on the right. The response is a survival object as returned by the `Surv` function (Therneau 2010). Here, `time` and `status` denote survival time and censoring indicator taking 1 (0) for uncensored (censored) observation; `x` denotes a fixed covariate and `id` denotes the subject identifier. The expression (**`1|id`**) specifies a random intercept model ((**`x|id`**) would specify a random slope model). The parameters `mord` and `dord` are the orders of Laplace approximations to fit the mean parameters (`mord`=0 or 1) and the dispersion parameters (`dord`= 1 or 2), respectively. The `Maxiter` parameter specifies the maximum number of iterations and `convergence` specifies the tolerance of the convergence criterion. If `varfixed` is specified as `TRUE` (or `FALSE`), the value of one or more of the variance terms for the frailties is fixed (or estimated) with starting value (e.g., 0.1) given in the `varinit`.

Previously, the frailty models have been implemented in several R functions such as the `coxph()` function in the **survival** package (Therneau 2010) and the `coxme()` function in the **coxme** package (Therneau 2011), based on the PPL, the `phmm()` function in the **phmm** package (Donohue and Xu 2012), based on a Monte Carlo EM (MCEM) method, and the `frailtyPenal()` function in the **frailtypack** package (Gonzalez et al. 2012), based on penalized marginal likelihood. The **phmm** package fits one-component frailty models, although it does allow for multivariate frailties. The `coxme()` function can also fit the multicomponent model as shown in Chap. 5. Results from the **`frailtyHL`** package are now compared with those from **`survival`**, **`coxme`**, and `phmm` packages.

4.3.4 Illustration

Example 4.1 (*Kidney infection data*) To demonstrate the differences among various estimation methods in small cluster size, we use the kidney data set in Sect. 1.2.1. The data consist of times until the first and second recurrences ($n_i = 2$) of kidney infection in 38 ($q = 38$) patients using a portable dialysis machine. The recorded information for the first three patients is as follows:

```
> library(frailtyHL)
> head(kidney)
  id time status age sex disease frail
1  1    8      1  28   1   Other   2.3
2  1   16      1  28   1   Other   2.3
3  2   23      1  48   2      GN   1.9
4  2   13      0  48   2      GN   1.9
5  3   22      1  32   1   Other   1.2
6  3   28      1  32   1   Other   1.2
```

Table 4.2 Comparison among different estimation methods for the kidney infection data

Method	Sex	Age	Patient
	$\hat{\beta}_1$ (SE)	$\hat{\beta}_2$ (SE)	$\hat{\alpha}$ (SE)
Cox's model			
	−0.821 (0.299)	0.002 (0.009)	– (–) – (–)
Log-normal model			
HL(0,1)	−1.380 (0.431)	0.005 (0.012)	0.535 (0.338)
HL(1,1)	−1.414 (0.432)	0.005 (0.012)	0.545 (0.340)
`coxph` (Breslow)	−1.388 (0.441)	0.005 (0.012)	0.551 (–)
`coxph` (Efron)	−1.411 (0.445)	0.005 (0.013)	0.569 (–)
`coxme` (Breslow)	−1.332 (0.414)	0.005 (0.012)	0.440 (–)
`coxme` (Efron)	−1.355 (0.417)	0.004 (0.012)	0.456 (–)
`phmm`	−1.329 (0.452)	0.004 (0.012)	0.378 (–)
Gamma model			
HL(0,2)	−1.691 (0.483)	0.007 (0.013)	0.561 (0.280)
HL(1,2)	−1.730 (0.485)	0.007 (0.013)	0.570 (0.281)
`coxph` (Breslow)	−1.557 (0.456)	0.005 (0.012)	0.398 (–)
`coxph` (Efron)	−1.587 (0.461)	0.005 (0.012)	0.412 (–)

The variable **`time`** contains the time until infection since an insertion of the catheter, and **`status`** is a censoring indicator (1 if infection has occurred and 0 otherwise). The covariates are **`age`**, **`sex`**, and **`disease`**: **`age`**, patient's age (in years); **`sex`**, 1 for male and 2 for female; **`disease`**, type of disease (GN, AN, PKD, or Other). The **`frail`** is the frailty estimate using the log-normal frailty model.

We fit the frailty models with two covariates (sex and age) using the functions, `frailtyHL()`, `coxph()`, `coxme()`, and `phmm()`. The results are summarized in Table 4.2. In the PPL procedures (`coxph()` and `coxme()`), the Breslow's method provides slightly smaller estimates for α than the Efron's method. With the log-normal frailty, the PREML procedures (`fraityHL()` and `coxph()`) give larger estimates for α than the ML (`phmm()`) and PML (`coxme()`) procedures. However, both MLE and PMLEs from `phmm()` and `coxme()` are somewhat different because the cluster size is small as $n_i = 2$ for all i. For the gamma frailty, `coxph()` uses the ML procedure, but it still gives smaller estimates for α than the PREML (h-likelihood) procedures. Compared with the h-likelihood methods, the PPL methods are computationally more efficient, but has larger biases (Ha et al. 2010). From Table 4.2, we see that the absolute magnitude and SEs of $\hat{\beta}_1$ and $\hat{\beta}_2$ in all frailty models are larger than those from the Cox model.

Example 4.2 (*Rat's tumorigenesis data (litter-matched rat data)*) The rat data set in Sect. 1.2.2 is based on a tumorigenesis study of 50 ($q = 50$) litters of female rats, with a litter size of $n_i = 3$. Event time (**time**) is time to development of tumor, measured in weeks.

We fit the frailty models with one covariate, **rx** (1 = drug; 0 = placebo), using `frailtyHL()`. Below, we present the R codes and results for the log-normal frailty model with HL(1,1). The resulting output shows that the effect of **rx** is significant (`t-value = 2.808 with p-value = 0.005`), implying that the **rx** group has a significantly higher risk than the control group. Here, the variance estimate of the frailty is $\hat{\alpha} = 0.427$ (with SE = 0.423). The difference in deviance (based on the partial REML) $-2p_{\beta,v}(h_p)$ between the Cox model without frailty and the log-normal frailty model is

$$364.15 - 362.56 = 1.59(< 2.71),$$

so that the corresponding p-value is $0.5P(\chi_1^2 > 1.59) = 0.104$. This indicates that the frailty effect is nonsignificant (i.e., $\alpha = 0$) at a 5% significance level. Thus, we can expect that the analyses from the frailty model and the Cox model should be similar. Note that the results from the Cox model without frailty are available by adding the two arguments **varfixed=TRUE** and `varinit=0` in the `frailtyHL` procedure (see below).

```
############### Fitting Cox's model ###############
> library(frailtyHL)
> data(rats)
> Cox<-frailtyHL(Surv(time,status)~rx+(1|litter),rats,
+                  varfixed=TRUE, varinit=0)
[1] "Results from the Cox model"
[1] "Model for conditional hazard:"
Surv(time, status)~rx + (1|litter)
[1]"Method : HL(0,1)"
[1]"Estimates from the mean model"
   Estimate Std. Error t-value  p-value
rx    0.8982      0.3174    2.83 0.004655
[1]"Estimates from the dispersion model"
         Estimate Std. Error
litter   "0"    "NULL"
         -2h0  -2*hp   -2*p_b,v(hp)
```

```
[1,] 363.69 363.69   364.15
       cAIC   pAIC   rAIC
[1,] 365.69 365.69 364.15


############# Fitting log-normal frailty model #############
> LNFM<-frailtyHL(Surv(time,status)~rx+(1|litter), rats,
+                    RandDist="Normal", mord=1, dord=1)
[1]"Results from the log-normal frailty model"
[1]"Model for conditional hazard:"
Surv(time, status) ~ rx + (1 |litter)
[1]"Method : HL(1,1)"
[1]"Estimates from the mean model"
   Estimate Std. Error t-value  p-value
[1]  0.9107     0.3226   2.823 0.004754
[1]"Estimates from the dispersion model"
       Estimate Std. Error
litter   0.4272     0.4232
       -2h0  -2*hp -2*p_v(hp) -2*p_b,v(hp)
[1,] 335.97 397.36     362.14     362.56
       cAIC   pAIC   rAIC
[1,] 362.22 366.14 364.56
```

The R codes and results for the gamma frailty model with HL(1,2) are presented below. The output shows that the results are similar to ones from the log-normal frailty model, particularly in estimation of β. The deviance difference (based on partial restricted likelihood) between the Cox model and gamma frailty model using the second-order approximation $-2s_{\beta,v}(h_p)$ is $364.15 - 362.12 = 2.03 (< 2.71)$, again indicating absence of the frailty effect (i.e., $\alpha = 0$) as shown in the log-normal frailty analysis.

```
############# Fitting gamma frailty model #############
> GFM<-frailtyHL(Surv(time,status)~rx+(1|litter), rats,
+                   RandDist="Gamma", mord=1, dord=2)
[1]"Results from the gamma frailty model"
[1]"Model for conditional hazard:"
Surv(time, status) ~ rx + (1|litter)
[1]"Method : HL(1,2)"
[1]"Estimates from the mean model"
   Estimate Std. Error t-value  p-value
rx   0.9126     0.3236    2.82 0.004806
[1]"Estimates from the dispersion model"
       Estimate Std. Error
litter   0.5757     0.5977
```

```
      -2h0   -2*hp -2*p_v(hp)   -2*s_v(hp)
[1,] 331.60 413.85    365.35    361.71
      -2*p_b,v(hp)   -2*s_b,v(hp)
       365.77         362.12
       cAIC    pAIC    rAIC
[1,] 365.30 365.71 364.12
```

Now we compare the results from **frailtyHL** and other packages. We consider three functions (**coxph**, **coxme**, and **phmm**) for the log-normal frailty model and the **coxph** function for the gamma frailty model. The codes of **coxph**, **coxme**, and **phmm** for fitting the log-normal frailty model are as follows:

```
> coxph(Surv(time, status)~rx+frailty(litter, dist="gauss"),
+       method = "breslow", rats)
> coxme(Surv(time, status)~ rx + (1|litter),
+       ties="breslow", rats)
> phmm(Surv(time, status)~ rx+(1|litter),rats,Gbs = 2000,Gbsvar = 3000,
+      VARSTART = 1, NINIT = 10, MAXSTEP = 200, CONVERG=90)
```

Table 4.3 Comparison among different estimation methods for the rat data

Method	Rx	Litter
	$\hat{\beta}$ (SE)	$\hat{\alpha}$ (SE)
Cox's model		
	0.898 (0.317)	– (–)
Log-normal model		
HL(0,1)	0.906 (0.323)	0.427 (0.423)
HL(1,1)	0.911 (0.323)	0.427 (0.423)
coxph (Breslow)	0.905 (0.322)	0.395 (–)
coxph (Efron)	0.913 (0.323)	0.412 (–)
coxme (Breslow)	0.905 (0.322)	0.406 (–)
coxme (Efron)	0.913 (0.323)	0.426 (–)
phmm	0.920 (0.326)	0.449 (–)
Gamma model		
HL(0,2)	0.908 (0.324)	0.575 (0.598)
HL(1,2)	0.913 (0.324)	0.576 (0.598)
coxph (Breslow)	0.906 (0.323)	0.474 (–)
coxph (Efron)	0.914 (0.323)	0.499 (–)

Table 4.3 summarizes the results. Even though the cluster size $n_i = 3$ is not large, the results are similar because the frailty effects are not significant ($\alpha = 0$). For example, the PMLE and MLE for α from **coxme** and **phmm** were somewhat different in Table 4.2, but they become similar in Table 4.3.

Next, an example of using **coxph** to fit the gamma frailty model is given below:

```
> coxph(Surv(time, status)~rx + frailty(litter,dist="gamma"),
+        method = "breslow", rats)
```

The results from **frailtyHL** (HL(0,2), HL(1,2)) and **coxph** with gamma frailty are also presented in Table 4.3. For estimation of β, both results from **frailtyHL** and **coxph** are similar, but they are somewhat different for α. That is, our PREMLEs from **frailtyHL** ($\hat{\alpha} = 0.575$ with HL(0,2) and $\hat{\alpha} = 0.576$ with HL(1,2)) are larger than the MLEs from **coxph** ($\hat{\alpha} = 0.474$ with Breslow's method and $\hat{\alpha} = 0.499$ with Efron's method).

4.4 Model Selection

In this section, we first review the basic concepts of the Akaike information. Then, we present three forms of Akaike information criterion (AIC) based on the partial h-likelihood for the frailty models (4.1).

4.4.1 Basic Concept of Akaike Information

Suppose that data $y = (y_1, \ldots, y_n)^T$ are generated from a true underlying distribution with density g, and that $f_\theta = f(\cdot|\theta)$ is a family of approximating models (or assumed models) with unknown parameters $\theta \in \Theta$. Akaike (1973) considered the Kullback–Leibler (1951) distance as a fundamental basis for model selection, defined by

$$I(f_\theta, g) = E_g\{\log g(y) - \log f_\theta(y)\},$$

where E_g denotes the expectation with respect to the true density g. Smaller values of $I(f_\theta, g)$ correspond to a better approximation of g by f_θ, and the minimum is obtained for some $\theta_0 \in \Theta$. If the true distribution g belongs to the fitted class of models $F = \{f_\theta, \theta \in \Theta\}$, then $f_{\theta_0} = g$ and $I(f_{\theta_0}, g) = 0$. In general, g may not be in F, so $I(f_\theta, g) \geq 0$. In practice θ should be estimated from the data y, so that $I(f_\theta, g)$ is approximated by $I(f_{\hat{\theta}}, g)$, where $\hat{\theta} = \hat{\theta}(y)$ is usually the MLE. The quality of the approximation of the true g by the class F is assessed, on average, by the quantity

$$E_g I(f_{\hat{\theta}}, g) = E_{g(y^*)} \log g(y^*) - E_{g(y)} E_{g(y^*)} \log f\{y^*|\hat{\theta}(y)\},$$

where y^* is another realization independent of y. When we are comparing different classes of models, the constant $E_{g(y^*)} \log g(y^*)$ can be ignored, and the relative fit of the competing models can be assessed using the Akaike information (AI), defined by

$$\mathrm{AI} = -2E_{g(y)}E_{g(y^*)} \log f\{y^*|\hat{\theta}(y)\}.$$

The AIC is an estimator of the AI, defined by

$$\mathrm{AIC} = -2 \log f\{y|\hat{\theta}(y)\} + 2K,$$

where K is the number of free parameters in the model F. When $\hat{\theta}(y)$ is the MLE and the approximating class of models F includes g,

$$E(\mathrm{AIC}) = \mathrm{AI} + o(1)$$

as the sample size $n \to \infty$; that is, the AIC is unbiased for the AI to a first order of n (Akaike 1973; Burnham and Anderson 2002).

4.4.2 Three AICs for the Frailty Models

To extend the AIC to the frailty models, we need to consider the random effects and nuisance parameters λ_0. For the semiparametric frailty models, we form the AICs using the partial h-likelihood h_p, which eliminates λ_0.

Throughout this book, for the frailty models we use three AICs (Ha et al. 2007a, 2012) based on h_p as follows: The conditional AIC (cAIC), based on ℓ_p in h_p (4.5), is defined by

$$\mathrm{cAIC} = -2\ell_p + 2\mathrm{df}_c, \tag{4.16}$$

where

$$\ell_p = \sum_{ij} \delta_{ij}(x_{ij}^T\widehat{\beta} + \widehat{v}_i) - \sum_k d_{(k)} \log\left\{ \sum_{(i,j)\in R_{(k)}} \exp(x_{ij}^T\widehat{\beta} + \widehat{v}_i) \right\},$$

and

$$\mathrm{df}_c = \mathrm{df}_c(\beta, v, \alpha) = \mathrm{trace}(H_p^{-1}H_p^*)$$

is an "effective degree of freedom adjustment" for estimating the fixed and random effects, computed using the Hessian matrices $H_p = -\partial^2 h_p/\partial\tau^2$ and $H_p^* = -\partial^2\ell_p/\partial\tau^2$ with $\tau = (\beta^T, v^T)^T$. In the Cox PH model without frailty, a degree of freedom df_c in (4.16) becomes the number of the fixed effects, p, i.e., the dimension of β. In general, df_c involves the fixed effects β, the random effects v and the frailty parameter α.

It can be shown that the cAIC in (4.16) is an extension of the AICs, defined in several existing models, to frailty models:
(i) For the Cox PH model without frailty it becomes the standard AIC (e.g., AIC in SAS PROC PHREG) using the Cox's partial likelihood.
(ii) For the LMMs with known variances, it yields the conditional AIC (Vaida and Blanchard 2005). Consider the LMMs with responses y and random effects v. Let y^* be an independently replicated outcome from the same distribution as y given v. Following Vaida and Blanchard (2005), the conditional Akaike information (cAI) is defined as

$$\text{cAI} = -2E_{g(y,v)}E_{g(y^*|v)}\ell_1(\hat{\beta}; y^*|v = \hat{v}),$$

where $\hat{\beta}$ and $\hat{v}$ are the MHLEs based on y. In the LMMs, Vaida and Blanchard (2005) showed that under some regular conditions,

$$E(\text{cAIC}) = \text{cAI} + o(1)$$

for large q and n_i; that is, the cAIC is an asymptotically unbiased estimator of cAI. Furthermore, Donohue et al. (2011) also showed that the asymptotic unbiasedness still holds in extended random-effect models such as the GLMMs and frailty models. In particular, under the frailty models, the corresponding cAI is defined by

$$\text{cAI} = -2E_{g(y_0,v)}E_{g(y_0^*|v)}\ell_1^*(\hat{\beta}; y_0^*|v = \hat{v}),$$

where $y_0 = (y, \delta)$ are generated under the frailty models (3.1) and y_0^* is another outcome which is independent of y_0.

Similarly, we define a partial marginal AIC (pAIC),

$$\text{pAIC} = -2m_p + 2\text{df}_p\,, \tag{4.17}$$

where m_p is the partial marginal likelihood in (4.8) and df_p is the number of fixed parameters (β, α). Since the computation of m_p is generally difficult, we use its Laplace approximation $p_v(h_p)$ or $s_v(h_p)$ (Ha et al. 2012). Xu et al. (2009) proposed a marginal AIC (mAIC),

$$\text{mAIC} = -2m^* + 2\text{df}_m, \tag{4.18}$$

where $m^* = m|_{\lambda_0=\hat{\lambda}_0}$ is a profile marginal likelihood after eliminating nuisance parameters λ_0. They used numerical approximations such as Laplace approximation based on $p_v(h)$ when it is difficult to obtain m. In the LMMs without λ_0, pAIC and mAIC become identical.

Similarly, the restricted AIC (rAIC) based on the partial restricted likelihood $p_{\beta,v}(h_p)$ in (4.10) is defined by

$$\text{rAIC} = -2p_{\beta,v}(h_p) + 2\text{df}_r\,, \tag{4.19}$$

where df_r is the number of dispersion parameters. It can be shown that in the LMMs, rAIC yields the AIC in `SAS PROC MIXED` (Wolfinger 1993) based upon the restricted likelihood for selecting a specific covariance structure, confirming that the quantity rAIC is an extension of the AIC in SAS. In the LMMs with an explicit form of the marginal likelihood, the mAIC is easily obtained by the standard statistical software, such as R function **`lme`** or **`SAS PROC MIXED`**.

For the pAIC and rAIC for the gamma frailty model using HL(0,2) or HL(1,2), we use the corresponding second-order approximations, defined by pAIC $= -2s_v(h_p) + 2\mathrm{df}_\mathrm{p}$ and rAIC $= -2s_{\beta,v}(h_p) + 2\mathrm{df}_\mathrm{r}$. The `frailtyHL` package provides three AICs, **`(cAIC, rAIC, pAICs).`**The cAIC is for model selection involving (v, β, α), the pAIC similarly involving (β, α) and rAIC involving α. Malfunction of cAIC and pAIC occur when $\hat{\alpha} = 0$ as will be seen.

In this book, we make model selection using the LRTs if models are nested. For non-nested models, we use the above AICs to select the final model.

Remark 4.3 The relative merits and disadvantages of those three AIC criteria are as follows:

- One can select a model that minimizes the AIC values. If the AIC difference is larger than 1, then the choice can be made (Sakamoto et al. 1986). However, if the difference is less than 1, a simpler model can be selected by a parsimony principal (Donohue et al. 2011).
- In the LMMs, Vaida and Blanchard (2005) demonstrated that the mAIC and its small sample correction are inappropriate when the interest is on clusters (Liang et al. 2008). Furthermore, Greven and Kneib (2010) showed that the mAIC is not an asymptotically unbiased estimator of the Akaike information due to the boundary problems regarding the random-effect variances, and that favors simpler models without random effects. Moreover, implementation of the mAIC is not straightforward because the marginal likelihood is hard to compute without having an explicit analytic form as in the GLMMs (Yu and Yau 2012). In this book, we use the Laplace approximations to compute the mAIC. Care is needed if $\hat{\alpha}$ is near 0, because it prefers a simpler model.
- For the cAIC, several authors have argued that ignoring the uncertainty in estimation of the random-effect covariance matrix can lead to a bias, so they have proposed using of the corrected AICs (Liang et al. 2008; Greven and Kneib 2010; Yu et al. 2013). However, Vaida and Blanchard (2005) and Donohue et al. (2011) recommended the cAIC without correction because under some regular conditions, the difference between cAIC and its corrected versions is asymptotically negligible. Ha et al. (2007a) illustrated that the cAIC selects a more complicated model when $\hat{\alpha} = 0$: see also Yu et al. (2013).
- The rAIC cannot be used to compare models with different fixed and random effects (β, v) and the mAIC cannot compare models with different random effects v. □

Below we illustrate how to use the three AICs for model selection among various models.

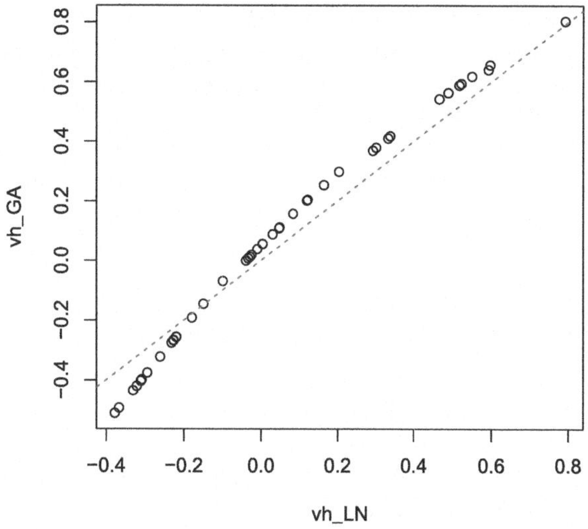

Fig. 4.3 Estimated log-frailty (vh-GA) in the gamma frailty model against estimated log-frailty (vh-LN) in the log-normal frailty model

Example 4.3 (*Illustration with the rat data used in Example* 4.2) For illustration of model selection, we consider the outputs of the frailty models presented in Example 4.2. With this data set, the Cox model gave cAIC = 365.69, pAIC = 365.69, and rAIC = 364.15, whereas the log-normal frailty model gave cAIC = 362.22, pAIC = 366.14, and rAIC = 364.56; and the gamma frailty model had cAIC = 365.30, pAIC = 365.71, and rAIC = 364.12. The LRT based on the partial restricted likelihood showed absence of the frailty effect ($\alpha = 0$). Thus, we may choose the Cox model as our final model parsimoniously. However, the cAIC selects the log-normal frailty model, indicating that this model could give a better conditional prediction. Thus, for subject-specific inference such as prediction of the frailties, we may prefer to use the log-normal frailty models. From Fig. 4.3, we see that the frailty estimates of $\hat{v}_i$ from the log-normal and gamma frailty models are somewhat different, which may explain the difference in cAICs between the two frailty models. In the log-normal frailty models, $\sum_i \hat{v}_i/q = 0$, while in the gamma frailty model $\sum_i \hat{u}_i/q = 1$, so in Fig. 4.3 we recentered them by putting a constraint $\sum_i \hat{v}_i/q = 0$.

4.5 Interval Estimation of the Frailty

It would be informative to investigate the heterogeneity among clusters (or centers) in order to understand and interpret the variability in event times. The semiparametric frailty models offer a flexible framework for modeling this heterogeneity. Such heterogeneity can be accounted for by the random cluster effects. In addition to estimation of the random effects, a measure of the uncertainty of these point estimates is necessary.

We now focus on interval estimation of the individual random effects. In multicenter clinical trials with a standardized protocol or a meta analysis combining multiple protocols, the treatment effect or the baseline risk may vary across the centers. To investigate and explain the source of such heterogeneity, interval estimation of a set of individual random effects for the centers has been studied using various methods such as EB, Full Bayesian and h-likelihood (HL) approaches (Gray 1994; Vaida and Xu 2000; Legrand et al. 2005; Ha et al. 2011, 2016b).

4.5.1 *Confidence Interval for the Frailty*

We now show how the Wald intervals for the random effects in Chap. 3 can be extended to the frailty models.

The individual $(1-\lambda)$-level HL confidence intervals (CIs) for the unidimensional components v_k of v have the form of

$$\hat{v}_k \pm z_{\alpha/2} \cdot \mathrm{SE}(\hat{v}_k - v_k), \tag{4.20}$$

where $\hat{v}_k$ maximizes h_p in (4.5) and $z_{\alpha/2}$ is the standard normal quantile with a probability of $\alpha/2$ in the right tail. Here, $\mathrm{SE}(\hat{v}_k - v_k)$ is $\sqrt{a_{kk}}$, where a_{kk} is the kth diagonal element of an approximated variance of $\hat{v} - v$, computed from the lower right-hand corner of the inverse of Hessian matrix $H_p = H(h_p; \beta, v)$ based on h_p:

$$\mathrm{var}(\hat{v} - v) \approx \{(Z^T W^* Z + Q) - (Z^T W^* X)(X^T W^* X)^{-1}(X^T W^* Z)\}^{-1}|_{\beta=\hat{\beta}, v=\hat{v}}.$$

Derivation of (4.20) is given in Appendix 4.7.6.

The Wald interval in (4.20) has been used for various random-effect models: see Paik et al. (2015). However, it gives null intervals when the variance-component estimates are zero (Ha et al. 2016b). Thus, it can lead to liberal intervals when the variance components or sample sizes are small. Following Ha et al. (2016b), we introduce a modification in order to overcome this shortcoming. Specifically, the partial restricted likelihood method based on $p_{\beta,v}(h_p)$ can give zero estimate for the frailty parameter α. This leads to the null CI for v in (4.20). This issue was recognized by Morris (2006) in the context of the LMMs. To extend the Morris' method (2006), we use a modification of the adjusted likelihood p_{adj}, defined as

$$p_{\text{adj}} = p_{\beta,v}(h_p) + \log \alpha. \tag{4.21}$$

Note that the last term in (4.21) is asymptotically negligible (i.e., the asymptotic property of the maximum p_{adj} estimator is asymptotically the same as that of the maximum $p_{\beta,v}(h_p)$ estimator). Furthermore,

$$\exp(p_{\text{adj}}) = \exp\{p_{\beta,v}(h_p)\}\alpha \geq 0,$$

and $\exp(p_{\text{adj}}) = 0$ only if $\alpha = 0$: see also Appendix of Li and Lahiri (2010). Thus, by adding the last term, we can effectively avoid zero estimate in the dispersion parameter. The adjusted likelihood p_{adj} is always defined, even when the original restricted likelihood based upon the marginal likelihood is hardly available. Ha et al. (2016b) showed via simulations that this correction generally works well.

Remark 4.4 Note that for the EB interval, $\text{SE}(\hat{v}_k - v_k)$ is simply $\sqrt{h_{kk}}$ (Vaida and Xu 2000; Othus and Li 2009). Here, h_{kk} is the kth diagonal element of the matrix $(-\partial^2 h/\partial v \partial v^T)^{-1}|_{\psi=\hat{\psi},v=\hat{v}} = (Z^T W_1 Z + Q)^{-1}|_{\psi=\hat{\psi},v=\hat{v}}$ which ignores the uncertainty caused by estimating $\psi = (\beta, \lambda_0)$ (Ha et al. 2011). Here, W_1 is a diagonal matrix defined in (4.35). Thus the EB method can underestimate the SE of $\hat{v} - v$, leading to a lower coverage probability of the CI than the nominal level (Chap. 3).

4.5.2 *Illustration*

Example 4.4 (Application to kidney infection data) We fit the univariate log-normal frailty model (4.1) with a single covariate, Sex (1 = male; 2 = female), using HL(0,1) in the **frailtyHL** package. The results show that the effect of Sex (Estimate = -1.353 and SE = 0.421) is highly significant (p-value = 0.001). That is, the female group has a significantly lower risk than the male group. Here, the variance estimate for the frailty is $\widehat{\sigma^2} = 0.478$ (with SE = 0.313). Following the asymptotic chi-square mixture distribution, $\chi^2_{0:1}$, the difference of partial restricted likelihood based on $-2p_{\beta,v}(h_p)$ between the Cox model without frailty and the log-normal frailty model is $369.96 - 364.68 = 5.28 (> \chi^2_{1,0.10} = 2.71)$, indicating that the frailty effect is significant (i.e., $\sigma^2 > 0$) at the 5% level. Thus we see that the frailty term is necessary for modeling the kidney data.

Below we present R codes for estimating the CIs and creating their plots for individual frailties.

```
############# Confidence intervals for frailties #############
> res<- frailtyHL(Surv(time,status) ~ sex+(1|id),data=kidney)
> p<- res$p; q<-res$q
> v_h<- res$v_h # estimates of log-frailties
> var<- diag(res$Hinv)[(p+1):(p+q)] # computation of var(v_h-v)
> SE<- sqrt(var) # SE(v_h-v)
> lb<- v_h -1.96*SE # lower bound
> ub<- v_h +1.96*SE # upper bound
> CI<- cbind(lb,ub)  # computation of CI
> patient<- 1:q
> plot(v_h ~ patient, ylim=c(min(lb)-0.5, max(ub)+0.5), ylab="Estimated
+  patient effects",xlab="Patient number",pch=20,type="o") # plot for CI
> abline(h=0)
> for (i in 1:q){
+     x1<- c(i,i)
+     y1<- c(lb[i],ub[i])
+    lines (y1~x1)
+     }
> text(21.8, v_h[21],21, cex=.8)
```

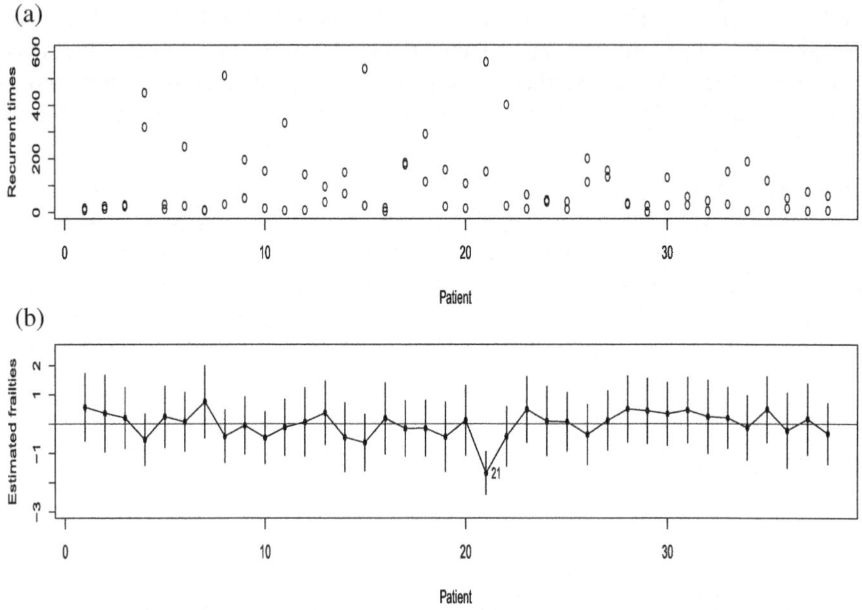

Fig. 4.4 **a** Recurrence times for 38 patients in the kidney infection data; **b** 95% confidence intervals of individual frailties of 38 patients, under the univariate log-normal frailty model

Figure 4.4a displays the recurrent infection times for 38 patients. Figure 4.4b shows the estimated frailties of 38 patients and their 95% CIs, which indicates that the patient's realized frailty effects on the recurrent times are heterogenous; in particular, the 21st patient has a very lower frailty (i.e., lower hazard) and the corresponding CI does not include zero. This is also confirmed from the fact that the 21st patient among all patients experienced the longest second infection time (i.e., 562 days) as shown in Fig. 4.4a. Thus we find that a graphical representation like Fig. 4.4b could be useful to identify the heterogeneity of a particular patient.

Example 4.5 (*Application to bladder cancer data*) The multicenter bladder cancer trial data in Chap. 1 are from 21 different centers in Europe (data set available in the **frailtyHL**: 'bladder0'). We are interested in investigating the heterogeneity of baseline risks across the centers. We consider the log-normal univariate frailty model in (4.1), $\lambda_{ij}(t|v) = \lambda_0(t)\exp(\eta_{ij})$, allowing a linear predictor

$$\eta_{ij} = \beta_1 x_{ij1} + \beta_2 x_{ij2} + v_{i0},$$

with $v_{i0} \sim N(0, \sigma_0^2)$. Here, x_{ij1} is CHEMO (0 = No, 1 = Yes) and x_{ij2} is TUSTAT (0 = Primary, 1 = Recurrent). The fitted estimates using the HL(0,1) method are as follows: $\hat{\beta}_1 = -0.695$ (SE = 0.175), $\hat{\beta}_2 = 0.544$ (SE = 0.149) and $\hat{\sigma}_0^2 = 0.070$ (SE = 0.058). Here, both fixed effects (β_j, $j = 1, 2$) are significant. In particular, the use of chemotherapy (CHEMO = 1) significantly prolongs time to first recurrence

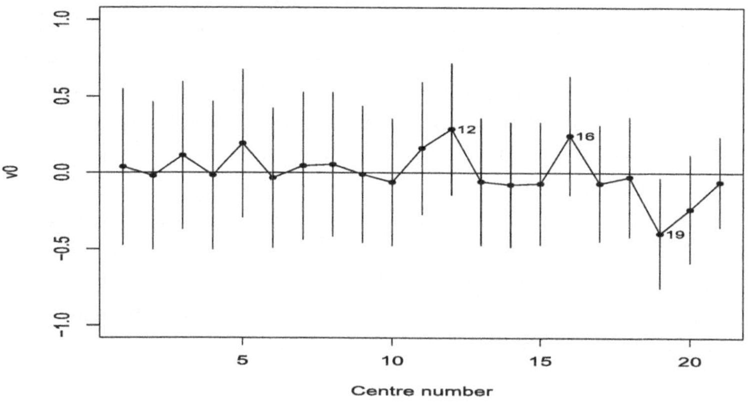

Fig. 4.5 Random center effects (v_{i0}) of 21 centers in the bladder cancer data and their 95% confidence intervals, under the univariate log-normal frailty model. Centers are sorted by the increasing order of number of patients

as compared to patients who do not receive chemotherapy (CHEMO = 0) (Legrand et al. 2005; Ha et al. 2011).

The corresponding random center effects (i.e., random baseline risks) and 95% CIs for the individual centers are plotted in Fig. 4.5. Here, the centers are sorted by the number of patients. It shows that substantial variations in the baseline risks across the centers. In particular, the centers (12,16) and 19 stand out as taking the highest and lowest baseline risks, respectively. But the 19th center has a significantly smaller baseline risk. Now we are also interested in testing the hypothesis $H_0 : \sigma_0^2 = 0$, i.e., no center effect, or no variation in random baseline risks. The difference in partial restricted likelihood ($-2p_{\beta,v}(h_p)$) between the Cox model and the univariate frailty model is 3.2(>2.71), indicating that the center effect is significant, i.e., $\sigma_0^2 > 0$, as in Example 4.4.

4.6 Discussion

We have presented various likelihood-based methods for the semiparametric frailty models. The correct model specification about the baseline hazard is crucial for parametric inference. If specified incorrectly, the regression parameter estimates suffer from potentially serious biases. Therefore, when the parametric frailty models are considered, model checking for the baseline hazard would be an important step. Thus, when the baseline hazard assumptions are uncertain, the semiparametric frailty model is recommended. The choice of a frailty distribution seems to have minimal effects on the regression parameter estimates unless the frailty variance is very large (Pickles and Crouchley 1995; Ng and Cook 2000; Ha and Lee 2003, 2005a, b; Ha et al. 2011).

The penalized maximum likelihood approach (Rondeau et al. 2008), which penalizes the baseline hazard $\lambda_0(t)$ in the marginal likelihood, has been proposed for inference on the parameters. However, it cannot be directly used for subject-specific inference involving the frailties, because it eliminates them by integration as in the standard marginal likelihood approach (Nielsen et al. 1992; Vaida and Xu 2000). Bayesian approaches (Legrand et al. 2005; Komarek et al. 2007) have been also suggested in the literature. Legrand et al. (2005) proposed a Bayesian approach using the partial h-likelihood h_p for the joint posterior density $\pi(v, \beta, \alpha|y, \delta)$ under the uniform priors and using a Laplace integration technique to approximate the marginal posterior density $\pi(\alpha|y, \delta)$. It can be shown that, under the uniform priors for (β, α) and the partial likelihood technique for λ_0,

$$\log\{\pi(v, \beta, \alpha|y, \delta)\} \propto h_p \text{ and } \log\{\pi(\alpha|y, \delta)\} \simeq p_{\beta,v}(h_p).$$

Thus, we see that the h-likelihood method based on the h_p is equivalent to Legrand et al.'s method under the uniform priors.

Copula models are also used to model dependence among multivariate survival data using a copula function (Oakes 1989; Shih and Louis 1995; Duchateau and Janssen 2008; Prenen et al. 2017). Following Sklar's (1959) theorem, a copula function expresses the joint distribution of random variables as a function of marginal distribution of each variable. However, frailty models and copula models are different types of models (Duchateau and Janssen 2008; Prenen et al. 2017) that take the association into account because the frailty model is a conditional modeling approach, whereas the copula model is a marginal modeling approach. Comparison of both models for the same data would be interesting.

Frailty models can also be used to model an unexplained heterogeneity or to introduce a non-PH in the independent survival data with cluster size $n_i = 1$: see Vaupel et al. (1979), Aalen (1988, 1992), Hougaard (1991), Henderson and Oman (1999), Kosorok et al. (2004), Barker and Henderson (2005) and Ha et al. (2010). In particular, Kosorok et al. (2004) studied robust inference under various frailty distributions.

As mentioned in Chap. 2, survival data can be left truncated when not all subjects in the data are observed from the time origin of interest, yielding both Left Truncation and Right Censoring (LTRC). The current h-likelihood procedure can be easily extended to the random-effect models with LTRC structure (Rondeau et al. 2003). In particular, as in the Cox PH models, the semiparametric frailty models under LTRC can be easily handled by replacing the risk set $R_{(k)} = \{(i, j) : y_{(k)} \leq y_{ij}\}$ by

$$R_{(k)} = \{(i, j) : a_{ij} \leq y_{(k)} \leq y_{ij}\}$$

where a_{ij} is the left truncation time, which is implemented as the **frailtyHL** R package (Ha et al. 2018). This LTRC technique is directly applied to semi-competing risk modeling in Chap. 10. For simplicity, we have assumed time-independent covariates, but the h-likelihood methods can be extended to the time-dependent covariates as in the Cox PH models (Therneau et al. 2016).

The semiparametric frailty models with nonparametric baseline hazards can be fitted via a Poisson HGLM (Ma et al. 2003; Ha and Lee 2005b) as in the Cox PH model in Remark 2.1. However, the number of nuisance parameters in the Poisson HGLM increases with sample size n, leading to a high-dimensional computation; for a practical use of this Poisson approach, development of an improved procedure is necessary.

The h-likelihood approaches can be extended to a general class of frailty models allowing various frailty structures such as nested or correlated frailties as will be shown in Chap. 5.

4.7 Appendix

4.7.1 *Proof of Remark 4.1*

The marginal survival function of T with covariates x, denoted by $S^{\mathrm{M}}(t; x)$, is given by

$$S^{\mathrm{M}}(t; x) = P(T > t; x) = \int S(t|u; x) f(u) du.$$

Since $S(t|u; x) = \exp\{-\Lambda(t|u; x)\} = \exp(-\Lambda_0(t) e^{x^T\beta} u)$, we see that $-dS(t|u; x)/dt = \lambda_0(t) \exp(x^T\beta) u S(t|u; x)$. Thus we have

$$\begin{aligned}
\lambda^{\mathrm{M}}(t; x) &= -\frac{dS^{\mathrm{M}}(t; x)/dt}{S^{\mathrm{M}}(t; x)} \\
&= \lambda_0(t) \exp(x^T\beta) \frac{\int u S(t|u; x) f(u) du}{S^{\mathrm{M}}(t; x)} \\
&= \lambda_0(t) \exp(x^T\beta) \int u f(u|T > t; x) du \\
&= \lambda_0(t) \exp(x^T\beta) E(U|T > t; x).
\end{aligned}$$

Note here that by Bayes' theorem,

$$f(u|T > t; x) = \frac{S(t|u; x) f(u)}{S^{\mathrm{M}}(t; x)},$$

which means the frailty density among the survivors at time t. This may be useful for computing the predictive distribution given that a subject is still alive up to just prior to time t (van Houwelingen and Putter 2012).

$E(U|T > t; x)$ can also be calculated from the Laplace transform. Define Laplace transform of the frailty U as $L(z) = E\{\exp(-zU)\}$. Then under (4.1) $S^{\mathrm{M}}(t; x)$ can be expressed as

$$S^{\mathrm{M}}(t; x) = L\{\Lambda(t; x)\}$$

where $\Lambda(t; x) = \Lambda_0(t)\exp(x^T\beta)$ is the cumulative hazard function. Let $L'(z)$ be the first derivative with respective to z. Then we obtain

$$-\frac{L'\{\Lambda(t; x)\}}{L\{\Lambda(t; x)\}} = \frac{\int uS(t|u; x)f(u)du}{S^{\mathrm{M}}(t; x)}$$
$$= E(U|T > t; x).$$

For the gamma frailty U with mean 1 and variance α, the Laplace transform has an explicit form, $L(z) = (1 + \alpha z)^{-1/\alpha}$, so that the computation is analytic, i.e., $E(U|T > t; x) = \{1 + \alpha\Lambda_0(t)\exp(x^T\beta)\}^{-1}$. However, for the log-normal frailty, a numerical integration is required because there is no explicit form of its Laplace transform. The marginal model, $\lambda^{\mathrm{M}}(t; x)$, is generally non-PH unless U follows a positive stable distribution (Hougaard 2000; Hsu et al. 2007; Ha and MacKenzie 2010). Exactly how the marginal model deviates from the proportionality is not known for many frailty distributions, including the log-normal distribution (Hougaard 2000, p. 245). This implies, though, that the frailty model is a more flexible model than the Cox PH model in that it can embrace various types of non-proportionalities. □

4.7.2 Derivation of the H-Likelihood for Frailty Model

We define the $n_i \times 1$ observed random vectors associated with the ith individual as $Y_i = (Y_{i1}, \ldots, Y_{in_i})^T$ and $\delta_i = (\delta_{i1}, \ldots, \delta_{in_i})^T$. The contribution, h_i, say, of the ith individual to the h-likelihood is given by the logarithm of the joint density of (Y_i, δ_i, V_i), where $V_i = \log(U_i)$:

$$h_i(\beta, \lambda_0, \alpha; y_i, \delta_i, v_i) = \log\{L_{1i}(\beta, \lambda_0; y_i, \delta_i|u_i)L_{2i}(\alpha; v_i)\}, \tag{4.22}$$

where L_{1i} is the conditional density of (Y_i, δ_i) given $U_i = u_i$ and L_{2i} is the density of V_i. By the conditional independence of $\{(T_{ij}, C_{ij}), j = 1, \ldots, n_i\}$ in Assumption 3, we have

$$L_{1i}(\beta, \lambda_0; y_i, \delta_i|u_i) = \prod_j L_{1ij}(\beta, \lambda_0; y_{ij}, \delta_{ij}|u_i), \tag{4.23}$$

where L_{1ij} is the conditional density of (Y_{ij}, δ_{ij}) given $U_i = u_i$. By the conditional independence of both T_{ij} and C_{ij} in Assumption 3 and the noninformativeness in Assumption 4, L_{1ij} in Eq. (4.23) becomes the ordinary likelihood for censored survival data given $U_i = u_i$:

$$L_{1ij} = \{\lambda(y_{ij}|u_i)\}^{\delta_{ij}}\exp\{-\Lambda(y_{ij}|u_i)\}, \tag{4.24}$$

where $\Lambda(\cdot|u_i)$ is the conditional cumulative hazard function of T_{ij} given $U_i = u_i$. Thus, from (4.1) and (4.22)–(4.24) we obtain

$$h_i = \sum_j \ell_{1ij} + \ell_{2i},$$

where $\ell_{1ij} = \log(L_{1ij}) = \delta_{ij}\{\log \lambda_0(y_{ij}) + \eta_{ij}\} - \{\Lambda_0(y_{ij})\exp(\eta_{ij})\}$ and $\ell_{2i} = \log(L_{2i})$. Therefore, the contribution from all individuals is given by

$$h = \sum_i h_i. \quad \square$$

4.7.3 Equivalence of Both Estimators of β Under the Gamma Frailty Model and the EM Estimating Equation of the Frailty Parameter α

For the gamma frailty model with $E(U_i) = 1$ and $\text{var}(U_i) = \alpha$, assume that the frailty parameter α is known. From Eq. (4.2), the h-likelihood is given by

$$h = \sum_{ij}\Big[\delta_{ij}\{\log \lambda_0(y_{ij}) + \eta_{ij}\} - \Lambda_0(y_{ij})\exp(\eta_{ij})\Big] + \sum_i \{\alpha^{-1}(v_i - u_i) + c(\alpha)\},$$

where $c(\alpha) = -\log \Gamma(\alpha^{-1}) - \alpha^{-1}\log\alpha$. Let $\widehat{\lambda}_{0k}(\tau)$ be the maximum hierarchical likelihood (MHL) estimator of $\lambda_{0k} = \lambda_0(y_{(k)})$ given $\tau = (\beta^T, v^T)^T$. Then we have

$$\begin{aligned}\frac{\partial h(\widehat{\lambda}_{0k}(\tau), \tau)}{\partial \tau} &= \frac{\partial h(\lambda_{0k}, \tau)}{\partial \tau}\bigg|_{\lambda_{0k}=\widehat{\lambda}_{0k}(\tau)} + \frac{\partial h(\lambda_{0k}, \tau)}{\partial \lambda_{0k}}\bigg|_{\lambda_{0k}=\widehat{\lambda}_{0k}(\tau)} \cdot \frac{\partial \widehat{\lambda}_{0k}(\tau)}{\partial \tau}\\ &= \frac{\partial h(\lambda_{0k}, \tau)}{\partial \tau}\bigg|_{\lambda_{0k}=\widehat{\lambda}_{0k}(\tau)}, \end{aligned} \tag{4.25}$$

since the second term is equal to zero in the score equations of the MHL estimator for λ_{0k}. Recall that the partial h-likelihood h_p is proportional to the profile h-likelihood h^*. Thus, given α the MHL score Eq. (4.25) for τ lead to the equations

$$\frac{\partial h_p}{\partial \tau} = \frac{\partial h^*}{\partial \tau} = \frac{\partial h}{\partial \tau}\bigg|_{\lambda_0=\widehat{\lambda}_0}. \tag{4.26}$$

By (4.26), the MHL equations for β_r $(r = 1, \ldots, p)$ become

$$\frac{\partial h}{\partial \beta_r}\bigg|_{\lambda_0=\widehat{\lambda}_0} = \sum_{ij}\left\{\delta_{ij} - \Lambda_0(y_{ij})\exp(x_{ij}^T\beta)u_i\right\} x_{ijr}\bigg|_{\lambda_0=\widehat{\lambda}_0} = 0. \tag{4.27}$$

From

$$\frac{\partial h}{\partial v_i} = \sum_j \left\{ \delta_{ij} - \Lambda_0(y_{ij}) \exp(x_{ij}^T \beta) u_i \right\} + \alpha^{-1} - \alpha^{-1} u_i = 0,$$

we have

$$\widehat{u}_i = \frac{\alpha^{-1} + \delta_{i+}}{\alpha^{-1} + \mu_{i+}}, \tag{4.28}$$

which also becomes $E(U_i|y_i, \delta_i)$ since the conditional distribution of U_i given the ith observed data (y_i, δ_i) is from a gamma distribution. Here, $\delta_{i+} = \sum_j \delta_{ij}$ is the total number of events in the ith individual, $\mu_{i+} = \mu_{i+}(\beta, \lambda_0) = \sum_j \mu_{ij}$ and $\mu_{ij} = \Lambda_0(y_{ij}) \exp(x_{ij}^T \beta)$ with $\Lambda_0(y_{ij}) = \sum_k \lambda_{0k} I(y_{(k)} \leq y_{ij})$.

On the other hand, from Eq. (4.7) the marginal likelihood m is given by

$$m = \sum_{ij} \delta_{ij} \{\log \lambda_0(y_{ij}) + x_{ij}^T \beta\} - \sum_i \{(\alpha^{-1} + \delta_{i+}) \log(\alpha^{-1} + \mu_{i+}) - f(\alpha)\}. \tag{4.29}$$

where $f(\alpha) = \log \Gamma(\alpha^{-1} + \delta_{i+}) + c(\alpha)$. The estimating equations for λ_{0k} are given by

$$\frac{\partial m}{\partial \lambda_{0k}} = \frac{d_{(k)}}{\lambda_{0k}} - \sum_{ij \in R(y_{(k)})} \exp(x_{ij}^T \beta) \widetilde{u}_i = 0,$$

where $\widetilde{u}_i = (\alpha^{-1} + \delta_{i+})/(\alpha^{-1} + \mu_{i+})$; this leads to the nonparametric MLEs,

$$\widetilde{\lambda}_{0k} = \frac{d_{(k)}}{\sum_{ij \in R(y_{(k)})} \exp(x_{ij}^T \beta) \widetilde{u}_i}. \tag{4.30}$$

Thus, we have the ML equations for β_r $(r = 1, \ldots, p)$

$$\frac{\partial m}{\partial \beta_r}\Big|_{\lambda_0 = \widetilde{\lambda}_0} = \sum_{ij} \left\{ \delta_{ij} - \Lambda_0(y_{ij}) \exp(x_{ij}^T \beta) \widetilde{u}_i \right\} x_{ijr}\Big|_{\lambda_0 = \widetilde{\lambda}_0} = 0, \tag{4.31}$$

which are equivalent to the MHL Eq. (4.27) with (4.28), and also become the EM equations for the ML estimator of β (Therneau and Grambsch 2000), i.e.,

$$E(\partial h_p / \partial \beta_r | \, y_i, \delta_i) = E(\partial h / \partial \beta_r | \, y_i, \delta_i, \widehat{\lambda}_0) = 0.$$

Accordingly, the MHL estimator for β given α is the same as the ML estimator.

Similar to the Poisson-gamma HGLM in Appendix 3.5, the ith component of adjustment term for $p_v(h)$ (4.9) is given by

$$H(h, v_i)|_{u_i=\hat{u}_i} = -\partial^2 h/\partial v_i^2|_{u_i=\hat{u}_i} = (\alpha^{-1} + \mu_{i+})\hat{u}_i = \alpha^{-1} + \delta_{i+} ,$$

which is free of (β, λ_0) but depends upon α. Since $H(h, v)|_{u=\hat{u}} = \text{diag}(\alpha^{-1} + \delta_{i+})$ is a $q \times q$ diagonal matrix, we have that

$$\begin{aligned} p_v(h) &= [h - \frac{1}{2}\log\det\{H(h, v)/(2\pi)\}]|_{u=\widehat{u}} \\ &= \sum_{ij}[\delta_{ij}\{x_{ij}^T\beta + \log\lambda_0(y_{ij})\}] + \sum_i\{-(\alpha^{-1} + \delta_{i+})\log(\alpha^{-1} + \mu_{i+}) \\ &+ (\alpha^{-1} + \delta_{i+})\log(\alpha^{-1} + \delta_{i+}) - (\alpha^{-1} + \delta_{i+}) - \log(\alpha^{-1} + \delta_{i+})/2 \\ &+ \log(2\pi)/2 + c(\alpha)\}. \end{aligned}$$

Thus, maximizing $p_v(h)$ given α is also equivalent to the marginal likelihood m.

Furthermore, the ML estimating equation of α is given by

$$\frac{\partial m}{\partial\alpha} = \sum_i\{\alpha^{-2}\log(\alpha^{-1} + \mu_{i+}) + \alpha^{-2}\widetilde{u}_i + f'(\alpha)\} = 0, \qquad (4.32)$$

where $f'(\alpha) = \partial f(\alpha)/\partial\alpha = -\alpha^{-2}\{dg(\alpha^{-1} + \delta_{i+}) - dg(\alpha^{-1}) - \log\alpha + 1\}$; this also becomes the EM equation for α, given by

$$E(\partial h/\partial\alpha|\ y_i, \delta_i, \widehat{\lambda}_0) = E(\partial\ell_2/\partial\alpha|\ y_i, \delta_i, \widehat{\lambda}_0) = 0.$$

Following Andersen et al. (1997), the ML estimators for (β, α) can be easily solved using the Newton–Raphson method, and the estimated SEs for β and α are also obtained from the inverse of the observed information matrix, $-\partial^2 m/\partial(\beta, \lambda_0, \alpha)^2$. □

4.7.4 Proof of Joint Score Equations in (4.12)

From (4.26) we have that

$$\begin{aligned} \partial h_p/\partial\beta &= \partial h/\partial\beta|_{\lambda_0=\widehat{\lambda}_0} = X^T(\delta - \mu)|_{\lambda_0=\widehat{\lambda}_0} \\ \partial h_p/\partial v &= \partial h/\partial v|_{\lambda_0=\widehat{\lambda}_0} = Z^T(\delta - \mu)|_{\lambda_0=\widehat{\lambda}_0} + \partial\ell_2/\partial v. \end{aligned}$$

The two equations above can be simply expressed as

$$\partial h_p/\partial\tau = \{E^T(\delta - \mu) + b\}|_{\lambda_0=\widehat{\lambda}_0} \qquad (4.33)$$

since $\partial h/\partial\tau = (\partial\eta/\partial\tau)(\partial h/\partial\eta)$ with $\eta = X\beta + Zv = E\tau$. Here, $E = (X, Z), b = (0^T, (\partial\ell_2/\partial v)^T)^T$, and δ and μ are the $n \times 1$ vectors of δ_{ij}'s, and μ_{ij}'s, respectively, where $\mu_{ij} = \Lambda_0(y_{ij})\exp(\eta_{ij})$ with $\Lambda_0(y_{ij}) = \sum_k \lambda_{0k} I(y_{(k)} \leq y_{ij})$.

Note that the vector μ can be written as a simple form by using a weighted risk indicator matrix M which contains the risk set $R_{(k)}$. Let L be the $n \times 1$ vector of L_{ij}'s with $L_{ij} = \Lambda_0(y_{ij})$. Since $\Lambda_0(y_{ij}) = \sum_k \lambda_{0k} I(y_{(k)} \le y_{ij})$, we have $L = MAJ$, where M is the $n \times D$ risk indicator matrix whose (ij, k)th element is $m_{ij,k}$ with $m_{ij,k} = I\{y_{ij} \ge y_{(k)}\}$, $A = \text{diag}(\lambda_{0k})$ is the $D \times D$ diagonal matrix and J is the $D \times 1$ vector with one. This gives $\mu = W_0(MAJ)$ with $W_0 = \text{diag}\{\exp(\eta_{ij})\}$.

Following Ha and Lee (2003), we have

$$-\frac{\partial^2 h^*}{\partial \tau^2} = \left\{ \left(\frac{-\partial^2 h}{\partial \tau^2}\right) - \left(\frac{-\partial^2 h}{\partial \tau \partial \lambda_0}\right)\left(\frac{-\partial^2 h}{\partial \lambda_0^2}\right)^{-1}\left(\frac{-\partial^2 h}{\partial \lambda_0 \partial \tau}\right)\right\}\Bigg|_{\lambda_0 = \widehat{\lambda}_0(\tau)}, \tag{4.34}$$

leading to

$$H_p = \begin{pmatrix} -\partial^2 h_p/\partial \beta^2 & -\partial^2 h_p/\partial \beta \partial v \\ -\partial^2 h_p/\partial v \partial \beta & -\partial^2 h_p/\partial v^2 \end{pmatrix} = \begin{pmatrix} X^T W^* X & X^T W^* Z \\ Z^T W^* X & Z^T W^* Z + Q \end{pmatrix}, \tag{4.35}$$

where $H_p = H(h_p; \beta, v) = -\partial^2 h_p/\partial(\beta, v)^2$, $W^* = W_1 - W_2$, $W_1 = \text{diag}(\mu)$, $W_2 = (W_0 M)C^{-1}(W_0 M)^T$ and $C = \text{diag}\{d_{(k)}/(\lambda_{0k})^2\}$ is the $D \times D$ diagonal matrix. That is, the Eq. (4.35) can be rewritten as

$$-\partial^2 h_p/\partial \tau^2 = E^T W^* E + F, \tag{4.36}$$

where $F = \text{BD}(0, Q)$ is a block diagonal matrix. Following Ha and Lee (2003), (4.33) and (4.36), we can show that given α, the MHL estimators of $\tau = (\beta^T, v^T)^T$ are obtained from the following score equations:

$$(E^T W^* E + F)\hat{\tau} = E^T w^* + g, \tag{4.37}$$

leading to (4.12), where $g = F\tau + b = (0^T, R^T)^T$. Note here that the terms λ_{0k} in W^* and w^* in (4.35)–(4.37) are evaluated at their estimates $\hat{\lambda}_{0k} = d_{(k)}/M_k^T \psi$, where M_k is the kth component vector of $M = (M_1, \ldots, M_D)$ and ψ is a vector of $\exp(\eta_{ij})$'s. This completes the proof. In addition, (4.37) can also be expressed as (4.13) because $(E^T W^* E + F) = \mathbf{P}^T \mathbf{V} \mathbf{P}$ and $E^T w^* + g = \mathbf{P}^T \mathbf{y}_0^*$. □

4.7.5 Computation of PREML Equation for Frailty Parameter α

The partial REML estimator for α is obtained by solving

$$\frac{\partial p_\tau(h_p)}{\partial \alpha} = 0. \tag{4.38}$$

Note here that

$$\frac{\partial p_\tau(h_p)}{\partial \alpha} = \frac{\partial \widehat{h}_p}{\partial \alpha} - \frac{1}{2}\text{tr}\left(\widehat{H}_p^{-1}\frac{\partial \widehat{H}_p}{\partial \alpha}\right), \tag{4.39}$$

where $\widehat{h}_p = h_p|_{\tau=\widehat{\tau}(\alpha)} = h_p(\widehat{\tau}(\alpha), \alpha)$ and $\widehat{H}_p = H(h_p; \tau)|_{\tau=\widehat{\tau}(\alpha)} = H_p(\widehat{\tau}(\alpha), \alpha)$, i.e.,

$$\hat{H}_p = \begin{pmatrix} X^T\hat{W}^*X & X^T\hat{W}^*Z \\ Z^T\hat{W}^*X & Z^T\hat{W}^*Z + \hat{Q} \end{pmatrix},$$

where $\hat{W}^* = W^*|_{\tau=\hat{\tau}(\alpha)} = W^*(\hat{\tau}(\alpha), \alpha)$ and and $\hat{Q} = Q(\hat{v}(\alpha), \alpha)$, and

$$\begin{aligned}\frac{\partial \hat{h}_p}{\partial \alpha} &= \left\{\left(\frac{\partial h_p}{\partial \alpha}\right) + \left(\frac{\partial h_p}{\partial \tau}\right)\left(\frac{\partial \hat{\tau}}{\partial \alpha}\right)\right\}\Bigg|_{\tau=\hat{\tau}} \\ &= \frac{\partial h_p}{\partial \alpha}\bigg|_{\tau=\hat{\tau}} = \left(\sum_i \frac{\partial \ell_{2i}}{\partial \alpha}\right)\bigg|_{v=\hat{v}},\end{aligned} \tag{4.40}$$

since $(\partial h_p/\partial \tau)|_{\tau=\hat{\tau}} = 0$ and h_p in (4.5) depends on ℓ_{2i} only with α. Following Lee and Nelder (2001a) and Ha and Lee (2003), we allow $\partial\hat{v}/\partial\alpha$ in implementing (4.38), but not $\partial\hat{\beta}/\partial\alpha$ (Lee et al. 2017b). Thus, the term $\partial\hat{H}_p/\partial\alpha$ in (4.39) becomes

$$\frac{\partial \hat{H}_p}{\partial \alpha} = \begin{pmatrix} X^T W' X & X^T W' Z \\ Z^T W' X & Z^T W' Z + Q' \end{pmatrix},$$

where $Q' = \partial Q/\partial\alpha$ and

$$W' = \frac{\partial \hat{W}^*}{\partial \alpha} = \left\{\left(\frac{\partial W^*}{\partial \alpha}\right) + \left(\frac{\partial W^*}{\partial v}\right)\left(\frac{\partial \hat{v}}{\partial \alpha}\right)\right\}\Bigg|_{v=\hat{v}}.$$

Note here that the term $\partial\hat{v}/\partial\alpha$ can be expressed (Lee and Nelder 1996) as

$$\frac{\partial \hat{v}}{\partial \alpha} = -\left(-\frac{\partial^2 h_p}{\partial v^2}\right)^{-1}\left(-\frac{\partial^2 h_p}{\partial v \partial \alpha}\right)\bigg|_{v=\hat{v}}. \tag{4.41}$$

Accordingly, the Eq. (4.38) is solved by using the Newton–Raphson method with the Hessian matrix, $-\partial^2 p_\tau(h_p)/\partial\alpha^2$ (Ha et al. 2011).

In particular, for the log-normal frailty with $v_i \sim N(0, \alpha)$, we have $\ell_{2i} = -\log(2\pi\alpha)/2 - v_i^2/(2\alpha)$. From (4.38)–(4.40), we have a simple partial REML estimator for α, given by

$$\hat{\alpha} = \frac{\hat{v}^T\hat{v}}{q - \gamma}, \tag{4.42}$$

where $\gamma = -\alpha\text{tr}\{\widehat{H}_p^{-1}(\partial\widehat{H}_p/\partial\alpha)\}$. □

4.7.6 Construction of CI of the Frailty in (4.20)

Since (β, λ_0, v) and $\alpha(=\text{var}(v_i))$ are asymptotically orthogonal as in the HGLMs, we only need to consider the Hessian matrix of v and $\psi = (\beta^T, \lambda_0^T)^T$. The definitions of the Hessian matrix of the fixed and random effects and the unconditional mean squared error (UMSE) of random effects in the HGLMs (Lee and Nelder 1996, 2009) are, respectively, extended to

$$H(h; \psi, v) \equiv -\partial^2 h/\partial(\psi, v)\partial(\psi, v)^{\top} \text{ and UMSE} \equiv E_{\psi}[\{\hat{v}(\hat{\psi}) - v\}\{\hat{v}(\hat{\psi}) - v\}^T]. \quad (4.43)$$

Here, $\hat{v}(\hat{\psi}) \equiv \hat{v}(\psi)|_{\psi=\hat{\psi}}$, where $\hat{v}(\psi)$ is the solution to $\partial h/\partial v = 0$ for a given ψ. Let y^* be a vector of observed data points $y^*_{ij} = (y_{ij}, \delta_{ij})$. Note that $\hat{v}(\psi) = E_{\psi}(v|y^*)$ asymptotically as $N = \min_{1\le i\le q} n_i \to \infty$. Following Lee and Nelder (1996) and Lee and Ha (2010), it can be shown that $H(h; \hat{\psi}, \hat{v})^{-1}$ gives the first-order approximation to the UMSE in (4.43), leading to a standard error (SE) for $\hat{v} - v$ and a Wald confidence interval for v.

In the semiparametric frailty models (4.1), the number of terms λ_{0k} in ψ increases with sample size n. Thus, $H(h; \hat{\psi}, \hat{v})^{-1}$ requires an inversion of a high-dimensional $(p + q + D)$ matrix. Following Ha et al. (2001), we use the partial HL h^* (i.e., h_p) that eliminates λ_0. Thus the covariance estimates for $\hat{v} - v$ are obtained from the lower right-hand corner of the inverse of $H(h_p; \beta, v)$ in (4.35). That is,

$$\begin{aligned} \text{var}(\hat{v} - v) &\approx \{(\frac{-\partial^2 h^*}{\partial v^2}) - (\frac{-\partial^2 h^*}{\partial v \partial \beta})(\frac{-\partial^2 h^*}{\partial \beta^2})^{-1}(\frac{-\partial^2 h^*}{\partial \beta \partial v})\}^{-1}|_{\tau=\hat{\tau}} \\ &= \{(Z^T W^* Z + Q) - (Z^T W^* X)(X^T W^* X)^{-1}(X^T W^* Z)\}^{-1}|_{\tau=\hat{\tau}}, \end{aligned}$$

where $\tau = (\beta^T, v^T)^T$ and $\hat{\tau} = (\hat{\beta}^T, \hat{v}^T)^T$. With h_p, though, we need to invert the $(p + q)$ matrix $H(h_p; \beta, v)$, leading to an efficient computation of the confidence interval for v given by (4.20). □

Chapter 5
Multicomponent Frailty Models

Time-to-event data (recurrent or multiple event times) are often observed on the same subject (or cluster), and the frailty models are useful for analysis of such data. In practice, the multicomponent frailty models are of interest with complicated frailty structures, nested or crossed. For example, in multicenter clinical trials, we may need the frailties for patients and hospitals (or centers), where patients are nested within a hospital. If the number of recurrences is large for each patient, we might need to accommodate autoregressive (AR) models for the frailties. In this Chapter, we present the multicomponent semiparametric frailty models with various frailty structures. The h-likelihood procedures in Chap. 4 can be easily extended to the multicomponent models with more than one random components.

5.1 Formulation of the Multicomponent Frailty Models

Consider a multicomponent frailty model, $\lambda(t|v) = \lambda_0(t)\exp(\eta)$ with

$$\eta = X\beta + Z_1 v^{(1)} + Z_2 v^{(2)} + \cdots + Z_k v^{(k)}. \tag{5.1}$$

Here, X is an $n \times p$ model matrix, Z_r $(r = 1, 2, \ldots, k)$ are $n \times q_r$ model matrices corresponding to the $q_r \times 1$ frailties $v^{(r)}$, and $v^{(r)}$ and $v^{(l)}$ are independent for all $r \neq l$. Let $Z = (Z_1, Z_2, \ldots, Z_k)$, $v = (v^{(1)T}, v^{(2)T}, \ldots, v^{(k)T})^T$, $\alpha = (\alpha_1^T, \ldots, \alpha_k^T)^T$, and $q = \sum_r q_r$. Here α are the dispersion parameters (or frailty parameters) in the frailty distribution. Suppose that

$$v^{(r)} \sim N(0, \Sigma_r) \text{ and } v \sim N(0, \Sigma), \tag{5.2}$$

where $\Sigma_r = \Sigma_r(\alpha_r)$, and $\Sigma = \Sigma(\alpha) = \text{BD}(\Sigma_1, \cdots, \Sigma_k)$ and BD denotes a block diagonal matrix. For identifiability, following Lee and Nelder (1996) as mentioned

I.D. Ha et al., *Statistical Modelling of Survival Data with Random Effects*,
Statistics for Biology and Health, https://doi.org/10.1007/978-981-10-6557-6_5

in Chap. 4, the frailties $v^{(r)}$ have constraints that $E(v^{(r)}) = 0$ for all r. Note that some component of α can be a vector, for example, $\alpha_3 = (\sigma^2, \rho)$ as in the AR(1) frailty in (5.6). Then, clearly, the multicomponent model (5.1) can be written as in one-component model (4.1) because (5.1) can be expressed as a one-component model

$$\eta = X\beta + Zv,$$

where $v \sim N(0, \Sigma)$. Thus, we may view a multicomponent model with $\Sigma_r = \Sigma_r(\alpha_r)$ as one-component model with the covariance matrix $\Sigma(\alpha)$ for the frailty. Thus, if we can allow a covariance structure for the frailties in one-component frailty model, the extension from one-component model to the multicomponent model is straightforward (Appendix 5.6.1). In the h-likelihood approach, allowing the covariance for the frailty is straightforward as we shall see in the next subsections.

5.1.1 Multilevel and Time-Dependent Frailties

The CGD data in Sect. 1.2.3 have a multilevel structure in which patients, nested within hospitals, have recurrent infection times. Below are various frailty structures which will be used for the data analysis later.

• **Multilevel frailties**

Let T_{ijk} be an infection time for the kth observation of the jth patient nested in the ith hospital. Let v_i be an unobserved log-frailty for the ith hospital and let v_{ij} be one for the jth patient in the i th hospital. For T_{ijk}, we consider a three-level frailty model, in which observations, patients, and hospitals are defined as the units at levels 1, 2 and 3, respectively:

$$\lambda_{ijk}(t|v_i, v_{ij}) = \lambda_0(t)\exp(\eta_{ijk}) \tag{5.3}$$

with

$$\eta_{ijk} = x_{ijk}^T\beta + v_i + v_{ij},$$

where $x_{ijk} = (x_{ijk1}, \ldots, x_{ijkp})^T$ is a vector of covariates, $\beta = (\beta_1, \ldots, \beta_p)^T$ is a vector of the fixed effects, and $v_i \sim N(0, \alpha_1)$ and $v_{ij} \sim N(0, \alpha_2)$ are mutually independent. The three-level model (5.3) can be also expressed as a multicomponent model (5.1) with $k = 2$:

$$\eta = X\beta + Z_1 v^{(1)} + Z_2 v^{(2)}, \tag{5.4}$$

$v^{(1)}$: hospital-frailty vector based on v_i's $\sim N(0, \Sigma_1 \equiv \alpha_1 I_{q_1})$,
$v^{(2)}$: patient-frailty vector based on v_{ij}'s $\sim N(0, \Sigma_2 \equiv \alpha_2 I_{q_2})$,

where I_{q_r} $(r = 1, 2)$ are the $q_r \times q_r$ identity matrices, and q_1 and q_2 are the number of hospitals and patients, respectively.

• Time-dependent frailties
Since T_{ijk} are the gap times, they may be serially correlated, so that the frailty of each patient in the same hospital may not be constant, but may vary stochastically over the gap times. Let $v_{ijk} \sim AR(1)$ be the unobserved AR(1) frailty on the kth gap time of the jth patient in the ith hospital, satisfying

$$v_{ijk} = \rho v_{ijk-1} + e_{ijk},$$

where $e_{ijk} \sim N(0, \sigma^2)$ and $|\rho| < 1$. Thus we can consider the following time-dependent AR(1) frailty model with

$$\eta_{ijk} = x_{ijk}^T \beta + v_{ijk}. \tag{5.5}$$

• Multilevel frailties with time-dependent structures
We consider

$$\eta_{ijk} = x_{ijk}^T \beta + v_i + v_{ij} + v_{ijk},$$

which can be written as the matrix form

$$\eta = X\beta + Z_1 v^{(1)} + Z_2 v^{(2)} + Z_3 v^{(3)}, \tag{5.6}$$

$$v^{(1)}\text{: hospital frailty} \sim N(0, \Sigma_1 \equiv \alpha_1 I_{q_1}),$$
$$v^{(2)}\text{: patient frailty} \sim N(0, \Sigma_2 \equiv \alpha_2 I_{q_2}),$$
$$v^{(3)}\text{: AR(1)-frailty vector based on } v_{ijk} \sim N(0, \Sigma_3 \equiv \sigma^2 A),$$

where $A = A(\rho) = (1 - \rho^2)^{-1} C(\rho)$ is an $n \times n$ symmetric matrix and $C(\rho)$ is an AR(1) correlation matrix whose (l, m)th element is given by $\text{corr}(v_l^{(3)}, v_m^{(3)}) = \rho^{|l-m|}$, where ρ is also a frailty parameter, and $|\rho| < 1$. Here $q_3 = n$ is the total number of observations.

5.1.2 *Correlated Frailties*

Correlated random effects are useful in investigating the heterogeneity of the random baseline risk and/or treatment effects across centers in multicenter clinical studies. Let T_{ij} be survival time for the jth observation in the ith cluster. Denote by $v_i = (v_{i0}, v_{i1}, \ldots, v_{i,m-1})^T$ a m-dimensional vector of unobserved log-frailties associated with the ith cluster. We allow for the correlation of random effects in one-random component v_i for the ith cluster. Given v_i, the conditional hazard function of T_{ij} is of the form

$$\lambda_{ij}(t|v_i) = \lambda_0(t) \exp(\eta_{ij}), \tag{5.7}$$

where

$$\eta_{ij} = x_{ij}^T \beta + z_{ij}^T v_i$$

is a linear predictor for the hazards, and $x_{ij} = (x_{ij1}, \ldots, x_{ijp})^T$ and $z_{ij} = (z_{ij1}, \ldots, z_{ijm})^T$ are $p \times 1$ and $m \times 1$ covariate vectors corresponding to fixed effects $\beta = (\beta_1, \ldots, \beta_p)^T$ and the log-frailties v_i, respectively. Here z_{ij} is often a subset of x_{ij}. Alternatively, it may be the constant (unity) representing a cluster effect on the baseline hazard. We assume the multivariate normal distribution for v_i:

$$v_i \sim N_m(0, \Sigma_1),$$

where the covariance matrix $\Sigma_1 = \Sigma_1(\alpha_1)$ depends on α_1, a vector of the unknown parameters.

• **Multicenter clinical study**

In multicenter randomized clinical trials, there may be two variations across centers; variations in the baseline risk and the treatment effect. Thus, we could model these variations using the correlated frailty models (5.7).

(i) Let v_{i0} be a random baseline intercept (i.e. a random center effect or a random baseline risk) and let v_{i1} be a random slope (i.e. a random treatment effect or a random treatment-by-center interaction). In model (5.7), if $z_{ij} = 1$ and $v_i = v_{i0}$ for all i, j, it becomes a random intercept (or univariate) model in (4.1) with

$$\eta_{ij} = x_{ij}^T \beta + v_{i0}, \tag{5.8}$$

where $v_{i0} \sim N(0, \sigma_0^2)$ for all i.

(ii) Let β_1 be the effect of the primary covariate x_{ij1} such as the main treatment effect and let β_l $(l = 2, \ldots, p)$ be the fixed effects corresponding to the covariates x_{ijl}. We can consider a bivariate frailty model,

$$\eta_{ij} = v_{i0} + (\beta_1 + v_{i1})x_{ij1} + \sum_{l=2}^{p} \beta_l x_{ijl}, \tag{5.9}$$

which can be written in the form (5.7) by taking $z_{ij} = (1, x_{ij1})^T$ and $v_i = (v_{i0}, v_{i1})^T$. Here,

$$\begin{pmatrix} v_{i0} \\ v_{i1} \end{pmatrix} \sim N\left\{ \begin{pmatrix} 0 \\ 0 \end{pmatrix}, \; \Sigma_1(\alpha_1) \equiv \begin{pmatrix} \alpha_{11} & \alpha_{12} \\ \alpha_{12} & \alpha_{22} \end{pmatrix} = \begin{pmatrix} \sigma_0^2 & \sigma_{01} \\ \sigma_{01} & \sigma_1^2 \end{pmatrix} \right\},$$

and the correlation between the two frailties becomes $\rho = \sigma_{01}/(\sigma_0 \sigma_1)$. By allowing a correlation between two random effects (v_{i0} and v_{i1}), the invariance of the model to parametrization of the treatment effect can be maintained (Rondeau et al. 2008; Lee et al. 2017b).

Remark 5.1 Below we present the interpretation of v_{i0} and v_{i1} in the multicenter clinical study. Consider model (5.9) with a single binary-treatment indicator, x_{ij},

$$\lambda_{ij}(t|v_{i0}, v_{i1}; x_{ij}) = \lambda_0(t)\exp\{v_{i0} + (\beta_1 + v_{i1})x_{ij}\}.$$

Then, the relative risk for treatment becomes

$$\psi_{ij}(t|x_{ij} = 1, x_{ij} = 0) = \frac{\lambda_0(t)\exp\{v_{i0} + (\beta_1 + v_{i1})\cdot 1\}}{\lambda_0(t)\exp\{v_{i0} + (\beta_1 + v_{i1})\cdot 0\}} = \exp(\beta_1 + v_{i1}),$$

which is free of time t and holds for all patients in center i. Here $\exp(\beta_1)$ is the usual expression for the relative risk in the standard PH model. Thus, $\psi_{ij}(t|x = 1, x = 0)$, represents a random multiplicative divergence from the standard relative risk under the PH model, homogeneous with respect to centers. Note that

$$\exp(\beta_1 + v_{i1}) = \exp(\beta_1)u_{i1},$$

where $u_{i1} = \exp(v_{i1})$ is often called the treatment hazard ratio in the ith center (Gray 1994; Yamaguchi and Ohashi 1999). We also have that

$$u_{i1} = \frac{\exp(\beta_1 + v_{i1})}{\exp(\beta_1)} = \exp(v_{i1}).$$

Thus, v_{i1} means the random deviation of the ith center from the overall treatment effect.

Similarly, in order to interpret v_{i0}, we consider the model without the covariate x_{ij} (i.e. $x_{ij} = 0$):

$$\lambda_{ij}(t|v_{i0}) = \lambda_0(t)\exp(v_{i0}).$$

Thus, the hazard ratio of v_{i0} against $v_{i0} = 0$ is given by

$$\phi_{ij}(t) = \frac{\lambda_0(t)\exp(v_{i0})}{\lambda_0(t)} = \exp(v_{i0}),$$

which is free of time t and holds for all patients in center i, and v_{i0} represents the random deviation of the ith center from the overall underlying baseline risk. □

5.2 H-Likelihood Procedures for the Multicomponent Models

We show how the h-likelihood estimation procedures for one-component (univariate) frailty models in Chap. 4 are extended to the multicomponent models (5.1). The h-likelihood for the multicomponent frailty models are of the form

$$h = h(\beta, \lambda_0, \alpha) = \ell_1 + \ell_2,$$

where $\ell_1 = \sum_{ij} \ell_{1ij}(\beta, \lambda_0; y_{ij}, \delta_{ij}|v) = \sum_{ij} \log\{f_{\beta,\lambda_0}(y_{ij}, \delta_{ij}|v)\}$, $\ell_2 = \ell_2(\alpha; v) = \log\{f_\alpha(v)\}$, and $f_{\beta,\lambda_0}(y, \delta|v)$ and $f_\alpha(v)$ denote the conditional density of (y, δ) given v and the density of v, respectively. Here $\ell_2(\alpha; v)$ becomes $\sum_r \ell_{2r}(\alpha_r; v^{(r)})$. Since we use $v^{(r)} \sim N(0, \Sigma_r)$ with $\Sigma_r = \Sigma_r(\alpha_r)$, we have

$$\ell_{2r} = \ell_{2r}(\alpha_r; v_r) = -\frac{1}{2}[\log \det\{2\pi \Sigma_r\}] - \frac{1}{2} v^{(r)T} \Sigma_r^{-1} v^{(r)}.$$

The nuisance parameters λ_0 in ℓ_1 are eliminated by the profiling method, so that we have

$$\ell_1^* = \ell_1|_{\lambda_0=\hat{\lambda}_0} = \ell_{1p} + \text{constant},$$

which does not depend on λ_0. The profile log-likelihood ℓ_1^* is again proportional to the partial log-likelihood ℓ_{1p}. Thus, if the distributions of $v^{(r)}$ are specified, the generalization of the h-likelihood procedure based on the partial h-likelihood

$$h_p = \ell_{1p} + \ell_2$$

is straightforward, as given in Appendix 5.6.1. The technical details of computation of $-\partial^2 p_\tau(h_p)/\partial\alpha^2$ with $\tau = (\beta, v)$ are also given in Appendix 5.6.2.

5.3 Examples

5.3.1 Mammary Tumor Data

Gail et al. (1980) presented data on multiple occurrences of mammary tumors for 48 female rats. The primary outcome of interest was time to development of a mammary tumor for 23 female rats in the treatment group and 25 female rats in the control group. Initially, 76 rats were injected with a carcinogen for mammary cancer at day zero, and then all rats were given retinyl acetate to prevent cancer for 60 days. After 60 days, forty-eight rats which remained tumor-free were randomly assigned to continue being treated with retinoid prophylaxis (treatment group) or to the control group receiving no further retinoid prophylaxis. Rats were palpated for tumors twice weekly and observation ended 182 days after the initial carcinogen injection. The main objective of the study was to evaluate treatment.

Our analysis is based on the original data set in Table 5.1 from Gail et al. (1980). We similarly define survival time as time to development of a mammary tumor from the initial carcinogen injection. The time origin is the day of the initial carcinogen injection. The inter-arrival (gap) time between the tumor recurrences T_{ij} $(j = 1, \ldots, n_i)$ is calculated as $T_{ij} = T_{i,j} - T_{i,j-1}$, where $T_{i,j}$ is the jth tumor occurrence time of the ith rat. Conventionally we set $T_{i,0} = 0$. Here the cluster size n_i ranges from 1 to 14. Some $T_{i,j}$ and $T_{i,j-1}$ are equal, leading to the gap time $T_{ij} = T_{i,j} - T_{i,j-1} = 0$.

Table 5.1 Model selection: three AICs for the mammary tumor data

Model	$-2p_{\beta,v}(h_p)$	df_r	rAIC	$-2\ell_p$	df_c	cAIC	$-2p_v(h_p)$	df_p	pAIC
M1 (Cox)	1946.8	0	7.0	1944.9	1	28.1	1944.9	1	6.1
M2 (R)	1939.0	1	1.2	1885.0	22.8	11.8	1937.8	2	1.0
M3 (S)	1946.8	1	9.0	1944.9	1.0	28.1	1944.9	2	8.1
M4 (AR(1))	1935.8	2	0	1820.4	49.2	0	1934.8	3	0
M5 (R+AR(1))	1935.8	3	2.0	1820.4	49.2	0.0	1934.8	4	2.0

R, rat frailty; AR(1), AR(1) frailty; S, saturated model with $\rho = 0$;
AIC, AIC differences where the smallest AIC is calibrated to be zero

For such cases, we added a small value (say 0.01) to all the observations (Fong Daniel et al. 2001; Ha et al. 2007a). Censoring (approximately 17%) occurred when no new tumor was found. Gap times on the same rat may be correlated due to the shared genetic or environmental effects and this correlation can be modeled by a shared (univariate) frailty. However, since T_{ij} are gap times of the same rat, they could be serially correlated. Thus, the frailty of each rat may not be constant, but can change stochastically over the gap times. Therefore, we consider several AR(1) frailty models for such dependency. Here we model the gap times T_{ij}, with a single fixed covariate x_{ij} (=1 for treatment and =0 for control). Let v_i be the unobserved shared frailty on the ith rat and let v_{ij} be the unobserved AR(1) frailty on the jth gap time of the ith rat. We consider the following five models, $\lambda_{ij}(t|v) = \lambda_0(t)\exp(\eta_{ij})$:

M1 (Cox): $\eta_{ij} = x_{ij}\beta$,
M2 (R): $\eta_{ij} = x_{ij}\beta + v_i$ with $v_i \sim N(0, \alpha_1)$,
M3 (S): $\eta_{ij} = x_{ij}\beta + v_{ij}$ with $v_{ij} \sim N(0, \alpha_2)$,
M4 (AR(1)): $\eta_{ij} = x_{ij}\beta + v_{ij}$ with $v_{ij} \sim$ AR(1),
M5 (R+AR(1)): $\eta_{ij} = x_{ij}\beta + v_i + v_{ij}$ with $v_i \sim N(0, \alpha_1)$ and $v_{ij} \sim$ AR(1).

Here $v_{ij} \sim AR(1)$ means that $v_{ij} = \rho v_{ij-1} + e_{ij}$, $e_{ij} \sim N(0, \alpha_2)$ and $|\rho| < 1$. The saturated model M3 can be AR(1) model with $\rho = 0$. Among these models, M5 is the full model and the rest of them are submodels by assuming the null components, i.e. M4 ($v_i = 0$), M3 ($v_i = 0$, $\rho = 0$), M2 ($v_{ij} = 0$) and M1 ($v_i = 0$, $v_{ij} = 0$). "S" stands for AR(1) model with $\rho = 0$, identical to the saturated frailty model.

If a model M1 is nested in M2, we denote it as "M1 $\subset$ M2". Here M5$\supset$M4$\supset$M3$\supset$M1 and M5$\supset$M2$\supset$M1. Thus, for the nested frailty models, we use the rAIC for model selection because they have a common linear predictor. However, {M4, M3} and M2 are not nested, so we may use the AICs to select a model. Between M5 and M4, the difference of the partial restricted likelihood is 0.0 (<2.71), so that the null hypothesis of $\alpha_1 = 0$ cannot be rejected. Between M4 and M3, the difference of the partial restricted likelihood is 11.0 (>3.84), so that we reject the null hypothesis of $\rho = 0$. Between M2 and M1, the difference of the partial restricted likelihood is 7.8 (>2.71), so again we reject the null hypothesis of $\alpha_1 = 0$. Therefore, in the presence of AR(1) frailty, the shared frailty v_i is not necessary and the LRT selects M4 as the final model.

Now we investigate the AIC values. For the ease of comparison and ranking of candidate models, we have set the smallest value of the AICs to be zero. In Table 5.1 we report the AIC differences, not the AIC values themselves. The cAICs of M3 and M1 are almost identical because $\widehat{\alpha}_2 \approx 0$. Similarly, those from M5 and M4 are almost identical because $\widehat{\alpha}_1 \approx 0$. In such cases, the cAIC prefers the complicated model (Yu et al. 2013), while the mAIC (therefore pAIC) prefers the simple model (Greven and Kneib 2010). The degree of freedom df_c for the cAIC does not reflect model complexity properly when the variance estimate of the frailties is zero (Ha et al. 2007a), so that between M4 and M5, we should choose M4 as the final model using the rAIC. Care is necessary for the cAIC and pAIC (mAIC) when the frailty variance estimate is near zero.

The final model M4 gives the parameter estimates: $\hat{\beta} = -0.927(SE = 0.236)$, $\hat{\rho} = 0.811$ and $\hat{\alpha} = 0.162$. That is, we see that the treatment group significantly reduces the tumor recurrences and that survival times have a large positive serial correlation. The estimation results from the final model (M4) are also presented in Sect. 5.4, together with R codes.

5.3.2 CGD Data

We consider the CGD data presented in Sect. 1.2.3. The gap time (inter-arrival time) between recurrent infection times, T_{ijk}, are calculated as $T_{ijk} = T_{ij,k} - T_{ij,k-1}$, where $T_{ij,k}$ ($T_{ij,0} = 0$) is the kth recurrent infection time of the jth patient nested in the ith hospital. From the data structure, we see that the survival time for a given patient would be correlated. However, since each patient belongs to one of the 13 hospitals, the correlation may also be due to a random hospital effect. Thus we may consider the multicomponent lognormal frailty models, in which infections, patients and hospitals are defined as level 1, level 2 and level 3 units, respectively. Let v_i be the frailty on the ith hospital and v_{ij} be that on the jth patient in the ith hospital. For T_{ijk}, we consider the following models based on $\lambda_{ijk}(t|v) = \lambda_0(t)\exp(\eta_{ijk})$:

M1 (Cox): $\eta_{ijk} = x_{ijk}^T\beta$,

M2 (H): $\eta_{ijk} = x_{ijk}^T\beta + v_i$, $v_i \sim N(0, \alpha_1)$,

M3 (P): $\eta_{ijk} = x_{ijk}^T\beta + v_{ij}$, $v_{ij} \sim N(0, \alpha_2)$,

M4 (H + P): $\eta_{ijk} = x_{ijk}^T\beta + v_i + v_{ij}$, $v_i \sim N(0, \alpha_1)$, $v_{ij} \sim N(0, \alpha_2)$,

where $x_{ijk} = (x_{ijk1}, \ldots, x_{ijk10})^T$ consist of 10 covariates as in Sect. 1.2.3. Here M4 is the full model and the other models are various submodels of M4, i.e. M3 ($v_i = 0$), M2 ($v_{ij} = 0$) and M1($v_i = 0$, $v_{ij} = 0$). Here $\hat{\alpha}_1 \approx 0$, so that df_c may not reflect the model complexity properly. From Table 5.2, we see that between M1 and M2 (M3 and M4), the pAIC prefers the simpler model M1 (M3).

Since T_{ijk}'s are gap times, they may be serially correlated, so that the frailty of each patient in the same hospital may vary stochastically over them. Let $v_{ijk} \sim AR(1)$ be the unobserved AR(1) frailty on the kth gap time of the jth patient in the ith

Table 5.2 Model selection: three AICs for the CGD data

Model	$-2p_{\beta,v}(h_p)$	df_r	rAIC	$-2\ell_p$	df_c	cAIC	$-2p_v(h_p)$	df_p	pAIC
M1 (Cox)	694.7	0	2.2	671.9	10	18.1	671.9	10	0
M2 (H)	694.7	1	4.2	671.9	10.0	18.1	671.9	11	2.0
M3 (P)	690.5	1	0	608.8	37.3	9.6	671.1	11	1.2
M4 (H+P)	690.5	2	2.0	608.8	37.3	9.6	671.1	12	3.2
M5 (S)	692.8	1	2.3	595.4	44.6	10.8	671.9	11	2.0
M6 (AR(1))	688.6	2	0.1	553.6	60.1	0	670.6	12	2.7
M7 (H+AR(1))	688.6	3	2.1	553.6	60.2	0.1	670.6	13	4.7
M8 (P+AR(1))	688.6	3	2.1	553.6	60.2	0.1	670.6	13	4.7
M9 (H+P+AR(1))	688.6	4	4.1	552.4	61.0	0.6	670.6	14	6.7

H, hospital frailty; P, patient frailty; S, saturate model with $\rho = 0$; AR(1), AR(1) frailty; AIC, AIC differences where the smallest AIC is calibrated to be zero

hospital, satisfying $v_{ijk} = \rho v_{ijk-1} + e_{ijk}$, $e_{ijk} \sim N(0, \alpha_3)$ and $|\rho| < 1$. We consider the following additional models:

M5 (S):$\eta_{ijk} = x_{ijk}^T\beta + v_{ijk}$ with $v_{ijk} \sim N(0, \alpha_3)$,
M6 (AR(1)): $\eta_{ijk} = x_{ijk}^T\beta + v_{ijk}$ with $v_{ijk} \sim$ AR(1),
M7 (H+AR(1)): $\eta_{ijk} = x_{ijk}^T\beta + v_i + v_{ijk}$ with $v_i \sim N(0, \alpha_1)$ and $v_{ijk} \sim$ AR(1),
M8 (P+AR(1)): $\eta_{ijk} = x_{ijk}^T\beta + v_{ij} + v_{ijk}$ with $v_{ij} \sim N(0, \alpha_2)$ and $v_{ijk} \sim$ AR(1),
M9 (H+P+AR(1)): $\eta_{ijk} = x_{ijk}^T\beta + v_i + v_{ij} + v_{ijk}$ with $v_i \sim N(0, \alpha_1)$, $v_{ij} \sim N(0, \alpha_2)$ and $v_{ijk} \sim$ AR(1).

Now, M9 is the full model which combines models M4 and M6 and the others are its various submodels: M8 ($v_i = 0$), M7 ($v_{ij} = 0$), and M6 ($v_i = 0$, $v_{ij} = 0$). First, consider the LRT for the nest models. From the partial restricted likelihoods of {M9, M8, M6, M5}, the LRT selects M6 as the best model. Between M6 and M5, the difference of the partial restricted likelihood is 4.2 (>3.84), so that the null hypothesis of $\rho = 0$ is not rejected. Now that {M2, M3, M5} are not the submodels of M6, the AICs may be used for model selection. We note that the rAIC selects either M3 or M6 as an adequate model because their difference is too small as 0.1, but the cAIC clearly selects M6. From the rAIC and cAIC, we select M6 as the final model. However, the pAIC selects M1 as the final model because it tends to select the simplest model (Greven and Kneib 2010). In this book, we select the final model by using the LRT and we will use various AICs if the LRT is not available. Between M1 and M6, the LRT based on the partial restricted likelihood selects M6, so that we choose M6 as the final model.

Table 5.3 shows the estimation results for the final model M6. Our main conclusions are as follows: (i) The Gamma-IFN is very significant, indicating that the new drug significantly reduces the infection rate in CGD patients; (ii) The longitudi-

Table 5.3 Parameter estimates of the final model (AR(1) frailty model) for the CGD data

Variable	Estimate	SE
Gamma-IFN	−1.239	0.364
Inheritance	−0.771	0.408
Age	−0.093	0.049
Height	0.100	0.015
Weight	0.009	0.021
Corticosteroids	2.201	0.956
Prophylactic	−0.714	0.489
Sex	−0.907	0.573
Hospital region	−0.759	0.432
Longitudinal	1.379	0.597
α_3	0.877	–
ρ	0.573	–

nal variable is positively significant, which implies an increasing risk of subsequent infection with elapsed time; (iii) The estimated AR(1) correlation is $\widehat{\rho} = 0.573$, indicating that there is a serial correlation effect among recurrent infection times.

In addition, the estimation results of the univariate model (M3), the multilevel model (M4) and the final model (M6) are presented in Sect. 5.4, together with R codes.

5.3.3 Bladder Cancer Data

We again consider the bladder cancer data, analyzed in Example 4.5 Let v_{i0} and v_{i1} be the random baseline risk (i.e. random center effect) and random treatment effect of the ith center, respectively. For the purpose of analysis, we consider the following five models, $\lambda_{ij}(t|v) = \lambda_0(t)\exp(\eta_{ij})$ with η_{ij} allowing several frailty structures M2-M5, where $(v_{i0}, v_{i1}) \sim BN$ means that $v_{i0} \sim N(0, \sigma_0^2)$, $v_{i1} \sim N(0, \sigma_1^2)$ and $\rho = \text{Corr}(v_{i0}, v_{i1})$, and $(v_{i0}, v_{i1}) \sim IN$ means BN with $\rho = 0$:

M1 (Cox): $\eta_{ij} = \beta_1 x_{ij1} + \beta_2 x_{ij2}$,

M2 (B): $\eta_{ij} = \beta_1 x_{ij1} + \beta_2 x_{ij2} + v_{i0}$, with $v_{i0} \sim N(0, \sigma_0^2)$,

M3 (T): $\eta_{ij} = \beta_1 x_{ij1} + \beta_2 x_{ij2} + v_{i1} x_{ij1}$, with $v_{i1} \sim N(0, \sigma_1^2)$,

M4 (Indep): $\eta_{ij} = (\beta_1 + v_{i1}) x_{ij1} + \beta_2 x_{ij2} + v_{i0}$, with $(v_{i0}, v_{i1}) \sim IN$,

M5 (Corr): $\eta_{ij} = (\beta_1 + v_{i1}) x_{ij1} + \beta_2 x_{ij2} + v_{i0}$, with $(v_{i0}, v_{i1}) \sim BN$.

Here x_{ij1} and x_{ij2} are the binary covariates which are already described in Example 4.5, and B and T denote the random baseline risk and the random treatment effect, respectively. Among M1-M5, M5 is the full model and the others are its various submodels by assuming the null components, i.e. M1 ($v_{i0} = 0$, $v_{i1} = 0$), M2 ($v_{i1} = 0$), M3 ($v_{i0} = 0$) and M4 ($\rho = 0$). In Table 5.4, we report the AIC differences, not the AIC values themselves.

Table 5.4 Model selection: three AICs for the bladder cancer data

Model	$-2\{p_{\beta,v}(h_p)\}$	df_r	rAIC	$-2\ell_p$	df_c	cAIC	$-2p_v(h_p)$	df_p	pAIC
M1 (Cox)	2196.2	0	1.2	2192.5	2	6.2	2192.5	2	1.2
M2 (B)	2193.0	1	0	2173.8	8.5	0.5	2189.3	3	0
M3 (T)	2194.2	1	1.2	2179.2	7.0	2.9	2190.6	3	1.3
M4 (Indep)	2193.0	2	2.0	2173.8	8.5	0.5	2189.3	4	2.0
M5 (Corr)	2192.7	3	3.7	2172.1	9.1	0	2189.3	5	4.0

B, random baseline risk (v_{i0}); T, random treatment effect (v_{i1})

Here M5⊃M4⊃M3⊃M1 and M5⊃M4⊃M2⊃M1. Among the models {M1, M2, M4, M5}, the LRT based on the partial restricted likelihood, $-2\{p_{\beta,v}(h_p)\}$, selects M2 as the final model. Now the final model should be decided between M2 and M3 which are non-nested. Both rAIC and pAIC selects M2 as the final model. The cAIC selects the most complicated model M5 as the final model. We choose the model M2 as the final model, not M5, due to the LRT principle. The estimation results for the final model M2 are given in Example 4.5.

5.4 Software and Examples Using R

5.4.1 Mammary Tumor Data: AR(1) Frailty Model

Below are the R codes and the results from fitting the AR(1) frailty model using the HL(0,1) method for the mammary tumor data in Sect. 5.3.1. In the `frailtyHL()` function in Sect. 4.3.3, the expression `(1|center)+(1|id)` specifies a multilevel frailty model and the inclusion of the option `RandDist="AR1"` allows to fit the AR(1) frailty model. The outputs are summarized in Table 5.1.

```
######################### AR(1) frailty model ####################
> data(ren, package="frailtyHL")
> res1<-frailtyHL(Surv(time,del)~gp+(1|rat),ren, RandDist="AR1",
+ Maxiter = 500)
########################### OUTPUT ###############################
iteration :
           302
convergence :
  9.749738e-07
[1] "converged"
[1] "Results from the lognormal frailty model with AR(1)"
[1] "Number of data :"
[1] 254
[1] "Number of event :"
[1] 212
[1] "Model for conditional hazard :"
```

```
Surv(time, del) ~ gp + (1 | rat)
[1] "Method : HL(0,1)"
[1] "Estimates from the mean model"
   Estimate Std. Error t-value   p-value
gp   -0.927     0.2357  -3.933 8.397e-05
[1] "Estimates from the dispersion model"
    Estimate
rat   0.1623
[1] "Estimates for rho in the AR(1) model"
      rho_h
[1,] 0.8114
       -2h0  -2*hp -2*p_b,v(hp)
[1,] 1820.4 1925.2       1935.8
       cAIC   pAIC   rAIC
[1,] 1918.8 1956.8 1939.8
```

5.4.2 CGD Data: Univariate, Multilevel and AR(1) Frailty Models

Below are the R codes and the results from fitting the univariate, multilevel and AR(1) frailty models using the HL(0,1) method for the CGD data. The outputs are summarized in Tables 5.2 and 5.3.

```
########################## Variable settings #####################
>  data(cgd, package="frailtyHL")
>  attach(cgd)
>  cgd$inherit=relevel(cgd$inherit,ref="autosomal")
>  hos=as.integer(hos.cat)
>  hospi<-ifelse(hos>=3,1,0)  ### 0=US, 1=Europe
>  L=ifelse(enum==1,0,tstart+1)
>  longi=L/365.25 ## longitudinal
######################## Univariate frailty model ###################
>  cgd_P<-frailtyHL(Surv(tstop-tstart,status)~
+   factor(treat)+ factor(inherit)+age+ height +weight
+   +factor(steroids)+factor(propylac)+factor(sex)+factor(hospi)+ longi
+   +(1|id),cgd,Maxiter = 500)
iteration :
        101
convergence :
  9.033824e-07
[1] "converged"
[1] "Results from the lognormal frailty model"
[1] "Number of data :"
[1] 203
[1] "Number of event :"
[1] 76
[1] "Model for conditional hazard :"
Surv(tstop - tstart, status) ~ factor(treat) + factor(inherit) +
    age + height + weight + factor(steroids) + factor(propylac) +
    factor(sex) + factor(hospi) + longi + (1 | id)
[1] "Method : HL(0,1)"
```

```
[1] "Estimates from the mean model"
                            Estimate Std. Error t-value  p-value
factor(treat)rIFN-g        -1.105033    0.33787 -3.2706 0.001073
factor(inherit)X-linked -0.658498    0.38189 -1.7243 0.084648
age                        -0.085951    0.04481 -1.9182 0.055091
height                      0.008576    0.01377  0.6229 0.533318
weight                      0.009931    0.02072  0.4793 0.631743
factor(steroids)1           1.991416    0.85668  2.3246 0.020094
factor(propylac)1          -0.690360    0.44859 -1.5390 0.123813
factor(sex)female          -0.758045    0.52845 -1.4345 0.151438
factor(hospi)1             -0.697510    0.39655 -1.7589 0.078590
longi                       0.794978    0.51075  1.5565 0.119591
[1] "Estimates from the dispersion model"
   Estimate Std. Error
id   0.7033     0.4357
        -2h0  -2*hp -2*p_b,v(hp)
[1,] 608.83 824.56       690.52
        cAIC   pAIC   rAIC
[1,] 683.37 694.63 692.52
>

#################### Multilevel frailty model ######################
> cgd_multi<-frailtyHL(Surv(tstop-tstart,status)~
+   factor(treat)+ factor(inherit)+age+ height +weight+ factor(steroids)
+   +factor(propylac)+factor(sex)+factor(hospi)+ longi
+   +(1|center)+(1|id),cgd,Maxiter = 500)
iteration :
         109
convergence :
  9.594358e-07
[1] "converged"
[1] "Results from the lognormal frailty model"
[1] "Number of data :"
[1] 203
[1] "Number of event :"
[1] 76
[1] "Model for conditional hazard :"
Surv(tstop - tstart, status) ~ factor(treat) + factor(inherit) +
    age + height + weight + factor(steroids) + factor(propylac) +
    factor(sex) + factor(hospi) + longi + (1|center) + (1|id)
[1] "Method : HL(0,1)"
[1] "Estimates from the mean model"
                            Estimate Std. Error t-value  p-value
factor(treat)rIFN-g        -1.105033    0.33787 -3.2706 0.001073
factor(inherit)X-linked -0.658498    0.38189 -1.7243 0.084648
age                        -0.085950    0.04481 -1.9182 0.055091
height                      0.008576    0.01377  0.6229 0.533317
weight                      0.009931    0.02072  0.4793 0.631744
factor(steroids)1           1.991417    0.85668  2.3246 0.020094
factor(propylac)1          -0.690360    0.44859 -1.5390 0.123814
factor(sex)female          -0.758045    0.52845 -1.4345 0.151438
factor(hospi)1             -0.697509    0.39655 -1.7589 0.078591
longi                       0.794977    0.51075  1.5565 0.119592
[1] "Estimates from the dispersion model"
       Estimate Std. Error
center   0.0000        NaN
id       0.7033     0.4357
```

```
        -2h0   -2*hp -2*p_b,v(hp)
[1,] 608.83 824.56       690.52
        cAIC    pAIC    rAIC
[1,] 683.37 696.63 694.52
>

######################## AR(1) frailty model ####################
> cgd_AR1<-frailtyHL(Surv(tstop-tstart,status)~
+   factor(treat)+ factor(inherit)+age+ height +weight+ factor(steroids)
+   +factor(propylac)+factor(sex)+factor(hospi)+ longi
+   +(1|id),cgd,RandDist="AR1",Maxiter = 500)
iteration :
         338
convergence :
  9.844568e-07
[1] "converged"
[1] "Results from the lognormal frailty model with AR(1)"
[1] "Number of data :"
[1] 203
[1] "Number of event :"
[1] 76
[1] "Model for conditional hazard :"
Surv(tstop - tstart, status) ~ factor(treat) + factor(inherit) +
    age + height + weight + factor(steroids) + factor(propylac) +
    factor(sex) + factor(hospi) + longi + (1 | id)
[1] "Method : HL(0,1)"
[1] "Estimates from the mean model"
                          Estimate Std. Error t-value   p-value
factor(treat)rIFN-g      -1.238621    0.36373 -3.4053 0.0006608
factor(inherit)X-linked -0.770533    0.40839 -1.8868 0.0591907
age                      -0.093005    0.04946 -1.8804 0.0600534
height                    0.009586    0.01479  0.6480 0.5169636
weight                    0.009197    0.02251  0.4087 0.6827812
factor(steroids)1         2.200806    0.95591  2.3023 0.0213177
factor(propylac)1        -0.714297    0.48942 -1.4595 0.1444324
factor(sex)female        -0.907046    0.57311 -1.5827 0.1134943
factor(hospi)1           -0.758608    0.43202 -1.7560 0.0790941
longi                     1.378750    0.59720  2.3087 0.0209614
[1] "Estimates from the dispersion model"
   Estimate
id   0.8768
[1] "Estimates for rho in the AR(1) model"
      rho_h
[1,] 0.5732
        -2h0   -2*hp -2*p_b,v(hp)
[1,] 553.55 999.79       688.59
        cAIC    pAIC    rAIC
[1,] 673.69 730.35 692.59
```

5.5 Discussion

The h-likelihood methods can be straightforwardly extended to the multicomponent semiparametric frailty models. Selecting a suitable model among a set of candidate models is very important in data analysis. The LRTs and AICs are useful in selecting the final model.

We have used the three AICs, i.e. cAIC, pAIC and rAIC, based on the partial h-likelihood h_p. Here the cAIC, pAIC and rAIC are model selection criteria for v, β, α, respectively. For example, the pAIC is useful for model selection for β given a frailty model on α. Thus, it may not work properly under different frailty models. The rAIC and cAIC are for selection of a frailty structure. By construction, the rAIC is more focussed on the selection of the frailty; the restricted likelihood is not appropriate for inference about an individual v_i because it integrates them out. The cAIC concerns a selection of the conditional model $y|v$ and its prediction, so that it seems to prefer a complicated random-effect model. For inference about the population, the rAIC may be preferred, while the cAIC may be preferred for a subject-specific model.

Extensions of the current multicomponent modeling approach to the frailty models allowing for spatial frailty structures (Henderson et al. 2002) or missing covariates (Herring et al. 2002) would be interested (Lee et al. 2017b). Furthermore, our h-likelihood approach can be extended to cure-rate modeling via frailty structures (Yau and Ng 2001; Xiang et al. 2011) or double HGLMs (Lee and Nelder 2006; Lee et al. 2017b) including structured dispersion on hazard (Noh et al. 2006; Burke and MacKenzie 2017).

5.6 Appendix

5.6.1 H-Likelihood Procedure in the Multicomponent Models

We show how the h-likelihood procedure of one-component models presented in Chap. 4 is extended to the multicomponent models (5.1). With the partial h-likelihood h_p, we estimate the fixed parameters (β, α) with $\alpha = (\alpha_1, \ldots, \alpha_k)^T$ and the random effects $v = (v^{(1)T}, v^{(2)T}, \ldots, v^{(k)T})^T$ as follows. Given α, estimation of $\tau = (\beta^T, v^T)^T$ is performed by solving

$$\frac{\partial h_p}{\partial \tau} = \left.\frac{\partial h}{\partial \tau}\right|_{\lambda_0=\widehat{\lambda}_0} = 0. \tag{5.10}$$

Here, the first partial derivatives, $\partial h/\partial \tau$, are given by the simple forms:

$$\frac{\partial h}{\partial \beta} = X^T(\delta - \mu) \text{ and } \frac{\partial h}{\partial v^{(r)}} = Z_r^T(\delta - \mu) - \Sigma_r^{-1} v^{(r)} \ (r = 1, \ldots, k),$$

where $\mu = \exp(\log \Lambda_0 + \eta)$ and $\Sigma_r = \Sigma_r(\alpha_r, \rho)$. Note that the asymptotic covariance of $\hat{\tau} - \tau$ in one-component model is obtained from the inverse of the information matrix $H_p = H(h_p; \tau)|_{\tau=\hat{\tau}(\alpha)} = -\partial^2 h_p/\partial \tau^2$, given by

$$H_p = \begin{pmatrix} X^T W^* X & X^T W^* Z \\ Z^T W^* X & Z^T W^* Z + Q \end{pmatrix}. \tag{5.11}$$

Let $Z = (Z_1, \ldots, Z_k)$ and $Q_r = -\partial^2 \ell_2 / \partial v^{(r)2} = \Sigma_r^{-1}$ $(r = 1, \ldots, k)$, and let $Q = \mathrm{BD}(Q_1, \ldots, Q_k) = \mathrm{BD}(\Sigma_1^{-1}, \ldots, \Sigma_k^{-1})$ be a $q \times q$ block diagonal matrix ($q = \sum_r q_r$). Then the information matrix (5.1) can be expressed as that of the multi-component models:

$$H_p = \begin{pmatrix} X^T W^* X & X^T W^* Z_1 & \ldots & X^T W^* Z_k \\ Z_1^T W^* X & Z_1^T W^* Z_1 + Q_1 & \ldots & Z_1^T W^* Z_k \\ \vdots & \vdots & \ddots & \vdots \\ Z_k^T W^* X & Z_k^T W^* Z_1 & \ldots & Z_k^T W^* Z_k + Q_k \end{pmatrix}.$$

We note that in the multicomponent models, the ILS equations of τ can be expressed as the same forms as (4.12) and (4.13), with $Z = (Z_1, \cdots, Z_k)$, $Q = \mathrm{BD}(Q_1, \ldots, Q_k)$, $\mathbf{V}^* = \mathrm{BD}(I_n, Q)$, $\mathbf{y}^* = (w^{*T}, \mathbf{0}^T)^T$ and $I_q = \mathrm{BD}(I_{q_1}, \ldots, I_{q_k})$.

Next, for estimation of the frailty parameters $\alpha = (\alpha_1, \ldots, \alpha_k)^T$, we use the adjusted partial h-likelihood (i.e. partial restricted likelihood) in (4.10), given by

$$p_\tau(h_p) = \left\{ h_p - \frac{1}{2} \log \det \left(\frac{H_p}{2\pi} \right) \right\} \Big|_{\tau = \hat{\tau}}, \tag{5.12}$$

where $\widehat{\tau} = \widehat{\tau}(\alpha) = (\widehat{\beta}^T(\alpha), \widehat{v}^T(\alpha))^T$. The partial REML estimator for α_r $(r = 1, \ldots, k)$ are obtained by solving iteratively

$$\frac{\partial p_\tau(h_p)}{\partial \alpha_r} = 0. \tag{5.13}$$

Note here that

$$\frac{\partial p_\tau(h_p)}{\partial \alpha_r} = -\frac{1}{2} \mathrm{tr}\left(\Sigma_r^{-1} \frac{\partial \Sigma_r}{\partial \alpha_r} \right) - \frac{1}{2} \widehat{v}^{(r)T} \left(\frac{\partial \Sigma_r^{-1}}{\partial \alpha_r} \right) \widehat{v}^{(r)} - \frac{1}{2} \mathrm{tr}\left(\widehat{H}_p^{-1} \frac{\partial \widehat{H}_p}{\partial \alpha_r} \right),$$

where $\widehat{H}_p = \widehat{H}_p(\alpha) = H(h_p; \tau)|_{\tau = \widehat{\tau}(\alpha)}$. Note also that in implementing (5.13), we still allow the term $\partial \widehat{v} / \partial \alpha_r$; details on the computation of the term $\partial \widehat{H}_p / \partial \alpha_r$ including $\partial \widehat{v} / \partial \alpha_r$ are given in Appendix 5.6.2. In particular, from (5.13) the partial REML estimator for α_r $(r = 1, 2)$ in the multilevel model (5.4) has a simple form

$$\widehat{\alpha}_r = \frac{\widehat{v}^{(r)T} \widehat{v}^{(r)}}{q_r - \gamma_r},$$

where $\gamma_r = -\alpha_r \mathrm{tr}\{\widehat{H}_p^{-1} (\partial \widehat{H}_p / \partial \alpha_r)\}$. Similarly, the partial REML estimator for σ^2 in the AR(1) model in (5.6) is given by

$$\widehat{\sigma^2} = \frac{\widehat{v}^{(3)T} A^{-1} \widehat{v}^{(3)}}{q_3 - \gamma_3},$$

where $\gamma_3 = -\sigma^2 \mathrm{tr}\{\widehat{H}_p^{-1}(\partial \widehat{H}_p / \partial \sigma^2)\}$. Note that the estimator for ρ in (5.5) and (5.6) is also obtained from (5.13): see also Yau and McGilchrist (1998). In fact, the estimator for ρ can also be easily obtained by maximizing $p_\tau(h_p)$ in (5.12) using the grid search method.

In summary, the estimates of τ and α are obtained by alternating between the two estimating Eqs. (5.10) and (5.13) until a convergence is achieved. Thus, the fitting algorithm of the simple frailty model in Sect. 4.3.2 can be straightforwardly applied to the multicomponent models.

5.6.2 Computation of $-\partial^2 P_\tau(h_p)/\partial\alpha^2$

The partial restricted likelihood in (5.12) can be expressed as

$$p_\tau(h_p) = \hat{h} - \frac{1}{2}\log\det(\hat{H}_p) + \frac{(p+q)}{2}\log(2\pi),$$

where $\tau = (\beta^T, v^T)^T$, $\hat{h}_p = h_p|_{\tau=\hat{\tau}(\alpha)} = h_p(\hat{\tau}(\alpha), \alpha)$ and $\hat{H}_p = H(h_p; \tau)|_{\tau=\hat{\tau}(\alpha)} = H_p(\hat{\tau}(\alpha), \alpha)$.

Since

$$\frac{\partial p_\tau(h_p)}{\partial \alpha_r} = \frac{\partial \hat{h}}{\partial \alpha_r} - \frac{1}{2}\mathrm{tr}\left(\hat{H}_p^{-1}\frac{\partial \hat{H}_p}{\partial \alpha_r}\right), \tag{5.14}$$

we have

$$-\frac{\partial^2 p_\tau(h_p)}{\partial \alpha_r \partial \alpha_s} = -\frac{\partial^2 \hat{h}}{\partial \alpha_r \partial \alpha_s} + \frac{1}{2}\mathrm{tr}\left(-\hat{H}_p^{-1}\frac{\partial \hat{H}_p}{\partial \alpha_r}\hat{H}_p^{-1}\frac{\partial \hat{H}_p}{\partial \alpha_s} + \hat{H}_p^{-1}\frac{\partial^2 \hat{H}_p}{\partial \alpha_r \partial \alpha_s}\right). \tag{5.15}$$

We now show how to compute the Eq. (5.14). Following Lee and Nelder (2001a) and Ha and Lee (2003), we allow $\partial \hat{v}/\partial \alpha_r$ in computing the two Eqs. (5.14) and (5.15), but not $\partial \hat{\beta}/\partial \alpha_r$. Then we have

$$\begin{aligned}\frac{\partial \hat{h}}{\partial \alpha_r} &= \left\{\left(\frac{\partial h_p}{\partial \alpha_r}\right) + \left(\frac{\partial h_p}{\partial v}\right)\left(\frac{\partial \hat{v}}{\partial \alpha_r}\right)\right\}\Bigg|_{\tau=\hat{\tau}} \\ &= \frac{\partial h_p}{\partial \alpha_r}\Bigg|_{\tau=\hat{\tau}}\end{aligned}$$

since $(\partial h_p/\partial v)|_{\tau=\hat{\tau}} = 0$. Along the lines of Appendix C of Lee and Nelder (1996), we can show that

$$\begin{aligned}\frac{\partial\hat{v}}{\partial\alpha_r} &= -\left(-\frac{\partial^2 h_p}{\partial v^2}\right)^{-1}\left(-\frac{\partial^2 h_p}{\partial v\partial\alpha_r}\right)\Bigg|_{\tau=\hat{\tau}} \\ &= -(Z^T\hat{W}^*Z+Q)^{-1}Q'_r\hat{v},\end{aligned}$$

where $\hat{W}^*$ is given in (5.17), $Q=\Sigma^{-1}$ and

$$Q'_r = \partial\Sigma^{-1}/\partial\alpha_r = -\Sigma^{-1}\,(\partial\Sigma/\partial\alpha_r)\Sigma^{-1}. \tag{5.16}$$

From these results, the first term on the right-hand side (RHS) of (5.15) becomes

$$\begin{aligned}-\frac{\partial^2\hat{h}}{\partial\alpha_r\partial\alpha_s} &= \left\{\left(-\frac{\partial^2 h_p}{\partial\alpha_r\partial\alpha_s}\right)-\left(\frac{\partial^2 h_p}{\partial\alpha_s\partial v}\right)\left(\frac{\partial\hat{v}}{\partial\alpha_r}\right)\right\}\Bigg|_{\tau=\hat{\tau}} \\ &= -\frac{1}{2}\mathrm{tr}\left(\Sigma^{-1}\frac{\partial\Sigma}{\partial\alpha_r}\Sigma^{-1}\frac{\partial\Sigma}{\partial\alpha_s}-\Sigma^{-1}\frac{\partial^2\Sigma}{\partial\alpha_r\partial\alpha_s}\right)+\frac{1}{2}\hat{v}^TQ''_{rs}\hat{v} \\ &\quad+\hat{v}^TQ'_s\left(\frac{\partial\hat{v}}{\partial\alpha_r}\right),\end{aligned}$$

where

$$\begin{aligned}Q''_{rs} = \partial^2Q/\partial\alpha_r\partial\alpha_s &= -Q'_s(\partial\Sigma/\partial\alpha_r)\Sigma^{-1}-\Sigma^{-1}(\partial\Sigma/\partial\alpha_r)Q'_s \\ &\quad -\Sigma^{-1}(\partial^2\Sigma/\partial\alpha_r\partial\alpha_s)\Sigma^{-1}.\end{aligned}$$

From (4.39) and (5.11), we have

$$\hat{H}_p = \begin{pmatrix} X^T\hat{W}^*X & X^T\hat{W}^*Z \\ Z^T\hat{W}^*X & Z^T\hat{W}^*Z+\hat{Q}\end{pmatrix}, \tag{5.17}$$

where $\hat{W}^* = W^*|_{\tau=\hat{\tau}(\alpha)} = W^*(\hat{\tau}(\alpha),\alpha)$, $\hat{Q}=Q(\hat{v}(\alpha),\alpha)$ and

$$W^* = W^*(\beta, v) = W_1 - W_2, \tag{5.18}$$

and the details are given in (4.35). Thus, the two derivatives in the second term on the RHS of (5.15) are computed as follows:

$$\frac{\partial\hat{H}_p}{\partial\alpha_r} = \begin{pmatrix} X^TW'_rX & X^TW'_rZ \\ Z^TW'_rX & Z^TW'_rZ+Q'_r\end{pmatrix}$$

and

$$\frac{\partial^2\hat{H}_p}{\partial\alpha_r\alpha_s} = \begin{pmatrix} X^TW''_{rs}X & X^TW''_{rs}Z \\ Z^TW''_{rs}X & Z^TW''_{rs}Z+Q''_{rs}\end{pmatrix}.$$

Here $W'_r = \partial \hat{W}^*/\partial \alpha_r$ and $W''_{rs} = \partial^2 \hat{W}^*/\partial \alpha_r \partial \alpha_s$ are calculated by the following procedures:

$$\frac{\partial \hat{W}^*}{\partial \alpha_r} = \left\{ \left(\frac{\partial W^*}{\partial \alpha_r} \right) + \left(\frac{\partial W^*}{\partial v} \right) \left(\frac{\partial \hat{v}}{\partial \alpha_r} \right) \right\} \Big|_{T=\hat{T}} = \left\{ \left(\frac{\partial W^*}{\partial v} \right) \left(\frac{\partial \hat{v}}{\partial \alpha_r} \right) \right\} \Big|_{T=\hat{T}}$$

since $\partial W^*/\partial \alpha_r = 0$, and

$$\frac{\partial^2 \hat{W}^*}{\partial \alpha_r \partial \alpha_s} = \left\{ \left(\frac{\partial \hat{v}}{\partial \alpha_r} \right) \left(word \frac{\partial^2 W^*}{\partial v^2} \right) \left(\frac{\partial \hat{v}}{\partial \alpha_s} \right) + \left(\frac{\partial W^*}{\partial v} \right) \left(\frac{\partial^2 \hat{v}}{\partial \alpha_r \partial \alpha_s} \right) \right\} \Big|_{T=\hat{T}},$$

where

$$\frac{\partial^2 \hat{v}}{\partial \alpha_r \partial \alpha_s} = -(Z^T \hat{W}^* Z + Q)^{-1} \left\{ (Z^T W'_r Z + Q'_r) \frac{\partial \hat{v}}{\partial \alpha_s} + Q'_s \frac{\partial \hat{v}}{\partial \alpha_r} + Q''_{rs} \hat{v} \right\},$$

and $\partial W^*/\partial v$ and $\partial^2 W^*/\partial v^2$ can be calculated by repeatedly differentiating (5.17) with respect to v.

Chapter 6
Competing Risks Frailty Models

Competing risks data arise when an occurrence of an event precludes other types of events from being observed. In this chapter, we extend the h-likelihood inference procedures for frailty models to competing risks models. We first review existing methods for competing risk models without the frailty.

6.1 Classical Competing-Risk Models

Competing risks data are encountered in various research areas such as medicine, engineering, econometrics, and sociology. In cancer studies, the primary outcome is often time to event and patients might experience multiple events, where the occurrence of one type of event hinders that of other types of events. For example, in breast cancer studies investigators may want to evaluate the effect of a new drug in terms of reducing the recurrence rate of the disease or mortality related to breast cancer. However, other types of primary cancers (i.e., competing events) that would shift the course of therapy to mask breast cancer recurrences or deaths due to other disease often preclude breast cancer-related deaths from being observed. Examples in other fields include failure of different components in a system in industrial reliability testing or time to part- or full-time employment in econometrics.

When a specific type of event is of primary interest, two useful summary measures could be considered. One is the cause-specific hazard, i.e., the conditional instantaneous event rate of a specific type given no prior event history and the other is the cumulative incidence function (CIF; Kalbfleisch and Prentice 1980), i.e., the cumulative probability of cause-specific events. For example, in a clinical cancer trial, the main event (Type 1 or cause 1) is time to recurrence, which makes time to death a competing event (Type 2 or cause 2). Then the event history can be diagrammatically displayed as in Fig. 6.1.

I.D. Ha et al., *Statistical Modelling of Survival Data with Random Effects*,
Statistics for Biology and Health, https://doi.org/10.1007/978-981-10-6557-6_6

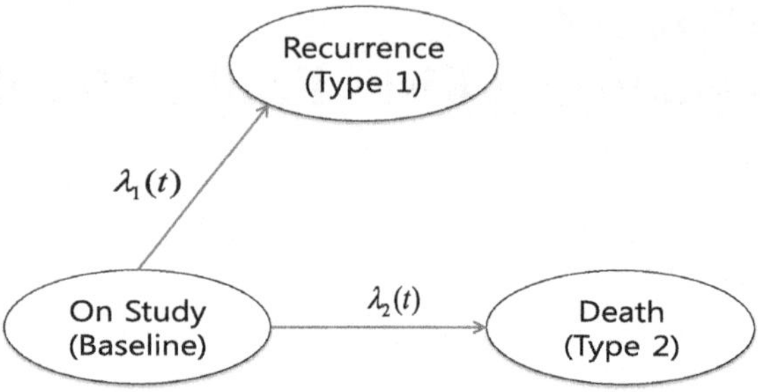

Fig. 6.1 Schematic display of competing risks model with cause-specific hazards $\lambda_k(t), k = 1, 2$

In this chapter, we are interested in clustered competing risks data where subjects within a cluster may experience more than (e.g., two events in Fig. 6.1) one type of event. In many applications involving competing risks data from multicenter randomized clinical trials, individual events within a cluster (center) may be correlated due to unobserved shared factors across individuals. Thus, we present the h-likelihood inference for correlated time-to-event data under competing risks, where the underlying cause-specific frailties might be correlated. First, we review existing methods for the independent survival data with competing risks.

6.1.1 Cause-Specific Hazard Function and Cumulative Incidence Function

Traditionally, two important quantities in analyzing competing risks data have been the cause-specific hazard function and the CIF. Suppose that the type of event is denoted by $k(k = 1, \ldots, K)$. Let T be time to the first event (i.e. the minimum of all of the latent event times) which is denoted by $T = \min(T_1, T_2, \ldots, T_K)$. Let $\epsilon \in \{1, 2, \ldots, K\}$ be the cause of event (or type of event). The cause-specific hazard function is the conditional intensity that a subject experiences an event type $\epsilon = k$ at a specific time point t, given that the individual has not experienced an event of any type up to just prior to time t:

$$\lambda_k(t) = \lim_{\Delta \to 0} \frac{Pr(t \le T < t + \Delta, \epsilon = k | T \ge t)}{\Delta}. \tag{6.1}$$

From the definition of the cause-specific hazard function above, we have

$$\begin{aligned} \lambda_k(t) &= \lim_{\Delta \to 0} \frac{Pr(t \le T < t + \Delta, \epsilon = k)}{\Delta Pr(T \ge t)} \\ &= \frac{1}{Pr(T \ge t)} \lim_{\Delta \to 0} \frac{Pr(t \le T < t + \Delta, \epsilon = k)}{\Delta} \\ &= \frac{f_k(t)}{S(t-)}, \end{aligned} \tag{6.2}$$

where $S(t-)$ is the survival function just prior to t for events from all causes and $f_k(t)$ is the subdistribution density function for Type k event. Thus, from (6.2) the CIF of event type k is defined by

$$F_k(t) \equiv Pr(T \leq t, \epsilon = k) = \int_0^t f_k(u)du = \int_0^t S(u-)d\Lambda_k(u), \qquad (6.3)$$

where $f_k(t) = dF_k(t)/dt$ and $\Lambda_k(u)$ is the cumulative hazard function of Type k event such that $d\Lambda_k(t)/dt = \lambda_k(t)$. In other words, the CIF $F_k(t)$ is the probability that a Type k event occurs at or before time t. The quantity $F_k(t)$ is also called as the subdistribution function because it is the cumulative joint probability of a Type k event, i.e., $F_k(t) = Pr(T \leq t, \epsilon = k)$. It is an improper distribution as

$$\lim_{t\to\infty} F_k(t) = \lim_{t\to\infty} Pr(T \leq t|\epsilon = k)Pr(\epsilon = k) = Pr(\epsilon = k).$$

Nonparametrically, the cause-specific hazard function can be estimated by a Nelson–Aalen type estimator $\hat{\lambda}_k(t) = d_{ik}/R_i$, after the competing events are censored at the time of their occurrences. Here d_{ik} is the number of Type k events at an ordered Type k event times $t_{(i)}$, $i = 1, \ldots, D_k$, and R_i is all subjects at risk at the time $t_{(i)}$. Therefore the CIF in (6.3) can be estimated by

$$\hat{F}_k(t) = \sum_{i:t_{(i)}\leq t} \hat{S}(t_{(i)}-)\frac{d_{ik}}{R_i},$$

where $\hat{S}(t_{(i)}-)$ is the Kaplan–Meier (KM) estimator of the all-cause survival function prior to $t_{(i)}$, i.e.,

$$\hat{S}(t_{(i)}-) = \prod_{j=1}^{i-1}\left\{1 - \frac{d_{jk} + e_{jc}}{R_j}\right\},$$

where e_{jc} is the number of competing events at time $t_{(j)}$. Without competing events the complement of the Kaplan–Meier estimator (denoted by $1 - KM$) is identical to the CIF since the $1 - KM$ for Type k events can be written as (Pintilie 2006)

$$1 - KM_k(t) = \sum_{i:t_{(i)}\leq t} KM_k(t_{(i)}-)\frac{d_{ik}}{R_i},$$

where

$$KM_k(t_{(i)}-) = \prod_{j=1}^{i-1}\left\{1 - \frac{d_{jk}}{R_j}\right\}.$$

Since $\hat{S}(t) \leq KM_k(t)$ for any t, from the above definitions of $\hat{F}_k(t)$ and $1 - KM_k(t)$, we have

$$\hat{F}_k(t) \leq 1 - KM_k(t).$$

Accordingly, we see that $(1 - KM)$ overestimates the true cumulative probability of the cause-specific event of interest, in the presence of competing risks. The CIF provides the correct estimate for the cause-specific cumulative probability (Gooley et al. 1999).

6.1.2 Subdistribution Hazard Function

The hazard function of a subdistribution (or subhazard function; Gray 1988) is defined by

$$\lambda_k^s(t) = \lim_{\Delta \to 0} \frac{Pr\{t \leq T < t + \Delta, \epsilon = k | T \geq t \text{ or } (T < t \cap \epsilon \neq k)\}}{\Delta}. \tag{6.4}$$

This is the instantaneous event rate at time t from cause k, given that an individual has not previously died from cause k. Since the risk set in this definition of the hazard function always includes those who have died from other causes before time t, the subhazard function is different from the cause-specific hazard in (6.1) in both definition and interpretation. Note that $\lambda_k^s(t)$ can be directly expressed using the CIF $F_k(t)$:

$$\begin{aligned} \lambda_k^s(t) &= \frac{f_k(t)}{1 - F_k(t)} \\ &= -\frac{d \log\{1 - F_k(t)\}}{dt}. \end{aligned} \tag{6.5}$$

6.1.3 Relationship Between Two Hazard Functions

The relationship between the cause-specific hazard function and the subhazard function is

$$\lambda_k(t) = \left\{ \frac{1 - F_k(t)}{S(t-)} \right\} \lambda_k^s(t)$$

since $f_k(t) = S(t-)\lambda_k(t) = \lambda_k^s(t)\{1 - F_k(t)\}$ by (6.2) and (6.5). This implies that when there are no competing events the two hazards become identical, but the cause-specific hazard is always larger than the subhazard under competing risks because $S(t-) \leq 1 - F_k(t)$ (Jeong 2014).

6.1.4 Regression Models Based on Two Hazard Functions

Two broad classes of regression models for analyzing the competing risks data have been developed based on the Cox PH model;

(a) Modeling the cause-specific hazard of each event type separately and
(b) Modeling the subhazard (i.e., the hazard function of a subdistribution) for a particular event of interest.

The cause-specific hazard regression model associates the covariates (e.g., confounding factors) with the cause-specific hazard function, whereas the subhazard regression model directly associates the covariates with the cumulative probability of cause-specific events through the subdistribution hazard. In other words, to adjust for the covariates x, we can use either (a) the cause-specific Cox PH model (Prentice et al. 1978),

$$\begin{aligned} \lambda_1(t; x) &= \lambda_{01}(t) \exp(x^T \beta_1), \\ &\vdots \\ \lambda_K(t; x) &= \lambda_{0K}(t) \exp(x^T \beta_K), \end{aligned} \tag{6.6}$$

where $\lambda_{0k}(\cdot)$ is an arbitrary baseline hazard for cause $k = 1, \ldots, K$, or (b) the subhazard regression model (Fine and Gray 1999), for a specific $k \in \{1, \ldots, K\}$

$$\lambda_k^s(t; x) = \lambda_{0k}^s(t) \exp(x^T \beta_k), \tag{6.7}$$

where $\lambda_{0k}^s(\cdot)$ is an arbitrary baseline subhazard for the cause k of interest.

The effect of a covariate on the cause-specific hazard function could be very different from the effect of the covariate on the corresponding subhazard function. Thus, the covariate effects β_k for cause k in both models can be different although we use the same notation. Since the subhazard model considers only one event type of interest, the subscript k will be suppressed. Under the cause-specific Cox model, the usual partial likelihood can be applied after the competing events are treated as censored at the time of occurrence, while under the subdistribution hazard regression, in principle, the competing events are replaced by the infinity (Gray 1988) but always contributed to the risk set.

The cause-specific hazard regression model (6.6) can be directly fitted via the R packages by treating competing events as censoring; the `coxph()` function in **survival** package or the `frailtyHL()` function in **frailtyHL** package. The subhazard regression model (6.7) can be fitted via the `crr()` function in **cmprsk** package. The two models can also be fitted via the `CSC()` and `FGR()` in **riskRegression** package, respectively.

Recently, these two modeling approaches have been extended for clustered competing risks data by using the frailties. Below we present those two extended models, i.e., cause-specific hazard frailty models and subhazard frailty models.

6.2 Cause-Specific Hazard Frailty Models

Suppose that there are $i = 1, \ldots, q$ clusters where each cluster has $j = 1, \ldots, n_i$ observations, so that the total sample size is $n = \sum_{i=1}^{q} n_i$. For a subject j in cluster i, let T_{ij} be time to the first event and let $\epsilon_{ij} \in \{1, 2, \ldots, K\}$ be the corresponding cause of the event. Let C_{ij} denote independent censoring time. Denote by U_i the frailty for cluster i. Assumptions 3 and 4 for the frailty models are extended to the competing-risks frailty models.

Assumption 5:
Given $U_i = u_i$, C_{ij} is conditionally independent of (T_{ij}, ϵ_{ij}) for $j = 1, \ldots, n_i$.

Assumption 6:
Given $U_i = u_i$, $\{C_{ij}, j = 1, \ldots, n_i\}$ are conditionally noninformative of (T_{ij}, ϵ_{ij}).

6.2.1 *Models*

In the competing risks frailty models, the event times within a cluster may be correlated. Under the Assumptions 5 and 6, we show how the h-likelihood procedures for the frailty models are extended to the competing-risks frailty models.

6.2.1.1 Univariate Frailty Models

The cause-specific hazard function conditional on the shared log-frailty v_i for the jth observation in the subject i who failed from cause k $(k = 1, \ldots, K)$ is

$$\lambda_{ijk}(t|v_i) = \lambda_{0k}(t) \exp(x_{ij}^T \beta_k + v_i), \tag{6.8}$$

where v_i is an unobserved random variable from a univariate distribution with parameter θ, $\lambda_{0k}(t)$ is the unspecified baseline hazard function for event type k and $\beta_k = (\beta_{k1}, \beta_{k2}, \ldots, \beta_{kp})^T$ is a $p \times 1$ vector of regression parameters for event type k, and $x_{ij} = (x_{ij1}, \ldots, x_{ijp})^T$ is a $p \times 1$ vector of fixed covariates corresponding to β_k. Let $\beta = (\beta_1^T, \beta_2^T, \ldots, \beta_K^T)^T$ be a $Kp \times 1$ vector of all the regression coefficients for all event types. Similarly, let $\lambda_0 = (\lambda_{01}(\cdot), \lambda_{02}(\cdot), \ldots, \lambda_{0K}(\cdot))^T$ denote the collection of all baseline hazard functions. If there is only one event type $K = 1$, then the cause-specific univariate frailty model (6.8) simply reduces to the standard univariate frailty model in the previous chapters.

6.2.1.2 Correlated (Multi-component or Multivariate) Frailty Models

The cause-specific univariate frailty model (6.8) has drawbacks. First, the model assumes that the common frailty v_i affects every event types within a cluster, even though there could be instances where, on average, subjects who experience one event type would be more frail than ones who experience another event type. Second, the model only allows a positive association within a cluster. If the true value of the log-frailty was greater than zero, then everyone in the cluster would experience an event early regardless of its event type. In practice, however, there may be cases where there is a negative association within a cluster, i.e., reducing the risk of dying from cancer could increase the risk of dying from cardiac disease.

To overcome these limitations, we can consider multivariate frailties for the competing risks events. Consider, the competing risks models with multivariate frailties $v_i = (v_{i1}, v_{i2}, \dots, v_{iK})^T$,

$$\lambda_{ijk}(t|v_i) = \lambda_{0k}(t)\exp(x_{ij}^T\beta_k + v_{ik}), \tag{6.9}$$

where v_{ik} $(k = 1, \dots, K)$ is the random effect for Type k event in cluster i. With this model, each cluster will have K random effects, one for each event type. When $K = 2$, the model (6.9) can account for a correlation between failure and informative censoring, where failure is an event type of main interest (Type 1) and informative censoring can be considered as the competing events (Type 2). Specifically, event times from cause 1 would follow a cause-specific PH model

$$\lambda_{ij1}(t|v_{i1}) = \lambda_{01}(t)\exp(x_{ij}^T\beta_1 + v_{i1}),$$

and event times from cause 2 would follow similarly a model

$$\lambda_{ij2}(t|v_{i2}) = \lambda_{02}(t)\exp(x_{ij}^T\beta_2 + v_{i2}).$$

Here v_{i1} and v_{i2} might be positively or negatively correlated. In the traditional cause-specific analysis, patients who failed from cause 2 are treated as censored for the analysis of Type 1 events, which ignores a potential correlation between v_{i1} and v_{i2}.

Remark 6.1 Suppose that there are two types of competing events, Type 1 and Type 2. It is well known that if only the first event (i.e., the event occurring first when Type 1 and Type 2 are competing) is observed, the joint distribution of times to those two types of events can be non-identifiable (Tsiatis 1975). In particular, for any joint distribution with arbitrary dependence between the two event times, there exists a different joint distribution with independent event times, which gives the same cause-specific hazards, leading to the same likelihoods. Thus, in the competing risks frailty models the joint distribution may not be identifiable if we observe only the first event time. Thus, the unobserved latent event times could not be predicted because it depends upon unidentifiable joint distribution.

Abbring and van den Berg (2003) have shown that the cause-specific hazard frailty models are identifiable under the following two assumptions about covariates and frailties: (i) variation of the observed regressors: $\{\exp(x_{ij}^T\beta_k), k = 1, 2\}$ contains a nonempty open set in R^2 and (ii) expectations of frailties: $\exp(v_{i1})$ and $\exp(v_{i2})$ are all finite. Accordingly, with the above competing risks PH frailty models, the joint distribution can be identifiable. □

For the cause-specific frailty models with bivariate frailties above, we may consider a "shared bivariate" frailty model,

$$v_{i1} = v_i \text{ and } v_{i2} = \gamma v_i,$$

where γ is a real-valued association parameter with a reference scale 0 that describes a dependency between Types 1 and 2 events; in this model, $\text{var}(v_{i2}) = \gamma^2\text{var}(v_{i1})$ and $\text{corr}(v_{i1}, v_{i2}) = \pm 1$. If $\gamma > 0$ [$\gamma < 0$], both event rates are positively [negatively] correlated; a cluster with higher frailty in Type 1 event will experience an earlier [delayed] Type 2 event, respectively. When $\gamma = 0$, Type 2 event rate $\lambda_{ij2}(t|v_i)$ does not depend on v_i and is noninformative for the Type 1 event rate $\lambda_{ij1}(t|v_i)$, so that the two rates are not associated (Liu et al. 2004; Rondeau et al. 2007).

The shared bivariate frailty model can be extended to the shared K-variate frailty model with

$$v_{i1} = v_i\ ,\ v_{i2} = \gamma_2 v_i, \ldots, v_{iK} = \gamma_K v_i,$$

which has K frailty parameters $\alpha = \text{var}(v_{i1}), \gamma_2, \ldots \gamma_K$.

A natural choice for the distribution of $(v_{i1}, v_{i2}, \ldots, v_{iK})$ is the multivariate normal distribution with mean 0 and $K \times K$ variance-covariance matrix Σ, characterized by Σ. An advantage of using the multivariate normal distribution is that the correlation is a natural measure of associations under normality. For Σ, we may consider independent (Σ_I), shared (Σ_S), exchangeable (Σ_E) and unstructured (Σ_U). For example, when $K = 3$,

$$\Sigma_I = \begin{pmatrix} \sigma_1^2 & 0 & 0 \\ 0 & \sigma_2^2 & 0 \\ 0 & 0 & \sigma_3^2 \end{pmatrix},$$

$$\Sigma_S = \begin{pmatrix} \sigma^2 & \gamma_2\sigma^2 & \gamma_3\sigma^2 \\ \gamma_2\sigma^2 & \gamma_2^2\sigma^2 & \gamma_2\gamma_3\sigma^2 \\ \gamma_3\sigma^2 & \gamma_2\gamma_3\sigma^2 & \gamma_3^2\sigma^2 \end{pmatrix},$$

$$\Sigma_E = \begin{pmatrix} \sigma_1^2 & \rho\sigma_1\sigma_2 & \rho\sigma_1\sigma_3 \\ \rho\sigma_1\sigma_2 & \sigma_2^2 & \rho\sigma_2\sigma_3 \\ \rho\sigma_1\sigma_3 & \rho\sigma_2\sigma_3 & \sigma_3^2 \end{pmatrix},$$

and

$$\Sigma_U = \begin{pmatrix} \sigma_1^2 & \rho_{12}\sigma_1\sigma_2 & \rho_{13}\sigma_1\sigma_3 \\ \rho_{12}\sigma_1\sigma_2 & \sigma_2^2 & \rho_{23}\sigma_2\sigma_3 \\ \rho_{13}\sigma_1\sigma_3 & \rho_{23}\sigma_2\sigma_3 & \sigma_3^2 \end{pmatrix},$$

respectively.

In Σ_I, Σ_S, Σ_E and Σ_U, we need K, K, $K+1$, and $K(K+1)/2$ frailty parameters, respectively. Thus, the K-variate shared frailty model is as parsimonious as the K-independent frailty model, while the number of the frailty parameters Σ_U in the unstructured frailty model increases rapidly with K. The exchangeable frailty model Σ_E is parsimonious, but it assumes the common correlation, while the shared frailty model Σ_S allows the correlation to change signs within a cluster.

Due to potential associations among the frailties, it is not recommended to model each event type separately as is normally done in the cause-specific analysis. Instead, the fixed and random effects for all event types need to be estimated jointly.

6.2.2 H-Likelihood Under the Cause-Specific Hazard Frailty Model

In this section, we present the h-likelihood-based inference under the cause-specific multivariate frailty model (6.9). For the jth observation in the ith cluster, let T_{ijk}, $k = 1, \ldots, K$, denote time to Type k event. Define time to the first event T_{ij} as

$$T_{ij} = \min(T_{ij1}, T_{ij2}, \ldots, T_{ijK}).$$

Then, the observed event time and the event indicator are, respectively, defined by

$$Y_{ij} = \min(T_{ij}, C_{ij}) \text{ and } \delta_{ijk} = I(Y_{ij} = T_{ijk}),$$

where $\delta_{ijk} = 1$ if Type k event occurs first (i.e., $Y_{ij} = T_{ijk}$) and 0 otherwise. Note that δ_{ijk} is often referred to as a cause-specific event indicator and that it can also be expressed as

$$\delta_{ijk} = I(T_{ij} \leq C_{ij})I(\epsilon_{ij} = k).$$

Under Assumptions 5 and 6, the conditional likelihood for cluster i under cause-specific frailty models (6.9) is defined by

$$L_i(\beta, \lambda_0|v_{ik}) = \prod_{k=1}^{K}\prod_{j=1}^{n_i} \left\{\lambda_{0k}(y_{ij})e^{x_{ij}^T\beta_k + v_{ik}}\right\}^{\delta_{ijk}} \exp\left\{-\Lambda_{0k}(y_{ij})e^{x_{ij}^T\beta_k + v_{ik}}\right\}.$$

To unify derivation of the h-likelihood procedures, we use $v_{ik} = z_{ij}^T v_k$, giving the linear predictor η_{ij},

$$\eta_{ij} = x_{ij}^T \beta_k + z_{ij}^T v_k,$$

where $z_{ij} = (z_{ij1}, z_{ij2}, \ldots, z_{ijq})^T$ is a $q \times 1$ cluster indicator vector such that $z_{ijm} = 1$ if $i = m$ and 0 otherwise, and $v_k = (v_{1k}, v_{2k}, \ldots, v_{qk})^T$ is a $q-$dimensional vector of the random effects from all clusters but only for event k.

We now show how to construct the h-likelihood function under a competing risks setting where, for simplicity, two types of events ($k = 1$ or 2) exist. The results can be easily generalized to K event types. From the h-likelihood construction of Appendix 4.7.1, the h-likelihood for cluster i under the cause-specific hazards frailty model (6.9) is given by

$$h_i = \log\{L_i(\beta_k, \lambda_{0k}|v_{ik})L_{2i}(\theta; v_i)\} = \log\left\{\prod_k \prod_j L_{1ijk}(\beta_k, \lambda_{0k}; y_{ij}, \delta_{ijk}|v_{ik})L_{2i}(\theta; v_i)\right\},$$

where $L_{1ijk}(\beta_k, \lambda_{0k}; y_{ij}, \delta_{ijk}|v_{ik})$ is the conditional likelihood of (y_{ij}, δ_{ijk}) given $V_{ik} = v_{ik}$ with parameters (β_k, λ_{0k}) and $L_{2i}(\theta; v_i)$ is the likelihood of $v_i = (v_{i1}, v_{i2})^T$ with parameter θ. Here, L_{1ijk} takes the form of

$$L_{1ijk}(\beta_k, \lambda_{0k}; y_{ij}, \delta_{ijk}|v_{ik}) = \left[\lambda_{0k}(y_{ij})e^{x_{ij}^T\beta_k + z_{ij}^T v_k}\right]^{\delta_{ijk}} \exp\left(-\Lambda_{0k}(y_{ij})e^{x_{ij}^T\beta_k + z_{ij}^T v_k}\right).$$

Assuming that $V_i = (V_{i1}, V_{i2})^T$ follows a bivariate normal distribution with mean 0 and covariance matrix Σ, its joint probability density is given by

$$f_i(v_i; \theta) = |2\pi\Sigma|^{-1/2} \exp\left(-\frac{1}{2}v_i^T \Sigma^{-1} v_i\right).$$

Let $v = (v_{11}, v_{21}, \ldots, v_{q1}, v_{12}, v_{22}, \ldots, v_{q2})^T$ be a $2q$-dimensional vector of all random effects, for all clusters and event times. Notice that the random effects are arranged by event type so that all of the random effects for the same event type are adjacent. Event times within a cluster are conditionally independent given the frailty $V_i = v_i$ and the frailties V_i are iid random variables. Thus, the h-likelihood for the cause-specific hazard frailty models (6.9) becomes

$$h(\beta, \lambda_0, v, \theta) = \sum_i h_i = \sum_{ijk} \ell_{1ijk}(\beta_k, \lambda_{0k}; y_{ij}, \delta_{ijk}|v_{ik}) + \sum_i \ell_{2i}(\theta; v_i) \quad (6.10)$$

where $\ell_{1ijk}(\cdot) = \log L_{1ijk}(\cdot)$ and $\ell_{2i}(\cdot) = \log L_{2i}(\cdot)$, and

$$\begin{aligned}\ell_{1ijk}(\beta_k, \lambda_{0k}; y_{ij}, \delta_{ijk}|v_{ik}) = \delta_{ijk}\left(\log\lambda_{0k}(y_{ij}) + x_{ij}^T\beta_k + z_{ij}^T v_k\right) \\ -\Lambda_{0k}(y_{ij})\exp\left(x_{ij}^T\beta_k + z_{ij}^T v_k\right)\end{aligned}$$

and

$$\ell_{2i}(\theta; v_i) = -\frac{1}{2}\log|2\pi\Sigma| - \frac{1}{2}v_i^T\Sigma^{-1}v_i.$$

6.2.3 *Partial H-Likelihood via Profiling*

To eliminate the high-dimensional baseline hazard function $\lambda_{0k}(y_{ij})$ in (6.10), again we apply the profiling method to the h-likelihood as in Chap. 4. First, define the cumulative baseline hazard function for event type k as a step function with jumps at observed event times

$$\Lambda_{0k}(t) = \sum_{r:\, y_{(kr)}\le t} \lambda_{0kr},$$

where $y_{(k1)} < y_{(k2)} < \cdots < y_{(kD_k)}$ denote the D_k ordered unique event times of type k and $\lambda_{0kr} = \lambda_{0k}(y_{(kr)})$. Let $d_{(kr)}$ be the number of events of Type k that occur at time $y_{(kr)}$,

$$D_{(kr)} = \{ij : \delta_{ijk} = 1 \text{ and } y_{ij} = y_{(kr)}\}$$

be a set of all individuals who have a Type k event at time $y_{(kr)}$,

$$s^T_{x(kr)} = \sum_{ij\in D_{(kr)}} x^T_{ij} \text{ and } s^T_{z(kr)} = \sum_{ij\in D_{(kr)}} z^T_{ij}$$

be the sums of the vectors x^T_{ij} and z^T_{ij} over $D_{(kr)}$, and

$$R_{(kr)} = \{ij : y_{ij} \ge y_{(kr)}\}$$

be the risk set at time $y_{(kr)}$. By using these notations, the h-likelihood (6.10) can be written as

$$h = \sum_{k=1}^{2}\left[\sum_{r=1}^{D_k} d_{(kr)}\log\lambda_{0kr} + s^T_{x(kr)}\beta_k + s^T_{z(kr)}v_k - \lambda_{0kr}\sum_{ij\in R_{(kr)}}\exp\left(x^T_{ij}\beta_k + z^T_{ij}v_k\right)\right] + \sum_{i=1}^{q}\ell_{2i}(\theta; v_i), \tag{6.11}$$

since λ_{0kr} only depends on the subscript k and r when the likelihood function is evaluated at the rth event time of type k. By replacing λ_{0kr} in (6.11) with the non-parametric MHLE, obtained from $\partial h/\partial\lambda_{0kr} = 0$,

$$\hat{\lambda}_{0kr} = \frac{d_{(kr)}}{\sum_{ij\in R_{(kr)}}\exp(x^T_{ij}\beta_k + z^T_{ij}v_k)},$$

the profile h-likelihood $h^* = h|_{\lambda_0=\hat{\lambda}_0}$ is given as a function of β, v, and θ only:

$$h^*(\beta, v, \theta) = \sum_{k=1}^{2}\left[\sum_{r=1}^{D_k} d_{(kr)} \log \hat{\lambda}_{0kr} + s_{x(kr)}^T\beta_k + s_{z(kr)}^T v_k - \hat{\lambda}_{0kr}\sum_{ij\in R_{(kr)}} \exp\left(x_{ij}^T\beta_k + z_{ij}^T v_k\right)\right] + \sum_{i=1}^{q} \ell_{2i}(\theta; v_i).$$

Then, h^* in the cause-specific frailty model becomes again proportional to the partial h-likelihood h_p, given by

$$h_p(\beta, v, \theta) = \sum_{ijk} \delta_{ijk}\left(x_{ij}^T\beta_k + z_{ij}^T v_k\right) - \sum_{kr} d_{(kr)} \log\left\{\sum_{ij\in R_{(kr)}} \exp\left(x_{ij}^T\beta_k + z_{ij}^T v_k\right)\right\} + \sum_{i=1}^{q} \ell_{2i}(\theta; v_i), \quad (6.12)$$

since

$$h^* = h_p + \sum_{kr} d_{(kr)}\{\log(d_{(kr)}) - 1\}$$

and the last term does not depend upon the unknowns (v, β, θ).

6.2.4 Fitting Procedure

Derivations of the gradient vector of $\tau = (\beta, v)$ given θ and the observed information matrix from the partial h-likelihood h_p, and those related to θ are provided in Appendix 6.7. In particular, below we derive the ILS equations and useful matrix forms for estimating (v, β, θ).

- **Matrix forms for estimating** β_k $(k = 1, 2)$ **and** v:

$$\frac{\partial h_p}{\partial \beta_k} = X^T(\delta_k - \mu_k),$$

$$\frac{\partial h_p}{\partial v} = \begin{pmatrix} Z^T(\delta_1 - \mu_1) \\ Z^T(\delta_2 - \mu_2) \end{pmatrix} - (\Sigma^{-1} \otimes I_q)v.$$

Here, Z is a $n \times q$ cluster indicator matrix whose ijth row is z_{ij}^T, δ_k is an $n \times 1$ Type k event indicator vector with ijth element δ_{ijk}, $\mu_k = \hat{\Lambda}_{0k} \exp(\eta_k)$ and $\otimes$ denotes the Kronecker product.

$$\text{Let } \mathbf{X} = \begin{pmatrix} X & \mathbf{0} \\ \mathbf{0} & X \end{pmatrix}, \quad \mathbf{Z} = \begin{pmatrix} Z & \mathbf{0} \\ \mathbf{0} & Z \end{pmatrix} \quad \text{and} \quad \mathbf{W}^* = \begin{pmatrix} W_1^* & \mathbf{0} \\ \mathbf{0} & W_2^* \end{pmatrix}.$$

Here $W_k^* = -\partial^2 h_p / \partial \eta_k \partial \eta_k^T$ with $\eta_k = X\beta_k + Zv_k$. Then, the ILS equations for (β, v) with $\beta = (\beta_1^T, \beta_2^T)^T$ are the same forms as in (4.12) with $R = 0$ due to the normality of the frailties, given by

$$\begin{pmatrix} \mathbf{X}^T\mathbf{W}^*\mathbf{X} & \mathbf{X}^T\mathbf{W}^*\mathbf{Z} \\ \mathbf{Z}^T\mathbf{W}^*\mathbf{X} & \mathbf{Z}^T\mathbf{W}^*\mathbf{Z} + Q \end{pmatrix} \begin{pmatrix} \widehat{\beta} \\ \widehat{v} \end{pmatrix} = \begin{pmatrix} \mathbf{X}^T\mathbf{w}^* \\ \mathbf{Z}^T\mathbf{w}^* \end{pmatrix}, \tag{6.13}$$

where $Q = -\partial^2 \ell_2 / \partial v^2 = \Sigma^{-1} \otimes I_q$ is a $Kq \times Kq$ matrix and $\mathbf{w}^* = (w_1^{*T}, w_2^{*T})^T$ with $w_k^* = W_k^* \eta_k + (\delta_k - \mu_k)$. The derivation of (6.13) is given in Appendix 6.7.1, including a simple univariate frailty case.

Let

$$\mathbf{P} = \begin{pmatrix} \mathbf{X} & \mathbf{Z} \\ \mathbf{0} & I_{k^*} \end{pmatrix} \text{ and } \mathbf{V} = \begin{pmatrix} \mathbf{W}^* & \mathbf{0} \\ \mathbf{0} & Q \end{pmatrix},$$

where $k^* \equiv K \times q = 2q$. Then the ILS Eq. (6.13) reduce to a simple matrix form

$$(\mathbf{P}^T\mathbf{V}\mathbf{P})\hat{\tau} = \mathbf{P}^T\mathbf{y}_0^*, \tag{6.14}$$

where $\mathbf{y}_0^* = (\mathbf{w}^{*T}, \mathbf{0}^T)^T$. □

• **Matrix forms for estimating** $\theta = (\sigma_{11}, \sigma_{22}, \sigma_{12})^T$:
The next step is to find the partial REMLE of frailty parameters θ by maximizing the partial restricted likelihood,

$$p_{\beta,v}(h_p) = \left[h_p - \frac{1}{2} \log\{\det(H_p/2\pi)\} \right] \Bigg|_{(\beta,\, v) = (\hat{\beta}(\theta),\, \hat{v}(\theta))}, \tag{6.15}$$

where

$$H_p \equiv H(h_p; \beta, v) = -\partial^2 h_p / \partial(\beta, v)^2 = \mathbf{P}^T\mathbf{V}\mathbf{P}$$

is a $K(p+q) \times K(p+q)$ observed information matrix from the partial h-likelihood h_p.

The partially restricted likelihood (6.15) can be rewritten as

$$p_{\beta,v}(h_p) = \hat{h}_p - \frac{1}{2} \log\{\det(\hat{H}_p)\} + \frac{K(p+q)}{2} \log(2\pi),$$

where $\hat{h}_p = h_p(\hat{\beta}(\theta), \hat{v}(\theta), \theta)$ and $\hat{H}_p = H_p(\hat{\beta}(\theta), \hat{v}(\theta), \theta)$ are the partial h-likelihood and observed information matrix evaluated at the current estimates of β and v, respectively. Derivations of the gradient vector of θ and the corresponding observed information matrix are also provided in Appendix 6.7.2. In particular, for the univariate cause-specific model (6.8) with the log-normal frailty, the partial REMLE of $\sigma^2 = \text{var}(v_i)$ has the same form as the standard univariate lognormal frailty model in (4.15).

6.3 Subdistribution Hazard Frailty Models

6.3.1 *Models*

In this section, we are interested in modeling the subhazard based on the CIF. Here, the observable random variables are expressed as

$$Y_{ij} = \min(T_{ij}, C_{ij}) \text{ and } \xi_{ij} = I(T_{ij} \le C_{ij})\epsilon_{ij},$$

where C_{ij} is an independent censoring and $\xi_{ij} \in \{0, 1, 2, \ldots, K\}$ because $\epsilon_{ij} \in \{1, 2, \ldots, K\}$. For simplicity, consider the two event types ($k = 1, 2$). Then, $\xi_{ij} \in \{0, 1, 2\}$, where 1 is for an event of interest, 2 for a competing event and 0 for censoring.

The CIF for events from cause 1 (i.e., $\epsilon_{ij} = 1$) is defined by

$$F_1(t) = Pr(T_{ij} \le t, \epsilon_{ij} = 1), \tag{6.16}$$

which represents the probability that an individual will experience a Type 1 event by time t. The corresponding hazard function of the subdistribution (subhazard function) is defined by

$$\lambda_1^s(t) = -\frac{d \log\{1 - F_1(t)\}}{dt}.$$

Fine and Gray (1999) first introduced this model to directly associate the effects of covariates with the CIF through the subhazard for a particular event type of interest (e.g., Type 1). Katsahian et al. (2006) and Christian (2011) have extended the Fine-Gray model to a subhazard frailty model with one random component to analyze multi-center competing risks data. Ha et al. (2016a) made a further extension to the general subhazard frailty model allowing the multicomponent random effects and their correlation. For example, in multicenter clinical trials, we may have a two-component model allowing random center and random treatment effects, where the random treatment effect means a random treatment-by-center interaction. In particular, a model allowing for the correlation between random center and random treatment effects can properly account for the heterogeneities from the treatment effects across centers as well as between-center variation.

Similar to Sect. 5.1.2, denote by $v_i = (v_{i0}, v_{i1}, \ldots, v_{i,m-1})^T$ an m-dimensional vector of the unobserved log-frailties associated with the ith ($i = 1, \ldots, q$) center.

Suppose that Assumptions 5 and 6 hold and we are interested in assessing the effects of covariates on the conditional CIF for cause 1 given the log-frailties v_i, defined by

$$F_1(t|v_i) = Pr(T_{ij} \leq t, \epsilon_{ij} = 1|v_i).$$

The conditional subhazard function for cause 1 given v_i is modeled as

$$\lambda^s_{ij1}(t|v_i) = \lambda^s_{01}(t)\exp(\eta_{ij}), \tag{6.17}$$

where $\lambda^s_{01}(\cdot)$ is an unknown baseline subhazard function,

$$\eta_{ij} = x^T_{ij}\beta + z^T_{ij}v_i$$

is a linear predictor for the log-hazard, and $x_{ij} = (x_{ij1}, \ldots, x_{ijp})^T$ and $z_{ij} = (z_{ij1}, \ldots, z_{ijm})^T$ are $p \times 1$ and $m \times 1$ covariate vectors corresponding to the fixed effects $\beta = (\beta_1, \ldots, \beta_p)^T$, and log-frailties v_i, respectively. We assume that the log-frailties v_i are independent and follow a multivariate normal distribution, i.e., $v_i \sim N_m(0, \Sigma_i(\theta))$, where the covariance matrix $\Sigma_i(\theta)$ depends on a vector of unknown parameters θ. Model (6.17) may include any frailty covariance structures mentioned in Sect. 6.2.1.

6.3.2 *H-Likelihood Under the Subhazard Frailty Model*

In this section, we first show how to construct the h-likelihood for the semiparametric subhazard frailty model (6.17).

6.3.2.1 Complete Data Case

For simplicity, we first outline the h-likelihood approach for competing risks data without censoring, i.e., when $\xi_{ij} = \epsilon_{ij} \in \{1, 2, \ldots, K\}$. Without the loss of generality, we assume $K = 2$. Let $t_{(r)}$ denotes the rth ($r = 1, \ldots, D$) smallest distinct event time of Type 1 among t_{ij}'s, where t_{ij} is the observed value of T_{ij} and D is the total number of distinct Type 1 events. Let $R_{0(r)}$ denote a risk set at $t_{(r)}$:

$$R_{0(r)} = R(t_{(r)}) = \{(i, j) : t_{ij} \geq t_{(r)} \text{ or } (t_{ij} \leq t_{(r)} \text{ and } \epsilon_{ij} \neq 1)\}.$$

In contrast to the Cox PH model, the risk set $R_{0(r)}$ includes not only individuals who have not failed by $t_{(r)}$ but also those who have previously failed from the competing causes. Since the functional form of the baseline subhazard function $\lambda^s_{01}(t)$ is unknown, at each t_{ij}, the baseline cumulative subhazard function $\Lambda^s_{01}(t)$ can be written as

$$\Lambda^s_{01}(t_{ij}) = \sum_r \lambda^s_{01r} I\{(i, j) \in R_{0(r)}\},$$

where $\lambda^s_{01r} = \lambda^s_{01}(t_{(r)})$ is the subhazard function for Type 1 events at $t_{(r)}$. Then, under Assumptions 5 and 6, the h-likelihood for the subhazard frailty models (6.17) is defined by

$$h = h(\beta, v, \lambda^s_{01}, \theta) = \sum_{ij} \ell_{1ij} + \sum_i \ell_{2i}, \tag{6.18}$$

where

$$\begin{aligned}
\sum_{ij} \ell_{1ij} &= \sum_{ij} I(\epsilon_{ij} = 1)\{\log \lambda^s_{01}(t_{ij}) + \eta_{ij}\} - \sum_{ij} \{\Lambda^s_{01}(t_{ij}) \exp(\eta_{ij})\} \\
&= \sum_r d_{0(r)} \log \lambda^s_{01r} + \sum_{ij} I(\epsilon_{ij} = 1)\eta_{ij} \\
&- \sum_{ij} \Big[\sum_r \lambda^s_{01r} I\{(i, j) \in R_{0(r)}\} \exp(\eta_{ij}) \Big]
\end{aligned}$$

is the sum of the logarithm of the conditional density function for T_{ij} and ϵ_{ij} given v_i, $\ell_{1ij} = \ell_{1ij}(\beta, \lambda^s_{01}; t_{ij}, \epsilon_{ij}|v_i)$, and

$$\ell_{2i} = \ell_{2i}(\theta; v_i) = -\frac{1}{2}[\log \det\{2\pi \Sigma_i(\theta)\}] - \frac{1}{2} v_i^T \Sigma_i(\theta)^{-1} v_i$$

is the logarithm of the density function for v_i with parameters $\theta = (\sigma_0^2, \sigma_1^2, \sigma_{01})^T$, i.e., the log-likelihood for v_i. Here, $v = (v_1^T, \ldots, v_q^T)^T$, v_i being a bivariate random vector because $K = 2$, $\lambda^s_{01} = (\lambda^s_{011}, \ldots, \lambda^s_{01D})^T$, and $d_{0(r)}$ is the number of Type 1 events at $t_{(r)}$.

- **Partial h-likelihood via profiling:** To eliminate the high-dimensional nuisance parameters λ^s_{01}, we use again the profile h-likelihood h^*, given by

$$h^* = h|_{\lambda^s_{01} = \widehat{\lambda}^s_{01}} = \sum_{ij} \ell^*_{1ij} + \sum_i \ell_{2i}, \tag{6.19}$$

where

$$\widehat{\lambda}^s_{01r}(\beta, v) = \frac{d_{0(r)}}{\sum_{(i,j) \in R_{0(r)}} \exp(\eta_{ij})}$$

are the solutions of the estimating equations, $\partial h / \partial \lambda^s_{01r} = 0$, for $r = 1, \ldots, D$. Since

$$\sum_{ij} \ell^*_{1ij} = \sum_{ij} \ell_{1ij}|_{\lambda^s_{01} = \widehat{\lambda}^s_{01}} = \sum_r d_{0(r)} \log \widehat{\lambda}^s_{01r} + \sum_{ij} I(\epsilon_{ij} = 1)\eta_{ij} - \sum_r d_{0(r)},$$

we see that the conditional profile likelihood $\sum_{ij} \ell^*_{1ij}$ is proportional to the conditional partial likelihood ℓ_p:

$$\sum_{ij} \ell^*_{1ij} = \ell_p + \sum_r d_{0(r)}\{\log d_{0(r)} - 1\},$$

where

$$\ell_p = \sum_{ij} I(\epsilon_{ij} = 1)\eta_{ij} - \sum_r d_{0(r)} \log\left\{ \sum_{(i,j)\in R_{0(r)}} \exp(\eta_{ij}) \right\},$$

which is the log conditional partial likelihood given v_i for complete data. Thus, the profile h-likelihood in (6.19) becomes again the partial h-likelihood

$$h_p = \ell_p + \sum_i \ell_{2i},$$

which is an extension of the Fine-Gray's partial likelihood to the subhazard frailty models without censoring.

6.3.2.2 Incomplete Data Case

Consider an incomplete data case with right censoring, where $\xi_{ij} \in \{0, 1, 2\}$. Let $R_{(r)}$ be the risk set at $y_{(r)}$, which is the rth smallest distinct event time of Type 1 event among the observed values y_{ij}'s, defined by

$$R_{(r)} = R(y_{(r)}) = \{(i, j) : y_{ij} \geq y_{(r)} \text{ or } (y_{ij} \leq y_{(r)} \text{ and } \xi_{ij} > 1)\}.$$

To define the h-likelihood for the incomplete data, we apply the inverse probability of censoring weighting (IPCW; Fine and Gray 1999) to the h-likelihood (6.18). The resulting weight is

$$w_{ij} = w_{ij}(y_{(r)}) = \frac{\hat{G}(y_{(r)})}{\hat{G}(y_{ij} \wedge y_{(r)})}$$

for a subject j in the cluster i at $y_{(r)}$, and $\hat{G}(\cdot)$ is the Kaplan–Meier estimate of the survival function for the censoring times. Here, $w_{ij} = 1$ as long as individuals have not failed by time $y_{(r)}$ (i.e., $y_{ij} \geq y_{(r)}$; the first condition of $R_{(r)}$), whereas $w_{ij} \leq 1$ and decreasing over time $y_{(r)}$ if they failed from Type 2 event before $y_{(r)}$ (i.e., $y_{ij} \leq y_{(r)}$ and $\xi_{ij} > 1$; the second condition of $R_{(r)}$) because the further the time point ($y_{(r)}$) moves away from Type 2 event (y_{ij}), the smaller the weight becomes (Pintilie 2006). We first define the weighted h-likelihood h_w based on the IPCW as

$$h_w = \sum_{ij} \ell_{w1ij} + \sum_i \ell_{2i}, \tag{6.20}$$

where

$$\sum_{ij} \ell_{w1ij} = \sum_r d_{(r)} \log \lambda^s_{01r} + \sum_{ij} \delta_{ij}\eta_{ij}$$
$$- \sum_{ij}\left[\sum_r \lambda^s_{01r} I\{(i,j) \in R_{0(r)}\} w_{ij} \exp(\eta_{ij})\right].$$

Here, $\delta_{ij} = I(\xi_{ij} = 1)$ is an event indicator representing whether subject j from cluster i experiences a Type 1 event, and $d_{(r)}$ is the number of Type 1 events at $y_{(r)}$.

• **Weighted partial h-likelihood via profiling:** The weighted profile h-likelihood h^*_w is defined by

$$h^*_w = h_w|_{\lambda^s_{01}=\widehat{\lambda}^w_{01}}, \tag{6.21}$$

where

$$\widehat{\lambda}^w_{01r}(\beta, v) = \frac{d_{(r)}}{\sum_{(i,j)\in R_{(r)}} w_{ij} \exp(\eta_{ij})}$$

are the solutions of the estimating equations, $\partial h_w/\partial \lambda^s_{01k} = 0$, for $r = 1, \ldots, D$.

Similarly to the previous section, it can shown that h^*_w is proportional to the weighted partial h-likelihood h_{pw}, given by

$$h^*_w = h_{pw} + \sum_r d_{(r)}\{\log d_{(r)} - 1\}. \tag{6.22}$$

Here

$$h_{pw} = \ell_{pw} + \sum_i \ell_{2i},$$

where $\ell_{pw} = \sum_{ij} \delta_{ij}\eta_{ij} - \sum_r d_{(r)} \log\left\{\sum_{(i,j)\in R_{(r)}} w_{ij} \exp(\eta_{ij})\right\}$ is the conditional partial likelihood.

In the absence of frailty, h_{pw} becomes the weighted partial likelihood of Fine and Gray (1999). Hereafter, we use the estimation procedure based on h_{pw} for model (6.17), which handles the general censoring case.

• **Fitting procedure**

The h-likelihood procedures for the correlated standard frailty model presented in Sect. 5.1.2 can be straightforwardly extended to the subhazard model (6.17) by using the weighted partial h-likelihood h_{pw}. That is, given the frailty parameters θ, the weighted MHLEs of $\tau = (\beta^T, v^T)^T$ are obtained by solving the score equations, $\partial h_{pw}/\partial \tau = 0$. It is shown that given θ, the score equations lead to the ILS equations for τ (Appendix 6.7.3):

$$\begin{pmatrix} X^T W^* X & X^T W^* Z \\ Z^T W^* X & Z^T W^* Z + Q \end{pmatrix} \begin{pmatrix} \widehat{\beta} \\ \widehat{v} \end{pmatrix} = \begin{pmatrix} X^T w^* \\ Z^T w^* \end{pmatrix}, \tag{6.23}$$

where X and Z are $n \times p$ and $n \times q^*$ ($q^* = Kq$) model matrices for the fixed effects β and the random effects v whose ijth row vectors are x_{ij}^T and z_{ij}^T, respectively, the form of W^* with the IPCW weight w_{ij} is the symmetric weight matrix given in (4.35) of Appendix 4.7.4, and $Q = -\partial^2 \ell_2 / \partial v^2 = \text{diag}(\Sigma_1^{-1}, \ldots, \Sigma_q^{-1})$ is a $q^* \times q^*$ matrix. Here

$$w^* = W^* \eta + (\delta - \mu)$$

with $\eta = X\beta + Zv$ and $\mu = \exp(\log w + \log \Lambda_{01}^s + \eta)$ where w is the IPCW weights. In the absence of frailty, the ILS equations in (6.23) reduces to the Fine and Gray (1999) estimating equation:

$$(X^T W^* X)\hat{\beta} = X^T w^*.$$

Note that the ILS Eq. (6.23) further reduce to a simple form

$$(\mathbf{P}^T \mathbf{V} \mathbf{P})\hat{\tau} = \mathbf{P}^T \mathbf{y}_0^*,$$

where

$$\mathbf{P} = \begin{pmatrix} X & Z \\ \mathbf{0} & I_{q^*} \end{pmatrix}, \quad \mathbf{V} = \begin{pmatrix} W^* & \mathbf{0} \\ \mathbf{0} & Q \end{pmatrix} \quad \text{and } \mathbf{y}_0^* = (w^{*T}, \mathbf{0}^T)^T.$$

For estimation of θ, we use the weighted partial restricted h-likelihood $p_\tau(h_{pw})$, given by

$$p_\tau(h_{pw}) = \left[h_{pw} - \frac{1}{2} \log \det \left\{ H_{pw}/(2\pi) \right\} \right] \Bigg|_{\tau = \hat{\tau}}, \tag{6.24}$$

where $\widehat{\tau} = \widehat{\tau}(\theta) = (\widehat{\beta}^T(\theta), \widehat{v}^T(\theta))^T$ and $H_{pw} = H(h_{pw}; \tau) = -\partial^2 h_{pw} / \partial \tau^2$ is an information matrix for τ. The weighted PREMLEs for θ are obtained by solving iteratively

$$\frac{\partial p_\tau(h_{pw})}{\partial \theta} = 0. \tag{6.25}$$

Note here that

$$\frac{\partial p_\tau(h_{pw})}{\partial \theta} = -\frac{1}{2} \text{tr}\left(\Sigma^{-1} \frac{\partial \Sigma}{\partial \theta} \right) - \frac{1}{2} \widehat{v}^T \left(\frac{\partial \Sigma^{-1}}{\partial \theta} \right) \widehat{v} - \frac{1}{2} \text{tr}\left(\widehat{H}_{pw}^{-1} \frac{\partial \widehat{H}_{pw}}{\partial \theta} \right),$$

where $\Sigma = \text{BD}(\Sigma_1, \ldots, \Sigma_q)$ is a $q^* \times q^*$ block diagonal matrix and $\widehat{H}_{pw} \widehat{H}_{pw}(\theta) = H(h_{pw}; \tau)|_{\tau = \widehat{\tau}(\theta)}$.

• **The estimated standard errors**
In this book, the SE estimates for $\hat{\tau} - \tau$ and $\hat{\theta}$ are, respectively, obtained from the inverses of the corresponding Hessian matrices,

$$H_{pw} = -\frac{\partial^2 h_{pw}}{\partial \tau^2} \text{ and } -\frac{\partial^2 p_\tau(h_{pw})}{\partial \theta^2}.$$

In particular, Fine and Gray (1999) proposed a robust/sandwich variance estimator to estimate $\text{var}(\hat{\beta})$ using empirical process theory because the martingale properties break due to the use of IPCW and thus the standard asymptotic theories are no longer valid. In the subhazard frailty model with one frailty term, Katsahian and Boudreau (2011) presented a sandwich variance estimator of $\hat{\beta}$ using Gray's (1992) method, estimated from

$$\text{var}_s(\hat{\tau}) = H_{pw}^{-1} H_1 H_{pw}^{-1},$$

where $\tau = (\beta^T, v^T)^T$, $H_1 = H(\ell_{1pw}; \tau) = -\partial^2 \ell_{1pw}/\partial \tau^2$ and ℓ_{pw} is the first term of h_{pw}. Thus,

$$\text{var}(\hat{\tau}) = H_{pw}^{-1} \geq H_{pw}^{-1} H_1 H_{pw}^{-1}, \qquad (6.26)$$

since $H_{pw} \geq H_1$. Ha et al. (2016a) showed via a simulation study that the SEs of $\hat{\beta}$ from $\text{var}(\hat{\tau}) = H_{pw}^{-1}$ in (6.26) perform better than those from $\text{var}_s(\hat{\tau}) = H_{pw}^{-1} H_1 H_{pw}^{-1}$ because the sandwich variance estimators often underestimate the true variances in the finite samples: see also Therneau et al. (2003).

6.4 Examples

In this section, we illustrate the h-likelihood approach for the competing risks frailty models with a breast cancer dataset, introduced in Sect. 1.2.6. First, we use the cause-specific hazard frailty to estimate the effect of tamoxifen on different types of failures when subjects can experience multiple events under competing risks. Second, the effect of tamoxifen on local or regional recurrence will be assessed adjusting for the possible center effects using the subhazard frailty model.

6.4.1 Cause-Specific Frailty Model for Breast Cancer Data

6.4.1.1 The Data

This analysis will use a high risk subset of patients from the B-14 study, with tumor size greater than 2.5 cm. In this subset, there were 731 women with follow-up (371

Table 6.1 Event type by treatment group for all observations including multiple observations from the same subject ($n = 731$ patients): breast cancer data

Types of Event	Placebo	Tamoxifen	Total
Type 1: Local or regional recurrence	73	40	113 (13.45%)
Type 2: Second primary in contralateral breast	32	32	64 (7.62%)
Type 3: Distance recurrence, other second primary or death	204	184	388 (46.19%)
No event (Censoring)	127	148	275 (32.74%)

placebo and 360 tamoxifen) who were eligible for the study. The median age for women on either placebo or treatment was 55 years. A series of multiple types of treatment failure were possible; local, regional, or distant recurrence of original cancer as well as a new second primary cancer or death because patients were followed as long as they did not withdraw their consents.

In this analysis, the types of failures will be divided into three event types:

Type 1: a local or regional recurrence,
Type 2: a new second primary cancer in the contralateral breast,
Type 3: a distant recurrence, other new second primary cancer or death.

We assume that these three types of events compete against each other because once a recurrence or second primary occurs, non-protocol therapies are often administered after the event, which would prohibit an accurate assessment of the effect of the treatment solely on that particular event type under consideration.

Table 6.1 gives the number of events by treatment group for all observations, including multiple observations from the same subject. The most common event type was Type 3. Subjects receiving placebo had more events in all types, except that both groups had exactly the same number of Type 2 events. The original B-14 manuscript reported that there was a significant reduction in new primary cancers in the contralateral breast as the first events for women receiving tamoxifen in the B-14 study. Based on the counts of Type 2 events in Table 6.1, it is clear that this analysis will not reach the same conclusion because it allows multiple events per patient. This difference is also due to the fact that this analysis used only a subset of the original data and would have much less power to detect a difference between treatment groups. About 57% of the 95 subjects who had multiple events experienced both Type 1 and Type 3 events and about 20% had Type 2 and Type 3 events.

6.4.1.2 Analyses from Cause-Specific Hazard Models

A cause-specific hazard frailty model with age and treatment as covariates was fitted assuming both univariate and exchangeable correlation structures among the random effects affecting different types of events. Here, the treatment is coded as 1 for tamoxifen and 0 for placebo. The regression coefficients and estimated variance of the random effects assuming a univariate normal distribution with only one random effect per subject are in the upper left-hand corner of Table 6.2. Adjusted for age, the relative risk of a Type 1 event for an individual on tamoxifen compared to the same individual being on placebo is $\exp(-0.742) = 0.48$ with a 95% confidence interval of $\exp(-0.742 \pm 1.96 \times 0.226) = (0.31, 0.74)$. The estimated variance of the random effects is 1.883, suggesting a fairly heterogeneous group of subjects. Ignoring the correlation among event types, fitting the standard frailty model for each event type by treating other events as independent censoring (Gorfine and Hsu 2011) results in smaller estimates of the treatment effect; fitting this naive model is equivalent to fitting a cause-specific hazard frailty model with an independent assumption among three random effects per subject (one random effect per event type; the upper right-hand corner of Table 6.2).

In Table 6.2, Exchangeable $\supset$ Indep, and Exchangeable $\supset$ Shared $\supset$ Univariate. Between Exchangeable and Indep models, the difference of the partial restricted likelihood based on $-2p_\tau(h_p)$ is 30.7 (> 3.84), so that the null hypothesis of $\rho = 0$ is rejected. Among three models of Univariate, Shared, and Exchangeable, the LRT selects the univariate model as the final model. Here, the rAIC also confirms this choice, even though the cAIC selects the shared model.

6.4.1.3 Predicted CIF and Frailty Effect

Figure 6.2 shows the predicted CIF curves of Type 1 event for a 55 year old (the median age) woman in each treatment group from the cause-specific univariate frailty model. To be brief, for a given set of the covariates x_0 and a known frailty value v_0, the predicted CIF for Type k events can be predicted from

$$\hat{F}_k(t|x_0, v_0) = \sum_{y_{ij} \le t} \hat{S}(y_{ij}|x_0, v_0)\hat{\lambda}_k(y_{ij}|x_0, v_0),$$

where $\hat{S}(y_{ij}|x_0, v_0) = \exp\{-\sum_k \hat{\Lambda}_k(y_{ij}|x_0, v_0)\}$, $\hat{\Lambda}_k(y_{ij}|x_0, v_0) = \hat{\Lambda}_{0k}(y_{ij}) \exp(x_0^T\hat{\beta}_k + v_0)$,

$$\hat{\Lambda}_{0k}(y_{ij}) = \sum_{r: y_{(kr)} \le y_{ij}} \frac{\delta_{ijk}}{\sum_{ij \in \mathcal{R}_{kr}} \exp(x_{ij}^T\hat{\beta}_k + \hat{v}_i)},$$

and hence

$$\hat{\lambda}_k(y_{ij}|x_0, v_0) = \frac{\delta_{ijk} \exp(x_0^T\hat{\beta}_k + v_0)}{\sum_{ij \in \mathcal{R}_{kr}} \exp(x_{ij}^T\hat{\beta}_k + \hat{v}_i)}.$$

The incidence of Type 1 events increases faster for the placebo group compared to the tamoxifen group. Ten years after surgery, an average women on tamoxifen has a 7% chance of experiencing a local or regional recurrence while a women on

Table 6.2 Estimates of the cause-specific hazard frailty models: univariate, independent, shared and exchangeable cases: breast cancer data

Event type	Effect	Univariate case with correlation		Independent case ignoring correlation	
		Estimate (SE)	95% CI	Estimate (SE)	95% CI
Type 1	Age	−0.015 (0.011)	(−0.036, 0.005)	−0.017(0.009)	(−0.035, 0.001)
	Trt	−0.742 (0.226)	(−1.185, −0.299)	−0.633(0.199)	(−1.023, −0.243)
Type 2	Age	−0.002 (0.014)	(−0.028, 0.025)	−0.001(0.013)	(−0.025, 0.024)
	Trt	−0.179 (0.274)	(−0.715, 0.358)	−0.041(0.250)	(−0.532, 0.449)
Type 3	Age	0.016 (0.007)	(0.001, 0.031)	0.017(0.005)	(0.007, 0.027)
	Trt	−0.261 (0.150)	(−0.555, 0.032)	−0.137 (0.102)	(−0.336, 0.063)
Random Effect	Variance	$\hat{\sigma}^2 = 1.883$	$\hat{\sigma}_I^2 = 0.116$, $\hat{\sigma}_{II}^2 = 0.001$, $\hat{\sigma}_{III}^2 = 0.001$		
$-2p_\tau(h_p)$		6958.5	6987.0		
AIC	rAIC = 6960.4, cAIC = 6733.6			rAIC = 6993.0, cAIC = 6977.9	
Shared case					
Event Type	Effect	Estimate	SE	95% CI	
Type 1	Age	−0.015	0.010	(−0.036, 0.005)	
	Trt	−0.744	0.226	(−1.187, −0.301)	
Type 2	Age	−0.002	0.013	(−0.028, 0.024)	
	Trt	−0.158	0.266	(−0.680, 0.364)	
Type 3	Age	0.016	0.008	(0.000, 0.032)	
	Trt	−0.289	0.165	(−0.612, 0.033)	
Random Effect	Variance	$\hat{\sigma}^2 = 1.954$			
Type 1, 2	Association	$\hat{\gamma}_2 = 0.820$			
Type 2, 3	Association	$\hat{\gamma}_3 = 1.169$			
$-2p_\tau(h_p)$		6957.2	(rAIC = 6963.2, cAIC = 6675.4)		

(continued)

Table 6.2 (continued)

Exchangeable case				
Event Type	Effect	Estimate	SE	95% CI
Type 1	Age	−0.017	0.010	(−0.036, 0.002)
	Trt	−0.684	0.211	(−1.057, −0.271)
Type 2	Age	−0.002	0.013	(−0.027, 0.024)
	Trt	−0.111	0.260	(−0.621, 0.399)
Type 3	Age	0.016	0.006	(0.004, 0.028)
	Trt	−0.203	0.125	(−0.448, 0.042)
Type 1	Variance	0.789		
Type 2	Variance	0.706		
Type 3	Variance	0.771		
Type1, 2, 3	Correlation	0.886		
$-2p_{\tau}(h_p)$		6956.3	(rAIC = 6964.3, cAIC = 6870.0)	

Trt, Treatment; CI, confidence interval; SE, standard error
σ^2, common variance for all event types
$\sigma_I^2, \sigma_{II}^2, \sigma_{III}^2$; independent variance for each event type

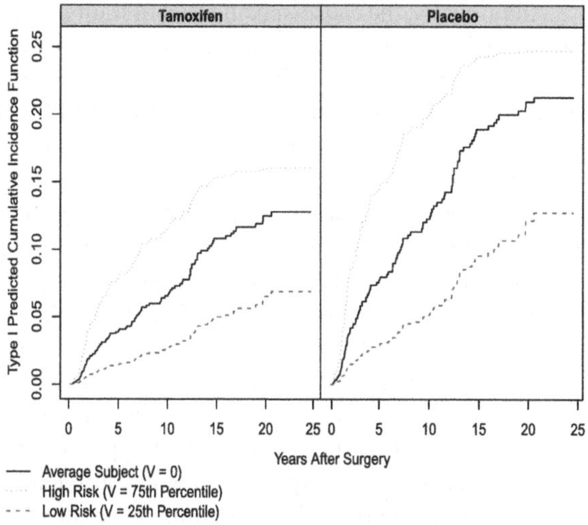

Fig. 6.2 Predicted cumulative incidence of Type 1 events, for an average subject $V = 0$, high-risk subject $V = 0.82$ (75th percentile), and low-risk subject $V = -1.07$ (25th percentile)

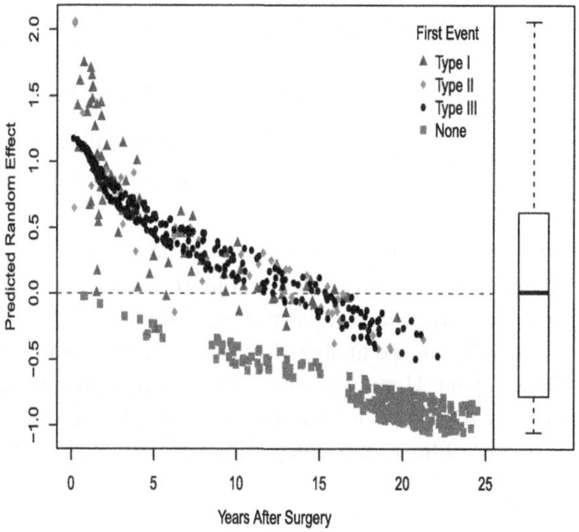

Fig. 6.3 Estimated cause-specific frailties versus the first observed event time for each subject; boxplot on the right-hand side is the distribution of the estimated cause-specific random effects

placebo has a 13% chance. This probability increases for women at higher risks (75th percentile of the estimated frailty distribution).

The estimated 25th and 75th percentiles of the frailties are -1.07 and 0.82, respectively. In Fig. 6.3, the estimated cause-specific frailties are larger in general for subjects who had an event early and decrease for later event times. Thus, those subjects who had an event early are more frail than those who survived longer, as expected. The estimated cause-specific random effects (i.e., log-frailties) for all subjects who did not have an event is less than 0 (similarly as in Fig. 4.1), which may be reasonable since there is no evidence from the observed data that these subjects should be at higher risk than an average person.

Additionally, the cause-specific hazard frailty model was fitted assuming a trivariate normal distribution with one random effect per event type (three random effects per subject) by using an exchangeable correlation structure. The estimated regression coefficients along with their standard errors and confidence intervals as well as the estimated variance components are given under the exchangeable case in Table 6.2. The estimated treatment effects for each event type are smaller than the corresponding estimates for the univariate case in Table 6.2, but the patterns were similar; patients on tamoxifen had a significantly lower risk of a Type 1 event compared to patients on placebo. Tamoxifen did not significantly lower the risk for other event types. The estimated variance of the random effects for each event type are all similar ranging from 0.706 to 0.789. There is also a strong positive correlation between the cause-specific random effects, indicating that patients who experienced a local or regional recurrence will also be at a greater risk for developing a second primary cancer in the contralateral breast as well as any of Type 3 events. This is because patients who have a larger frailty for Type 1 event would also tend to have a larger frailty for Type 2 and Type 3 events, and larger frailties would increase the risk of failure for an individual or a cluster.

We also fitted the cause-specific "shared" frailty model with three types of events:

$$\begin{aligned}\lambda_{ij1}(t|v_i) &= \lambda_{01}(t)\exp(x_{ij}^T\beta_1 + v_i),\\ \lambda_{ij2}(t|v_i) &= \lambda_{02}(t)\exp(x_{ij}^T\beta_2 + \gamma_2 v_i),\\ \lambda_{ij3}(t|v_i) &= \lambda_{03}(t)\exp(x_{ij}^T\beta_3 + \gamma_3 v_i),\end{aligned}$$

where the shared log-frailties $v_i \sim N(0, \sigma^2)$ $(i = 1, \ldots, q)$. The fitted results of the model above are also given in Table 6.2, but they are similar to the results from the cause-specific univariate frailty model with $\gamma_2 = \gamma_3 = 1$, since $\hat{\gamma}_2$ and $\hat{\gamma}_3$ are near 1.

In addition, we analyzed the data set after combining Types 2 and 3 events into Type 2 event. Here, we considered four models with univariate, independent, shared and BN frailties. The fitted results are all given in Table 6.3, which shows similar estimates of the regression parameters. In Table 6.3, BN $\supset$ Shared $\supset$ Univariate, and

Table 6.3 Estimation results of cause-specific hazard frailty models: breast cancer data

Model	Event	Age(SE)	Trt(SE)	$\hat{\sigma}_1^2$	$\hat{\sigma}_2^2$	$\hat{\sigma}_{12}$	$-2p_\tau(h_p)$
Univ	Type 1	−0.016(0.010)	−0.741(0.226)	$\hat{\sigma}^2$			
	Type 2	0.014(0.007)	−0.249(0.144)	1.87	–	–	6954.7
Indep	Type 1	−0.017(0.009)	−0.633(0.199)	0.12	–		
	Type 2	0.014(0.005)	−0.123(0.094)	–	0.001	–	6992.4
Shared	Type 1	−0.016(0.010)	−0.738(0.223)	$\hat{\sigma}^2$	$\hat{\gamma}_2$		
	Type 2	0.013(0.007)	−0.259(0.149)	1.74	1.10	–	6952.9
BN	Type 1	−0.016(0.010)	−0.734(0.223)				
	Type 2	0.013(0.007)	−0.246(0.143)	1.69	1.79	1.74	6883.1

Univ, Univariate frailty model

Table 6.4 First observed event types by two treatment arms ($n = 2817$ patients)

Types of Event	Placebo	Tamoxifen	Total
Type 1: Local or regional recurrence	205	109	314 (11.15%)
Type 2: Distance recurrence, second primary, or death	671	632	1303 (46.25%)
No event (Censoring)	537	663	1200 (42.60%)

BN $\supset$ Indep. Between the Indep and BN models, the LRT selects the BN model since the difference is 69.8($>$ 3.84). Between the Shared and BN models, it also selects the BN model. Thus, we chooses the BN model as the final model.

6.4.2 Subhazard Frailty Model for Breast Cancer Data

6.4.2.1 The Data

In the breast cancer data, a total of 2,817 eligible patients from 167 distinct centers were followed up for about 20 years since randomization. The number of patients per center varied from 1 to 241, with the mean of 16.9 and the median of 8. The patients were randomized to one of two treatment arms, tamoxifen (1413 patients) or placebo (1404 patients). The average age was 55 and the average tumor size was about 2 cm.

The aim of this analysis is to investigate the effect of treatment on local or regional recurrence. Here we consider two event types:

Type 1: local or regional recurrence,
Type 2: a new primary cancer, distant recurrence or death.

Only the event that occurs first is of interest in this analysis, so that the repeated event times are not considered. Table 6.4 shows the distribution of first observed event types by two treatment arms; Type 1 is an event of interest (314 patients, 11.15%), Type 2 is a competing event (1,303 patients, 46.25%), and the remaining patients are censored at the last follow-up (1,200 patients, 42.60%).

Figure 6.4 presents the estimated CIFs for the two treatment arms. The tamoxifen group has lower CIFs compared to placebo group for both Type 1 and Type 2 events. For Type 1 events, the difference in the CIFs between two arms seems to be more noticeable than for Type 2 events. In particular, the estimated probability that a patient in tamoxifen group would experience Type 1 event within 10 years after surgery is 5%, while for a patient in the placebo group it is 10%.

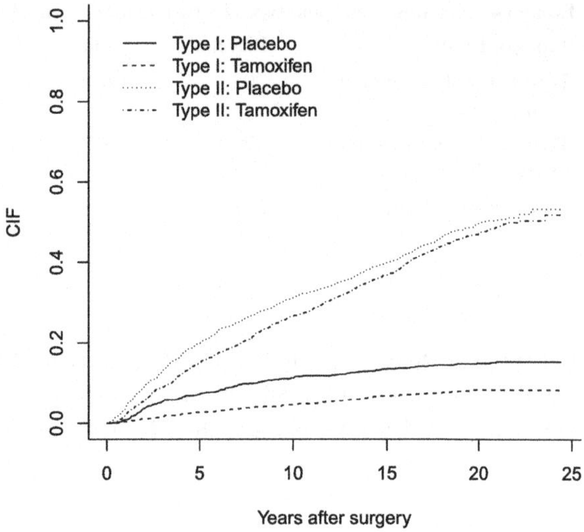

Fig. 6.4 Estimated CIFs for tamoxifen vs placebo for the two types of events in the breast cancer data

6.4.2.2 Analyses Using the Subhazard Models

As mentioned in Chap. 5, in multicenter randomized clinical trials the treatment effects or baseline risks may vary among centers. This situation can also be applied to the current multicenter competing risks data. For data analysis, we consider three covariates of interest: treatment ($x_{ij1} = 1$ for tamoxifen and 0 for placebo), age (x_{ij2}), and tumor size (x_{ij3}) as continuous covariates. Let v_{i0} and v_{i1} be the random center effects and the random treatment effects (i.e., random treatment-by-center interaction), respectively. The event type of interest is Type 1, and three models are considered, including the subhazards model without random effects and two subhazard frailty models,

$$\lambda^s_{1ij}(t|v) = \lambda^s_{01}(t)\exp(\eta_{ij}),$$

where η_{ij} is the linear predictor:

M1 (F-G): $\eta_{ij} = \beta_1 x_{ij1} + \beta_2 x_{ij2} + \beta_3 x_{ij3}$;
M2 (Center): $\eta_{ij} = v_{i0} + \beta_1 x_{ij1} + \beta_2 x_{ij2} + \beta_3 x_{ij3}$, with $v_{i0} \sim N(0, \sigma_0^2)$;
M3 (Corr): $\eta_{ij} = v_{i0} + (\beta_1 + v_{i1})x_{ij1} + \beta_2 x_{ij2} + \beta_3 x_{ij3}$, with $(v_{i0}, v_{i1}) \sim BN$,

where "F-G", "Center", and "Corr" indicate Fine-Gray model without frailties, the subhazard frailty model with a random center effect v_{i0} and the subhazard correlated frailty model with $v_{i0} \sim N(0, \sigma_0^2)$, $v_{i1} \sim N(0, \sigma_1^2)$ and $\rho = \text{Corr}(v_{i0}, v_{i1})$, respectively. Here M3 ($\sigma_0^2 \geq 0, \sigma_1^2 \geq 0, \rho \neq 0$) is our full model and the others are its sub-models by assuming the null components, i.e., M1 ($v_{i0} = 0, v_{i1} = 0; \sigma_0^2 = 0, \sigma_1^2 = 0$) and M2 ($v_{i1} = 0; \sigma_0^2 \geq 0, \sigma_1^2 = 0$). The estimation results are listed in Table 6.5.

Table 6.5 Results for fitting the three subhazard frailty models to Type 1 event of the breast cancer data

Model	$\hat{\beta}_1$(SE)	$\hat{\beta}_2$(SE)	$\hat{\beta}_3$(SE)	$\hat{\sigma}_0^2$(SE)	$\hat{\sigma}_1^2$(SE)	$\hat{\sigma}_{01}$(SE) [$\hat{\rho}$]	$-2p_{\tau}(h_{pw})$
M1	−0.667	−0.026	0.082	–	–	–	4870.5
(F-G)	(0.119)	(0.005)	(0.042)				
M2	−0.672	−0.026	0.081	0.043	–	–	4869.4
(Center)	(0.119)	(0.005)	(0.042)	(0.051)			
M3	−0.658	−0.026	0.079	0.091	0.249	−0.108	4865.7
(Corr)	(0.137)	(0.005)	(0.043)	(0.026)	(0.073)	(0.037) [−0.721]	

β_1, Treatment effect; β_2, Age effect; β_3, Tumor-size effect

Note that M3 ⊃ M2 ⊃ M1. First, the null hypothesis $H_0 : \sigma_0^2 = 0$ (i.e., no center effect) lies on the boundary of the parameter space. From Table 6.5, we obtain the difference of 1.1 in $-2p_{\beta,v}(h_{pw})$ between M1 and M2 (with p-value = $0.5P(\chi_1^2 > 1.1) = 0.147$) based on the asymptotic statistic of $\chi_{0:1}^2$, as shown in Sect. 4.3.2, indicating that the random center effect is not significant. Furthermore, the difference between M2 and M3 is 3.7 (with p-value = 0.106) based on the asymptotic statistic of $\chi_{1:2}^2$ (Verbeke and Molenberghs 2003), leading to acceptance of the null hypothesis of $\sigma_1^2 = 0$; the p-value (Sect. 4.3.2) of 0.106 is calculated as

$$\begin{aligned} p &= P\{\chi_{1:2}^2 > 3.7\} \\ &= \frac{1}{2}P(\chi_1^2 > 3.7) + \frac{1}{2}P(\chi_2^2 > 3.7) \\ &= 0.10582. \end{aligned}$$

Thus, the LRTs support the simplest model M1. Later, we show how to confirm the heterogeneity (i.e., M1).

In all three subhazard models, only two fixed effects (β_j, $j = 1, 2$) are significant, except β_3. In particular, tamoxifen significantly reduces the risk of local or regional recurrence (Type 1 event) as compared to patients who receive placebo. We also observe that overall there is no substantial change in the fixed-effects estimates, although the effect of the main treatment (β_1) slightly decreases due to the increased standard error when both random components and their correlation are included. In M2 and M3, the variance components (σ_0^2 and σ_1^2) indicate the amount of variation between centers in the baseline risk (i.e., center effect) and in the treatment effect, respectively. Here, the estimate of σ_1^2 and its SE are relatively larger than those of σ_0^2, which is also confirmed in Fig. 6.5. Furthermore, the correlated model M3 explains the degree of dependency between the two random components (i.e., the random center effect v_0 and the random treatment-by-center interaction v_1). The estimate of ρ gives a negative value ($\hat{\rho} = -0.721$), indicating that the two predicted random components ($\hat{v}_0$ and $\hat{v}_1$) have a negative correlation. The negative correlation leads

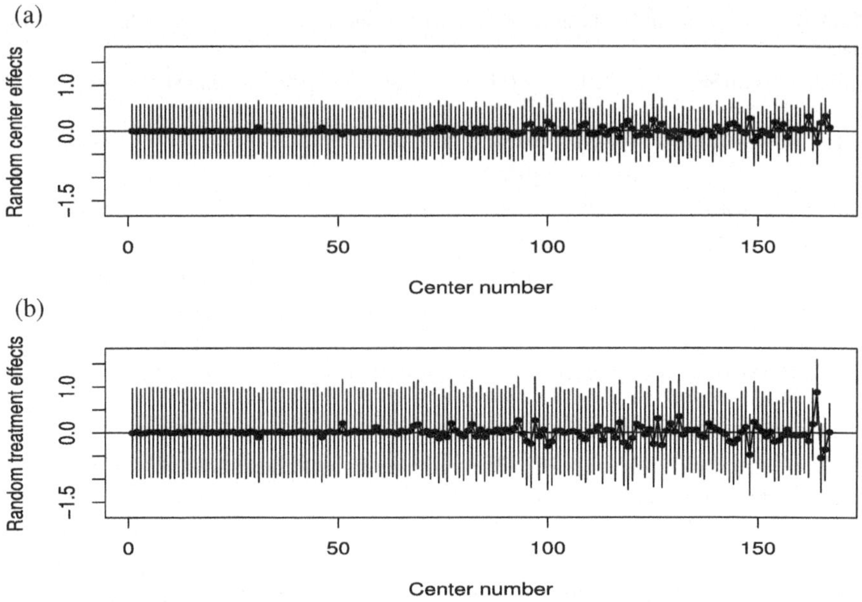

Fig. 6.5 Random effects of 167 centers in the breast cancer data (event of interest is Type 1) and their 95% confidence intervals, under subhazard correlated frailty model (M3); **a** random center effects (v_{i0}); **b** random treatment-by-center interaction (v_{i1}); Centers are sorted in the increasing order of number of patients

to a conclusion that treatment confers more benefit in centers with a higher baseline risk. This is consistent with the findings by Rondeau et al. (2008) in the context of meta-analysis and by Ha et al. (2011) in the context of of multicenter trials.

6.4.2.3 Investigating and Testing for Heterogeneity in Treatment Effects

We demonstrate how to investigate heterogeneity in treatment effects over centers using the Wald confidence intervals presented in Chap. 4 for the frailties of the individual centers: for more discussions about heterogeneity in treatment effects, see Lee (2002). Note that the standard intervals using $p_{\tau}(h_{pw})$ in (6.24) can be null due to zero estimation of the variance components, especially for small sample sizes or small variance components. To avoid the null frailty variance estimator, we use an adjusted likelihood (4.21), defined by

$$p_{\text{adj}} = p_{\tau}(h_{pw}) + \log\det(\Sigma_i),$$

which leads to positive frailty variance estimators. The individual $(1-\alpha)$-level h-likelihood confidence intervals for the uni-dimensional components v_k of the random effects v are of the form

$$\hat{v}_k \pm z_{\alpha/2} \cdot \text{SE}(\hat{v}_k - v_k),$$

where $\hat{v}$ maximizes the weighted partial h-likelihood h_{pw}, $z_{\alpha/2}$ is the normal quantile with probability of $\alpha/2$ in the right tail, and $\text{SE}(\hat{v}_k - v_k)$ are obtained from $H(h_{pw}, \hat{\beta}, \hat{v})^{-1}$. Figure 6.5 shows the estimated random effects and their 95% confidence intervals for the 167 centers using the subhazard correlated model M3. Here, centers are ordered by the number of patients accrued. Figure 6.5a, b give the confidence intervals for the random center effect (v_{i0}) and the random treatment-by-center interaction (v_{i1}), respectively. Overall, the lengths of the intervals are seen to decrease as the number of patients per center increases.

Figure 6.5a indicates overall homogeneity in the baseline risk across 167 centers and Fig. 6.5b shows that there is no substantial variation in the treatment effects across centers although three centers (148, 164, and 165) among 167 centers noticeably stand out. Note here that the centers (148, 165) and 164 provide the lowest and the highest treatment-by-center interactions, respectively, but that the corresponding three intervals include zero; this indicates that the homogeneity of treatment effects also extends to these three centers. Thus, in this multicenter trial, there is little variation in treatment effects across centers and the treatment is shown to be effective, These results suggest that the treatment effect may be generalized to a broader patient population.

6.5 Software and Examples Using R

6.5.1 A Simulated Data Set

For an illustration, we consider a simulated data set.

• **Simulation scheme**

A data set for the cause-specific hazard frailty model assuming a bivariate normal distribution is generated using a technique similar to Beyersmann et al. (2009) and Christian et al. (2016). Let there be two event types, Types 1 and 2, as well as independent censoring. We considered a sample size, $n = 100$ with $(q, n_i) = (50, 3)$. Data were generated with two covariates (x_{ij1}, x_{ij2}), where x_{ij1} follows a standard normal distribution and x_{ij2} is a Bernoulli random variable with probability 0.5. The random effects are from bivariate normal with

$$\begin{pmatrix} v_{i1} \\ v_{i2} \end{pmatrix} \sim N\left(\begin{pmatrix} 0 \\ 0 \end{pmatrix}, \begin{pmatrix} \sigma_1^2 & \sigma_{12} \\ \sigma_{21} & \sigma_2^2 \end{pmatrix}\right)$$

where $\theta = (\sigma_{11}, \sigma_{22}, \sigma_{12}) = (1, 1, -0.5)$. The conditional cause-specific hazard rates for each event type are,

$$\lambda_{ij1}(t|x_{ij}, v_{i1}) = 2\exp(0.6x_{ij1} - 0.4x_{ij2} + v_{i1}),$$
$$\lambda_{ij2}(t|x_{ij}, v_{i2}) = 0.5\exp(-0.3x_{ij1} + 0.7x_{ij2} + v_{i2}).$$

That is, $\beta_1 = (\beta_{11}, \beta_{12}) = (0.6, -0.4)$ and $\beta_2 = (\beta_{21}, \beta_{22}) = (-0.3, 0.7)$. Censoring times are generated from a Uniform(0, 1.3) distribution. Under this scenario, with 25.2% censoring, the proportions of Type 1 and Type 2 events are 53.2% and 21.6%, respectively.

• Comparison of cause-specific hazards and subhazard models

For the model fitting, we consider the following two types of competing-risks frailty models:

(1) Cause-specific hazard frailty models:

$$\lambda_{ij1}(t|v_{i1}) = \lambda_{01}(t)\exp(x_{ij1}\beta_{11} + x_{ij2}\beta_{12} + v_i),$$
$$\lambda_{ij2}(t|v_{i2}) = \lambda_{02}(t)\exp(x_{ij1}\beta_{21} + x_{ij2}\beta_{22} + v_{i2}),$$

For the random effects (v_{i1}, v_{i2}) we consider the three cases with univariate (Univ), shared and bivariate normal (BN):

(i) Univ: $v_{i1} = v_i$ and $v_{i2} = v_i$ with $v_i \sim N(0, \sigma^2)$,
(ii) Shared: $v_{i1} = v_i$ and $v_{i2} = \gamma_2 v_i$ with $v_i \sim N(0, \sigma^2)$,
(iii) BN: $(v_{i1}, v_{i2})^T \sim BN(\sigma_1^2, \sigma_2^2, \sigma_{12})$.

(2) Subhazard frailty models:
We consider the subhazard frailty model for Type 1 events,

$$\lambda^s_{ij1}(t|v_i) = \lambda^s_{01}(t)\exp(x_{ij1}\beta_{11} + x_{ij2}\beta_{12} + v_i),$$

where $v_i \sim N(0, \sigma^2)$, and consider the separate model for Type 2 events,

$$\lambda^s_{ij2}(t|v_i) = \lambda^s_{02}(t)\exp(x_{ij1}\beta_{21} + x_{ij2}\beta_{22} + v_i),$$

where $v_i \sim N(0, \sigma^2)$.

We note that for different types of events, the cause-specific frailty models are jointly fitted via a common frailty or correlated frailties, whereas the subhazard frailty models fit each event type separately.

Below are the R codes and outputs from fitting the two competing-risks frailty models for a simulated dataset.

• R codes and outputs

```
> data(test, package="frailtyHL")
> head(test)
  obs id       time status          x1 x2
1   1  1 0.17317534      1 -0.75313807  1
2   2  1 0.38924435      0 -0.10980574  1
```

```
3    3  1 0.08503643      1 -0.01271682  1
4    4  1 0.15189636      1 -1.16308531  0
5    5  1 0.58214961      0 -0.26338994  1
6    6  2 0.02019265      2  1.58529452  1
############### 1) Cause-specific Cox PH models ################
> cs1<-frailtyHL(Surv(time,status==1)~x1+x2+(1|id),varfixed=T,
+ varinit=0,data=test)
> cs2<-frailtyHL(Surv(time,status==2)~x1+x2+(1|id),varfixed=T,
+ varinit=0,data=test)

#cs1<-coxph(Surv(time,status==1)~x1+x2,ties="breslow", data=test)
#cs2<-coxph(Surv(time,status==2)~x1+x2,ties="breslow", data=test)

########### 2)  Cause-specific frailty models (Shared) ###########
> data_conti<-test
> data_surv<-data_conti
> jm1<-jointmodeling(Model="mean",RespDist="FM",Link="log",
+ LinPred=Surv(time,status==1)~x1+x2+(1|id),RandDist="gaussian")
> jm2<-jointmodeling(Model="mean",RespDist="FM",Link="log",
+ LinPred=Surv(time,status==2)~x1+x2+(1|id),RandDist="gaussian")
> res<-jmfit(jm1,jm2,data_conti,data_surv,Maxiter=200)
[1] "iterations : "
[1] 74
      beta_h    se_beh   t_value       p_value
  0.5069570 0.1039914  4.874987 1.088155e-06 #Type1
 -0.3800849 0.1976954 -1.922579 5.453299e-02 #Type1
 -0.3003151 0.1788708 -1.678949 9.316193e-02 #Type2
  1.3378959 0.3196841  4.185056 2.850961e-05 #Type2
       alpha_h       rho_h
[1,] 1.024705 -0.9487815

######## 3) Cause-specific frailty models (Univariate) ##########
>  beta.init <-c(sapply(1:2, function(k) coxph(Surv(time,status==k)
+  ~x1+x2,data=test)$coef))
> theta.init = 0.05
> q = length(unique(test$id))
> v.init=rep(0,q) #v.init = rnorm(q,0,1)
> CSFM <-hlike.frailty(formula=CmpRsk(time,status)~x1+x2+cluster(id),
+ data=test,frailty.cov="none",inits=list(beta=beta.init,
+ theta=theta.init,v=v.init),order=1, MAX.ITER=500, TOL=1E-5)
> summary(CSFM)
  Type Effect    Estimate         SE        2.5%      97.5%
1    1      x1  0.5228740 0.1056892  0.3157270 0.7300211
2    1      x2 -0.1456480 0.1925659 -0.5230703 0.2317742
3    2      x1 -0.1816378 0.1580546 -0.4914192 0.1281436
4    2      x2  0.8366907 0.3002129  0.2482843 1.4250972
  Var.Comp  Estimate
1 Sigma.11 0.3998222
 Successfully Converged

########### 4) Cause-specific hazard frailty models (BN) ###########
> theta.init <-matrix(c(1,0.5,0.5,1),nrow=2)
> v.init<-matrix(0,q,2)
> CSFM_BN <-hlike.frailty(formula=CmpRsk(time,status)~x1+x2+cluster(id),
+ data=test,frailty.cov="unstructured",inits=list(beta=beta.init,
+ theta=theta.init,v=v.init), order=1, MAX.ITER=500, TOL=1E-5)
> CSFM_BN
```

```
  Type Effect   Estimate        SE       2.5%       97.5%
1    1     x1  0.5292916 0.1064490  0.3206553 0.737927872
2    1     x2 -0.3889682 0.2013525 -0.7836118 0.005675494
3    2     x1 -0.3253187 0.1941188 -0.7057844 0.055147114
4    2     x2  1.4377490 0.3416088  0.7682080 2.107290009
  Var.Comp   Estimate
1 Sigma.11  1.0122354
2 Sigma.21 -0.5947998
3 Sigma.22  1.3884678
 Successfully Converged

########### 5) Subhazard without frailty (Fine-Gray) ################
> library(cmprsk)
> attach(data_conti)
> da<-data_conti
>  x=cbind(da$x1,da$x2)
>   SH1=crr(time,status, x, failcode=1)  #Type 1
>  summary(SH1)
Competing Risks Regression
Call:
crr(ftime = time, fstatus = status, cov1 = x, failcode = 1)
     coef exp(coef) se(coef)     z p-value
x1  0.457     1.579   0.0953  4.79 1.7e-06
x2 -0.309     0.734   0.1726 -1.79 7.3e-02
   exp(coef) exp(-coef)  2.5% 97.5%
x1     1.579      0.633 1.310  1.90
x2     0.734      1.363 0.523  1.03
Num. cases = 250
Pseudo Log-likelihood = -656
Pseudo likelihood ratio test = 26.1  on 2 df,
>
>  SH2=crr(time,status, x, failcode=2)  #Type 2
>  summary(SH2)
Competing Risks Regression
Call:
crr(ftime = time, fstatus = status, cov1 = x, failcode = 2)
     coef exp(coef) se(coef)     z p-value
x1 -0.458     0.633    0.155 -2.96   0.003
x2  0.864     2.372    0.279  3.09   0.002
   exp(coef) exp(-coef)  2.5% 97.5%
x1     0.633      1.581 0.467 0.856
x2     2.372      0.422 1.373 4.099
Num. cases = 250
Pseudo Log-likelihood = -275
Pseudo likelihood ratio test = 18.9  on 2 df,

############ 6) Subhazard frailty models #################
>  # Subhazard Frailty (Type 1)
> beta.init<-c(0,0); v.init=rep(0,q); theta.init = 0.05
> SHFM1<-hlike.frailty(CmpRsk(time,status)~x1+x2+cluster(id),
+ data=data_conti,inits=list(beta=beta.init,theta=theta.init,
+ v=v.init),order=1,frailty.cov="none",subHazard=T,MAX.ITER=300)
> summary(SHFM1)
   Type Effect   Estimate        SE       2.5%      97.5%
x1    1     x1  0.4974055 0.1048294  0.2919436  0.7028673
x2    1     x2 -0.6164925 0.1987410 -1.0060177 -0.2269673
  Var.Comp  Estimate
```

```
1 Sigma.11 0.8836048
 Successfully Converged
status_2<-ifelse(da$status==2,1,2*da$status) #Transformation of status
>  SHFM2<-hlike.frailty(CmpRsk(time,status_2)~x1+x2+cluster(id),
+ data=da,inits=list(beta=beta.init,theta=theta.init,v=v.init),
+ order=1,frailty.cov="none",subHazard=T,MAX.ITER=500,TOL=1E-6)
> summary(SHFM2)
   Type Effect    Estimate         SE        2.5%       97.5%
x1    1      x1 -0.5658455 0.1899055 -0.9380534 -0.1936376
x2    1      x2  1.2966654 0.3294090  0.6510355  1.9422952
  Var.Comp Estimate
1 Sigma.11 1.595583
 Successfully Converged
```

The R outputs from the cause-specific hazard frailty model show that for example, the effect of x_1 is significant for Type 1 event (`t-value = 4.875` with p-value = 0.000), but it is not significant for Type 2 event (`t-value = -1.679` with p-value = 0.093). The estimated association parameter $\widehat{\gamma}_2 = -0.949$ shows a negative association between the risks of these two events, which reflects a true negative correlation, -0.5, under the true BN model. The fitted results are summarized in Table 6.6. The resulting estimates from the shared and univariate cause-specific frailty models are different because the estimated association parameter $\hat{\gamma}_2 = -0.949$ is not close to $\gamma_2 = 1$. We also fitted the cause-specific model with bivariate normal frailties, using `frailty.cov="unstructured"` in this package. The results confirm the negative association from the shared model. This means that reducing the risk of dying from Type 1 event increases the risk of dying from Type 2 event. We observe the results from the subhazard model show similar trends to those from the cause-specific models.

In Table 6.6, we see that in the cause-specific models, BN $\supset$ Shared $\supset$ Univ $\supset$ Cox. Between BN and Shared models, the difference in $-2p_{\tau}(h_p)$ is 4.6, so that the true BN model is selected as the final model. We also observe that the two AICs (rAIC and cAIC) select the BN model as the best model too.

Next, for the subhazard models, we should select a proper model in each type. For Type 1, the difference in $-2p_{\tau}(h_{pw})$ between Fine-Gray model without frailty and the shared model is $1130.0 - 1122.3 = 9.7$ (> 2.71), so that the null hypothesis of $\sigma^2 = 0$ is rejected. Thus, we select the shared model as the final model. Similarly, for Type 2 we also select the shared model. This indicates there is a correlation among survival times in each type.

There are R packages to fit the subhazard models without frailty terms, e.g., the function `crr()` in the **cmprsk** package to fit the subhazard model (Fine and Gray 1999) with univariate competing risks data, and the function `crrc()` in the **crrSC** package to fit a marginal subhazard model (Zhou et al. 2012; Zhou and Latouche 2015) for clustered competing risks data. The outlines of R codes used with the simulated data are as follows.

Table 6.6 Comparison of competing-risks frailty models with cause-specific hazard (CSH) and subhazard (SH): a simulated data set

Model	Event	$\hat{\beta}_1$(SE)	$\hat{\beta}_2$(SE)	$\hat{\sigma}^2$	Association
CSH	Type 1	0.462(0.098)	−0.038(0.177)	–	–
(Cox)	Type 2	−0.217(0.155)	0.843(0.291)	–	–
$-2p_{\tau}(h_p)$		1765.3	(rAIC = 1765.3, cAIC = 1766.2)		
CSH	Type 1	0.523(0.106)	−0.146(0.193)	0.400	–
(Univ)	Type 2	−0.182(0.158)	0.837(0.300)		
$-2p_{\tau}(h_p)$		1749.9	(rAIC = 1751.9, cAIC = 1735.8)		
CSH	Type 1	0.507(0.104)	−0.380(0.198)	1.025	$\hat{\gamma}_2 = -0.949$
(Shared)	Type 2	−0.300(0.179)	1.338(0.320)		
$-2p_{\tau}(h_p)$		1717.2	(rAIC = 1721.2, cAIC = 1683.2)		
CSH	Type 1	0.529(0.106)	−0.389(0.201)	$\hat{\sigma}_1^2 = 1.012$	$\hat{\sigma}_{12} = -0.595$
(BN)	Type 2	−0.325(0.194)	1.438(0.342)	$\hat{\sigma}_2^2 = 1.388$	–
$-2p_{\tau}(h_p)$		1712.6	(rAIC = 1718.6, cAIC = 1677.2)		
SH(F-G)	Type 1	0.457(0.095)	−0.309(0.173)	–	–
$-2p_{\tau}(h_{pw})$		1316.2	(rAIC = 1316.2, cAIC = 1315.7)		
	Type 2	−0.458(0.155)	0.864(0.279)	–	–
$-2p_{\tau}(h_{pw})$		551.6	(rAIC = 551.6, cAIC = 553.0)		
SH(Shared)	Type 1	0.497(0.105)	−0.616(0.199)	0.884	–
$-2p_{\tau}(h_{pw})$		1284.3	(rAIC = 1286.3, cAIC = 1261.2)		
	Type 2	−0.566(0.190)	1.297(0.329)	1.596	–
$-2p_{\tau}(h_{pw})$		528.4	(rAIC = 530.4, cAIC = 517.1)		

F-G, Fine-Gray subhazard model without frailty

```
##### Subhazard models without frailty term #####
> x=cbind(da$x1,da$x2)
> SH1<-crr(time, status, x, failcode=1)
> SH2<-crr(time, status, x, failcode=2)
> SHF1<-crrc(time, status, x, failcode=1,cluster=id)
> SHF2<-crrc(time, status, x, failcode=2,cluster=id)
```

6.5.2 Bladder Cancer Data

We consider an extension (data set available in the **frailtyHL**: ‘bladder’) of the bladder cancer data introduced in Sect. 1.2.4. Here we consider 396 patients with bladder cancer from 21 centers, focusing on two competing endpoints, i.e, time to first bladder recurrence (an event of interest; Type 1 event) and time to death prior to recurrence (competing event; Type 2 event). Of 396 patients, 200 (50.51%) had recurrence of bladder cancer and 81 (20.45%) died prior to recurrence. One hundred and fifteen patients (29.04%) who were still alive without recurrence were censored at the date of the last available follow-up. The numbers of patients per center varied from 3 to 78, with the mean of 18.9 and the median of 14. Two covariates are considered: Chemo (no, yes) and Age ($\leq$65 years, >65 years). The corresponding R codes and outputs are provided below.

```
############ 1) Cause-specific frailty models (Shared) ##############
> data(bladder, package="frailtyHL")
> data_conti <- bladder
> data_surv<-data_conti
> jm1<-jointmodeling(Model="mean",RespDist="FM", Link="log",
+  LinPred=Surv(surtime,status==1)~CHEMO+AGE+(1|center),
+  RandDist="gaussian")
> jm2<-jointmodeling(Model="mean",RespDist="FM",Link="log",
+  LinPred=Surv(surtime,status==2)~CHEMO+AGE+(1|center),
+  RandDist="gaussian")
> res<-jmfit(jm1,jm2,data_conti,data_surv, Maxiter=200)
[1] "iterations : "
[1] 56
           beta_h    se_beh    t_value      p_value
CHEMO   -0.6662422 0.1745319 -3.8173084 0.0001349155
AGE     -0.1496644 0.1436548 -1.0418339 0.2974886732
CHEMO    0.1141791 0.3816191  0.2991965 0.7647901300
AGE      0.6889314 0.2656665  2.5932189 0.0095082234
       alpha_h    rho_h
[1,] 0.07255925 1.135538
> res$V.Est # log-frailty estimates
 [1]  0.151408743  0.184588705 -0.002456431 -0.034954163  0.002932653
 [6] -0.142895085  0.193996286  0.031210209 -0.052866610  0.094447854
[11]  0.047394088  0.108819042 -0.090623066  0.044104201 -0.278417345
[16] -0.360868161  0.107422797  0.090266872  0.229043252 -0.368074147
[21]  0.045520307
######### 2) Cause-specific frailty models (Univariate) ############
> q = length(unique(data_conti$center))
> beta.init <- c(sapply(1:2, function(k) coxph(Surv(surtime, status==k)
+ ~CHEMO + AGE, data=data_conti)$coef))
> theta.init = 0.05
> v.init = rnorm(q,0,1) #v.init=rep(0,q)
> CSFM<-hlike.frailty(formula=CmpRsk(surtime,status)~CHEMO+AGE
+ +cluster(center),data=data_conti,frailty.cov="none",inits=list(beta=
+ beta.init,theta=theta.init,v=v.init),order=1,MAX.ITER=200,TOL=1E-5)
> CSFM
  Type  Effect   Estimate        SE       2.5%       97.5%
1    1   CHEMO -0.6679760 0.1747679 -1.0105148 -0.3254372
2    1     AGE -0.1488389 0.1437895 -0.4306613  0.1329834
3    2   CHEMO  0.1194893 0.3809705 -0.6271992  0.8661778
4    2     AGE  0.6772935 0.2652078  0.1574957  1.1970913
  Var.Comp   Estimate
1 Sigma.11 0.07742706
 Successfully Converged
> CSFM$iterations
[1] 72
> unique(CSFM$v) # log-frailty estimates
########## 3) Cause-specific frailty models (BN) ###########
```

```
theta.init <-matrix(c(1,0.5,0.5,1),nrow=2)
v.init<-MASS::mvrnorm(q,mu=rep(0,2),Sigma=theta.init)#or v.init=matrix(0,q,2)
> CSFM_BN<-hlike.frailty(formula=CmpRsk(surtime,status)~CHEMO+AGE
+  +cluster(center),data=data_conti,frailty.cov="unstructured",
+  inits=list(beta=beta.init, theta=theta.init, v=v.init),
+  order=1, MAX.ITER=500, TOL=1E-5)
> CSFM_BN
  Type Effect   Estimate        SE       2.5%      97.5%
1    1  CHEMO -0.6674387 0.1747014 -1.0098471 -0.3250303
2    1    AGE -0.1484962 0.1437680 -0.4302763  0.1332840
3    2  CHEMO  0.1199023 0.3821346 -0.6290678  0.8688723
4    2    AGE  0.6865932 0.2660755  0.1650949  1.2080915
  Var.Comp   Estimate
1 Sigma.11 0.07356526
2 Sigma.21 0.07632592
3 Sigma.22 0.09516952
 Successfully Converged
> CSFM$iteration
[1] 377
############ 4) Subhazard frailty models ############
> # Subhazard Frailty (Type 1)
>  beta.init<-c(0,0); v.init=rep(0,q); theta.init = 0.05
> SHFM1<-hlike.frailty(CmpRsk(surtime,status)~CHEMO+AGE+cluster(center),
+  data=data_conti,inits=list(beta=beta.init,theta=theta.init,v=v.init),
+  order=1,frailty.cov="none",subHazard=TRUE, MAX.ITER=300)
>  SHFM1
        Type  Effect   Estimate        SE      2.5%       97.5%
CHEMO      1   CHEMO -0.7004651 0.1751228 -1.043699 -0.35723077
AGE        1     AGE -0.2154904 0.1443744 -0.498459  0.06747824
  Var.Comp   Estimate
1 Sigma.11 0.06347167
 Successfully Converged
> unique(SHFM1$v) # log-frailty estimates
># Subhazard Frailty (Type 2)
> da<-data_conti
> status_2<-ifelse(da$status==2,1,2*da$status) #Transformation of status
> theta.init = 0.001
> SHFM2<-hlike.frailty(CmpRsk(surtime,status_2)~CHEMO+AGE+cluster(center),
+   data=da, inits=list(beta=beta.init, theta=theta.init, v=v.init),
+   order=1, frailty.cov="none", subHazard=TRUE, MAX.ITER=300, TOL=1E-6)
>  SHFM2
        Type  Effect  Estimate        SE        2.5%    97.5%
CHEMO      1   CHEMO 0.6375418 0.3732185 -0.09395313 1.369037
AGE        1     AGE 0.9289147 0.2581870  0.42287737 1.434952
  Var.Comp     Estimate
1 Sigma.11 0.0009959938
 Successfully Converged
############ 5) Subhazard without frailty (Fine-Gray) ################
#library(cmprsk)
> attach(data_conti)
> x=cbind(CHEMO,AGE)
> SH1=crr(surtime,status, x, failcode=1)  #Type 1
> summary(SH1)
Competing Risks Regression
Call:
crr(ftime = surtime, fstatus = status, cov1 = x, failcode = 1)
        coef exp(coef) se(coef)     z p-value
CHEMO -0.673     0.510    0.178 -3.77 0.00016
```

```
AGE    -0.228       0.796     0.143 -1.60 0.11000
       exp(coef) exp(-coef)  2.5% 97.5%
CHEMO      0.510        1.96 0.360 0.724
AGE        0.796        1.26 0.602 1.053
Num. cases = 396
Pseudo Log-likelihood = -1099
Pseudo likelihood ratio test = 16.8  on 2 df,
> SH2=crr(surtime,status, x, failcode=2)  #Type 2
> summary(SH2)
Competing Risks Regression
Call:
crr(ftime = surtime, fstatus = status, cov1 = x, failcode = 2)
       coef exp(coef) se(coef)     z p-value
CHEMO 0.637      1.89    0.344 1.85 0.06400
AGE   0.930      2.53    0.245 3.80 0.00015
      exp(coef) exp(-coef)  2.5% 97.5%
CHEMO      1.89      0.529 0.963  3.71
AGE        2.53      0.395 1.568  4.09
Num. cases = 396
Pseudo Log-likelihood = -408
```

Table 6.7 Comparison of competing-risks frailty models with cause-specific hazard (CSH) and subhazard (SH): bladder cancer data; Age = I(age at diagnosis > 65)

Model	Event	Chemo(SE)	Age(SE)	$\hat{\sigma}^2$	Association
CSH	Type 1	−0.626(0.172)	−0.164(0.142)	–	–
(Cox)	Type 2	0.180(0.378)	0.556(0.264)	–	–
$-2p_\tau(h_p)$		2819.6	(rAIC = 2819.6, cAIC = 2822.8)		
CSH	Type 1	−0.668(0.175)	−0.149(0.144)	0.077	–
(Univ)	Type 2	0.119(0.381)	0.677(0.265)		
$-2p_\tau(h_p)$		2812.5	(rAIC = 2814.5, cAIC = 2812.1)		
CSH	Type 1	−0.666(0.175)	−0.150(0.144)	0.073	$\hat{\gamma}_2 = 1.136$
(Shared)	Type 2	0.114(0.382)	0.689(0.266)		
$-2p_\tau(h_p)$		2812.5	(rAIC = 2816.5, cAIC = 2812.1)		
CSH	Type 1	−0.667(0.175)	−0.148(0.144)	$\hat{\sigma}_1^2 = 0.074$	$\hat{\sigma}_{12} = 0.076$
(BN)	Type 2	0.120(0.382)	0.687(0.266)	$\hat{\sigma}_2^2 = 0.095$	–
$-2p_\tau(h_p)$		2812.0	(rAIC = 2818.0, cAIC = 2812.1)		
SH(F-G)	Type 1	−0.673(0.178)	−0.228(0.143)	–	–
$-2p_\tau(h_{pw})$		2201.6	(rAIC = 2201.6, cAIC = 2201.8)		
	Type 2	0.637(0.344)	0.930(0.245)	–	–
$-2p_\tau(h_{pw})$		816.7	(rAIC = 816.7, cAIC = 819.7)		
SH(Shared)	Type 1	−0.700(0.175)	−0.215(0.144)	0.063	–
$-2p_\tau(h_{pw})$		2199.2	(rAIC = 2201.2, cAIC = 2197.4)		
	Type 2	0.637(0.373)	0.929(0.258)	0.000	–
$-2p_\tau(h_{pw})$		816.7	(rAIC = 818.7, cAIC = 819.7)		

F-G, Fine-Gray subhazard model without frailty

The fitted results are summarized in Table 6.7. It may be an interesting comprehensive analysis to compare the cause-specific hazard and subhazard models. We first observe that the trends under the two models are also similar.

For the cause-specific models in Table 6.7, we have BN $\supset$ Shared $\supset$ Univ $\supset$ Cox. The three models (BN, Shared and Univ models) have almost the same value of $-2p_{\tau}(h_p)$. Between the Univ and Cox models, the likelihood difference is 7.1 (> 2.71), so that the null hypothesis of $\sigma^2 = 0$ is rejected. Thus, the LRT selects the univariate model as the final model. Note that the rAIC selects the univariate model as the best model, but the cAIC does not because the frailty variances are near zero as in Sect. 5.3. Next, for the subhazard models, we select a proper model in each event type. For Type 1, the difference in $-2p_{\tau}(h_{pw})$ between the Fine-Gray model and the shared model is 2.4 (< 2.71), so that the null hypothesis of $\sigma^2 = 0$ is not rejected. For Type 2, the two models has practically the same value of $-2p_{\tau}(h_{pw})$. Thus, we see that in the subhazard models, the frailty term is necessary for neither event type. In fact, Table 6.7 presents that the two subhazard models give very similar estimates for Type 1 and Type 2.

6.6 Discussion

The h-likelihood procedures are applied to fit the cause-specific frailty models as well as the subhazard frailty models. The competing risks models with correlated frailties provides systematically more informative results for the analysis of multicenter competing risk data. We also demonstrate how to investigate the heterogeneity in treatment effect over centers and how to test such heterogeneity.

The cause-specific frailty model can take into account the correlation among events of interest and competing events via frailties, while the subhazard frailty models can not, assuming that the frailty effects on both types of events are independent. Therefore, the cause-specific frailty model would be more appropriate when a dependency between different types of events or informative censoring is present. The subhazard model is useful for direct statistical inference about the CIF of the particular event type of interest. Developing an extended frailty modeling approach under the subhazard frailty model to allow a correlation between different event types would be an interesting topic for future work.

The present work only considered the lognormal frailty distribution. It may be also interesting to consider other distributions when competing risks are present, in particular the gamma frailty distribution.

Even if in this chapter the h-likelihood procedures have been presented for clustered competing-risks data, they can be applied to univariate competing-risks data without clustering. For example, with the h-likelihood the cause-specific PH model (6.9) allowing for the BN frailties would be applied to dependent competing risks or informative censoring problem under the univariate competing-risks structure (Huang and Zhang 2008; Chen 2010).

We have analyzed the competing risks data where the observation time ends upon occurrence of the first failure. However, a subject may experience a nonterminal event (e.g., disease recurrence) and/or a terminal event (e.g., death), where the terminal event censors the nonterminal event but not vice versa. This is called as semi-competing risks data (Fine et al. 2001), which is an extension of competing risks data. We will present the frailty models for semi-competing risks data in Chap. 10.

6.7 Appendix

6.7.1 Calculation of the Gradient Vector and Elements for the Information Matrix from the Partial Likelihood

Let us define $\tau = (\beta^T, v^T)^T$. Since the partial h-likelihood involves the sample estimate of the cause-specific baseline (cumulative) hazard function, following Appendix 2 of Ha et al. (2001), the MHL score equation for τ for fixed θ is given by

$$\partial h_p/\partial\tau = (\partial h/\partial\tau)|_{\Lambda_{0k}=\hat{\Lambda}_{0k}},$$

where the h-likelihood h was defined in (6.18).

First, from (6.19) the elements of the gradient vector $(\partial h_p/\partial\beta, \partial h_p/\partial v)^T$ are calculated. The k th element of $\partial h_p/\partial\beta = (\partial h_p/\partial\beta_1, \partial h_p/\partial\beta_2)^T$ is the derivative of h_p with respect to the regression coefficients for event k,

$$\frac{\partial h_p}{\partial\beta_k} = \sum_{ij} x_{ij}\delta_{ijk} - x_{ij}\hat{\Lambda}_{0k}(y_{ij})\exp\left(x_{ij}^T\beta_k + z_{ij}^T v_k\right) \tag{6.27}$$

and $\partial h_p/\partial v = (\partial h_p/\partial v_1, \partial h_p/\partial v_2)^T$ is the derivative of h_p with respect to the random effects for each event Type k,

$$\frac{\partial h_p}{\partial v_k} = \sum_{ij} z_{ij}\delta_{ijk} - z_{ij}\hat{\Lambda}_{0k}(y_{ij})\exp\left(x_{ij}^T\beta_k + z_{ij}^T v_k\right) - \sum_{i=1}^{q} v_i \bullet (\sigma^{kk}, \sigma^{12}) \tag{6.28}$$

where $\bullet$ denotes the inner product of two vectors, and σ^{kk} and σ^{12} are elements of the precision matrix Σ^{-1}.

The following matrices and notation are used for the remainder of this section. Let $R_k = (R_1, R_2, \ldots, R_{D_k})$ be an $n \times D_k$ at risk indicator matrix where the ijth element in column r is one if $y_{ij} \geq y_{(kr)}$ and zero otherwise. Define δ_k as an $n \times 1$ vector of Type k event indicators with its ijth element being δ_{ijk}. Let μ_k be an $n \times n$

diagonal matrix with elements $\hat{\Lambda}_{0k}(y_{ij})\exp(x_{ij}^T\beta_k + z_{ij}^T v_k)$, and let N_k be an $n \times n$ diagonal matrix with elements $\exp(x_{ij}^T\beta_k + z_{ij}^T v_k)$. Finally, let I_q be a $q \times q$ identity matrix and let $\otimes$ denote the Kronecker product. Recall that X is an $n \times p$ matrix of p covariates and Z is an $n \times q$ matrix of cluster indicators.

Using this notation, Eq. (6.27) can be expressed as

$$\frac{\partial h_p}{\partial \beta_k} = X^T(\delta_k - \mu_k) \tag{6.29}$$

and the derivative (6.28) of h_p with respect to all random effects v is,

$$\frac{\partial h_p}{\partial v} = \begin{pmatrix} Z^T(\delta_1 - \mu_1) \\ Z^T(\delta_2 - \mu_2) \end{pmatrix} - (\Sigma^{-1} \otimes I_q)v. \tag{6.30}$$

Next, the observed information matrix H of β and v from the profile h-likelihood for fixed θ is calculated. Again because the partial h-likelihood includes the sample estimate of the cause-specific baseline (cumulative) hazard function, Ha and Lee (2003, Appendix B) showed that

$$\partial^2 h_p/\partial\tau^2 = (H_1 - H_2)|_{\Lambda_{0k}=\hat{\Lambda}_{0k}}, \tag{6.31}$$

where $H_1 = \partial^2 h/\partial\tau^2$ and

$$H_2 = (-\partial^2 h/\partial\tau\partial\lambda_0)(-\partial^2 h/\partial\lambda_0^2)^{-1}(-\partial^2 h/\partial\lambda_0\partial\tau).$$

Denoting C_k for a diagonal $D_k \times D_k$ matrix where the rth element is $d_{(kr)}/\hat{\lambda}_{0kr}^2$, let us define $\mathbf{X}$, $\mathbf{Z}$ and $\mathbf{W}^*$ as block diagonal matrices such that,

$$\mathbf{X} = \begin{pmatrix} X & \mathbf{0} \\ \mathbf{0} & X \end{pmatrix}, \quad \mathbf{Z} = \begin{pmatrix} Z & \mathbf{0} \\ \mathbf{0} & Z \end{pmatrix} \quad \text{and} \quad \mathbf{W}^* = \begin{pmatrix} W_1^* & \mathbf{0} \\ \mathbf{0} & W_2^* \end{pmatrix} \tag{6.32}$$

where $\mathbf{0}$ is a conformable matrix of zeros and $W_k^* = W_k^*(\beta_k, v_k) = \mu_k - N_k R_k C_k^{-1}$ $(R_k N_k)^T$ for $k = 1, 2$. Then the observed information matrix H is a $K(p+q) \times K(p+q)$ matrix,

$$H_p = H_p(\beta, v, \theta) = \begin{pmatrix} \mathbf{X}^T\mathbf{W}^*\mathbf{X} & \mathbf{X}^T\mathbf{W}^*\mathbf{Z} \\ \mathbf{Z}^T\mathbf{W}^*\mathbf{X} & \mathbf{Z}^T\mathbf{W}^*\mathbf{Z} + Q \end{pmatrix}. \tag{6.33}$$

where Q is a $Kq \times Kq$ matrix that is the negative second derivative of the log of the joint density function for all random effects with respect to the vector v,

$$Q(v, \theta) = -\frac{\partial^2}{\partial v^2}\sum_{i=1}^{q} \ell_{2i}(\theta; v_i) = \Sigma^{-1} \otimes I_q. \tag{6.34}$$

In addition, consider the univariate frailty case with $v \sim N(0, \sigma^2 I_q)$. Then, the estimating equations of β_k are identical to (6.29), while those of v are slightly different from (6.30) because they are given by

$$\frac{\partial h_p}{\partial v} = \sum_k Z^T(\delta_k - \mu_k) - \sigma^{-2} v.$$

This leads to the observed information for (β_k, v) with $k = 1, 2$:

$$H_p = H_p(\beta, v, \sigma^2) = \begin{pmatrix} X^T W_1^* X & 0 & X^T W_1^* Z \\ 0 & X^T W_2^* Z & X^T W_2^* Z \\ Z^T W_1^* X & Z^T W_2^* X & \sum_k Z^T W_k^* Z + Q \end{pmatrix},$$

where $Q = \sigma^{-2} I_q$. □

6.7.2 Derivation of the Gradient Vector and Elements for the Information Matrix from the Partial Restricted Likelihood

Let θ_r and θ_s denote the rth and sth components of $\theta = (\sigma_{11}, \sigma_{22}, \sigma_{12})$ for $r, s = 1, 2, 3$. To evaluate the derivatives of the adjusted profile h-likelihood (5.12), the following two identities from matrix calculus will be used here (Searle et al. 1992, Appendix M): For a matrix A and a scalar x,

$$\frac{\partial}{\partial x} \log(\det(A)) = \text{trace}\left(A^{-1} \frac{\partial A}{\partial x}\right)$$

and

$$\frac{\partial A^{-1}}{\partial x} = -A^{-1} \frac{\partial A}{\partial x} A^{-1}.$$

By using these results, the rth component of the gradient vector of the restricted partial h-likelihood $p_\tau(h_p)$, i.e., $\partial p_\tau(h_p)/\partial \theta$ where $\tau = (\beta^T, v^T)^T$, can be written as

$$\frac{\partial p_\tau(h_p)}{\partial \theta_r} = \frac{\partial \hat{h}_p}{\partial \theta_r} - \frac{1}{2} \text{trace}\left(\hat{H}_p^{-1} \frac{\partial \hat{H}_p}{\partial \theta_r}\right). \tag{6.35}$$

Furthermore, the element in row r and column s of the 3×3 observed information matrix $\partial^2 p_\tau(h_p)/\partial \theta^2$ for the frailty parameter θ is given by

$$-\frac{\partial^2 p_\tau(h_p)}{\partial\theta_r\partial\theta_s} = -\frac{\partial^2 \hat{h}_p}{\partial\theta_r\partial\theta_s} + \frac{1}{2}\text{tr}\left(-\hat{H}_p^{-1}\frac{\partial \hat{H}_p}{\partial\theta_r}\hat{H}_p^{-1}\frac{\partial \hat{H}_p}{\partial\theta_s} + \hat{H}_p^{-1}\frac{\partial^2 \hat{H}_p}{\partial\theta_r\partial\theta_s}\right) \tag{6.36}$$

Since $\hat{\beta}(\theta)$ and $\hat{v}(\theta)$ are functions of θ, it is not appropriate to just use the partial derivatives in (6.35) and (6.36). Instead the total derivative should be used. The total derivative of $p_\tau(h_p)$ with respect to θ_r is,

$$\frac{\partial p_\tau(h_p)}{\partial\theta_r} = \frac{\partial p_\tau(h_p)}{\partial\theta_r} + \left(\left.\frac{\partial p_\tau(h_p)}{\partial\beta}\right|_{\beta=\hat{\beta}}\right)\frac{\partial\hat{\beta}}{\partial\theta_r} + \left(\left.\frac{\partial p_\tau(h_p)}{\partial v}\right|_{v=\hat{v}}\right)\frac{\partial\hat{v}}{\partial\theta_r}. \tag{6.37}$$

Note, in general, that when $x = x(t)$ and $y = y(t)$ are functions of t, the total derivative of $f(t, x, y)$ with respect to t is defined as

$$\frac{df}{dt} = \frac{\partial f}{\partial t} + \frac{\partial f}{\partial x}\frac{dx}{dt} + \frac{\partial f}{\partial y}\frac{dy}{dt}.$$

The total derivative allows the other arguments of h_A, $\hat{\beta}(\theta_r)$ and $\hat{v}(\theta_r)$ to depend on θ_r, which should not be overlooked in our conditionally iterative optimization procedure. Originally, however, Lee and Nelder (1996) and Ha et al. (2001) ignored $\partial\hat{\beta}/\partial\theta_r$ and $\partial\hat{v}/\partial\theta_r$ when differentiating $\hat{h}_p$ and $\hat{H}$ with respect to θ. Following Ha and Lee (2003), in our derivation $\partial\hat{\beta}/\partial\theta$ is ignored because there is no direct dependency between $\hat{\beta}$ and θ_r whereas $\partial\hat{v}/\partial\theta_r$ is included because there exists a direct dependency between $\hat{v}$ and θ_r, which is clear from (6.29) and (6.30).

Now first, we calculate the derivatives in (6.35). Let $\hat{\tau} = \hat{\tau}(\theta) = (\hat{\beta}(\theta), \hat{v}(\theta))$. Since $\partial h_p/\partial v|_{\tau=\hat{\tau}} = 0$, the total derivative of the first term $\partial\hat{h}_p/\partial\theta_r$ is

$$\begin{aligned}
\frac{\partial\hat{h}_p}{\partial\theta_r} &= \left.\frac{\partial h_p}{\partial\theta_r}\right|_{\tau=\hat{\tau}} + \left(\left.\frac{\partial h_p}{\partial v}\right|_{\tau=\hat{\tau}}\right)\left(\frac{\partial\hat{v}}{\partial\theta_r}\right) \\
&= \left.\frac{\partial h_p}{\partial\theta_r}\right|_{\tau=\hat{\tau}} \\
&= \sum_{i=1}^{q}\frac{\partial\ell_{2i}(\theta;\hat{v}_i)}{\partial\theta_r} \\
&= \sum_{i=1}^{q} -\frac{1}{2}\text{trace}\left(\Sigma^{-1}\Sigma'_r\right) + \frac{1}{2}\hat{v}_i^T\left(\Sigma^{-1}\Sigma'_r\Sigma^{-1}\right)\hat{v}_i,
\end{aligned}$$

where $\Sigma'_r = \partial\Sigma/\partial\theta_r$.

The derivative of the second term, $\partial\hat{H}_p/\partial\theta_r$, in (6.35) is given by the following form:

$$\frac{\partial \hat{H}_{pi}}{\partial \theta_r} = \left.\frac{\partial H_{pi}}{\partial \theta_r}\right|_{T=\hat{T}} + \left(\left.\frac{\partial H_{pi}}{\partial v}\right|_{T=\hat{T}}\right)\left(\frac{\partial \hat{v}}{\partial \theta_r}\right),$$

where $\hat{H}_{pi}$ and H_{pi} are the (i, i)th components of $\hat{H}_p = \mathbf{P}^T\hat{\mathbf{V}}\mathbf{P}$ and $H_p = \mathbf{P}^T\mathbf{V}\mathbf{P}$, respectively. Here $\hat{\mathbf{V}} = \text{BD}(\hat{\mathbf{W}}^*, Q)$: see also Appendix A.2. The term $\partial\hat{v}/\partial\theta_r$ is calculated following Lee et al. (2017b). From h_p, given θ_r, let $\hat{v}(\theta_r)$ be the solution to $g(\theta_r) = \partial h_p/\partial v|_{T=\hat{T}} = 0$. Then,

$$\frac{\partial g(\theta_r)}{\partial \theta_r} = \left.\frac{\partial^2 h_p}{\partial v \partial \theta_r}\right|_{T=\hat{T}} + \left(\left.\frac{\partial^2 h_p}{\partial v^2}\right|_{T=\hat{T}}\right)\left(\frac{\partial \hat{v}}{\partial \theta_r}\right) = 0.$$

Solving for $\partial\hat{v}/\partial\theta_r$ gives a $2q \times 1$ vector,

$$\begin{aligned}\frac{\partial \hat{v}}{\partial \theta_r} &= \left(-\left.\frac{\partial^2 h_p}{\partial v^2}\right|_{T=\hat{T}}\right)^{-1}\left(\left.\frac{\partial^2 h_p}{\partial v \partial \theta_r}\right|_{T=\hat{T}}\right)\\ &= \left(\mathbf{Z}^T\hat{\mathbf{W}}^*\mathbf{Z} + Q\right)^{-1}\left(\left[(\Sigma^{-1}\Sigma_r'\Sigma^{-1}) \otimes I_q\right]\hat{v}\right),\end{aligned}$$

where $\hat{\mathbf{W}}^*$ is $\mathbf{W}^*$ evaluated at $(\hat{\beta}, \hat{v}, \theta)$, that is, when $W_k^* = W_k^*(\hat{\beta}_k, \hat{v}_k, \theta) = \hat{W}_k^*$. Now, since $\mathbf{X}$ and $\mathbf{Z}$ are constant matrices that do not depend on θ, following Appendix 4.7.5, the total derivative of $\partial\hat{H}_p/\partial\theta_r$ is given by

$$\frac{\partial \hat{H}_p}{\partial \theta_r} = \begin{pmatrix} \mathbf{X}^T\hat{\mathbf{W}}_r'\mathbf{X} & \mathbf{X}^T\hat{\mathbf{W}}_r'\mathbf{Z} \\ \mathbf{Z}^T\hat{\mathbf{W}}_r'\mathbf{X} & \mathbf{Z}^T\hat{\mathbf{W}}_r'\mathbf{Z} + Q_r' \end{pmatrix},$$

where $\hat{\mathbf{W}}_r' = \partial\hat{\mathbf{W}}^*/\partial\theta_r$ and $Q_r' = \partial Q/\partial\theta_r$. Since $\hat{W}_k^*$ does not depend on θ, the total derivative of $\hat{W}_k^*$ is

$$\hat{W}_{kr}' = \frac{\partial \hat{W}_k^*}{\partial \theta_r} = \left.\frac{\partial W_k^*}{\partial \theta_r}\right|_{T=\hat{T}} + \left(\left.\frac{\partial W_k^*}{\partial v_k}\right|_{T=\hat{T}}\right)\left(\frac{\partial \hat{v}_k}{\partial \theta_r}\right) = \left(\left.\frac{\partial W_k^*}{\partial v_k}\right|_{T=\hat{T}}\right)\left(\frac{\partial \hat{v}_k}{\partial \theta_r}\right).$$

The derivative $\partial W_k^*/\partial v_k$ is found by differentiating $W_k^*(\beta_k, v_k) = \mu_k - N_k R_k C_k^{-1} (R_k N_k)^T$ with respect to v_k. Given the structure of v defined earlier, $\partial\hat{v}_1/\partial\theta_r$ is the first q elements of the vector $\partial\hat{v}/\partial\theta_r$ and $\partial\hat{v}_2/\partial\theta_r$ are the last q elements. Since Q does not depend on v, the total derivative is not needed to find $\partial Q/\partial\theta_r$ so,

$$Q_r' = \frac{\partial Q}{\partial \theta_r} = (-\Sigma^{-1}\Sigma_r'\Sigma^{-1}) \otimes I_q. \tag{6.38}$$

Using (6.38) there is a slightly simpler expression for $\partial\hat{v}/\partial\theta_r$,

$$\partial\hat{v}/\partial\theta_r = -\left(\mathbf{Z}^T\hat{\mathbf{W}}^*\mathbf{Z} + Q\right)^{-1}(Q_r'\hat{v}).$$

The next step is to calculate the terms in the observed information (6.36). First, we have

$$\begin{aligned}
-\frac{\partial^2 \hat{h}_p}{\partial\theta_r\partial\theta_s} &= -\left.\frac{\partial^2 h_p}{\partial\theta_r\partial\theta_s}\right|_{\tau=\hat{\tau}} - \left(\left.\frac{\partial^2 h_p}{\partial v\partial\theta_r}\right|_{\tau=\hat{\tau}}\right)\left(\frac{\partial\hat{v}}{\partial\theta_s}\right) \\
&= \sum_{i=1}^{q} -\frac{\partial^2 \ell_{2i}(\theta;\hat{v}_i)}{\partial\theta_r\partial\theta_s} + (Q'_r\hat{v})\left(\frac{\partial\hat{v}}{\partial\theta_s}\right) \\
&= \sum_{i=1}^{q}\left\{-\frac{1}{2}\text{tr}\left(\Sigma^{-1}\Sigma'_s\Sigma^{-1}\Sigma'_r + \Sigma^{-1}\Sigma''_{rs}\right) + \frac{1}{2}\hat{v}_i^T S_{rs}\hat{v}_i\right\} + Q'_r\hat{v}\left(\frac{\partial\hat{v}}{\partial\theta_s}\right),
\end{aligned}$$

where $S_{rs} = \partial(-\Sigma^{-1}\Sigma'_r\Sigma^{-1})/\partial\theta_s = (\Sigma^{-1}\Sigma'_s\Sigma^{-1}\Sigma'_r\Sigma^{-1}) + (\Sigma^{-1}\Sigma'_r\Sigma^{-1}\Sigma'_s\Sigma^{-1}) - (\Sigma^{-1}\Sigma''_{rs}\Sigma^{-1})$ and $\Sigma''_{rs} = \partial^2\Sigma/\partial\theta_r\partial\theta_s$. The last term needed to calculate (6.36) is

$$\frac{\partial^2 \hat{H}_p}{\partial\theta_r\partial\theta_s} = \begin{pmatrix} \mathbf{X}^T\hat{\mathbf{W}}''_{rs}\mathbf{X} & \mathbf{X}^T\hat{\mathbf{W}}''_{rs}\mathbf{Z} \\ \mathbf{Z}^T\hat{\mathbf{W}}''_{rs}\mathbf{X} & \mathbf{Z}^T\hat{\mathbf{W}}''_{rs}\mathbf{Z} + Q''_{rs} \end{pmatrix},$$

where

$$Q''_{rs} = \frac{\partial^2 Q}{\partial\theta_r\partial\theta_s} = S_{rs}\otimes I_q$$

and $\hat{\mathbf{W}}''_{rs} = \partial\hat{\mathbf{W}}'_r/\partial\theta_s$. Like earlier, $\hat{\mathbf{W}}''_{rs}$ can be found by evaluating $\partial^2\hat{W}^*_k/\partial\theta_r\partial\theta_s$ for $k = 1, 2$,

$$\hat{W}''_{krs} = \frac{\partial^2\hat{W}^*_k}{\partial\theta_r\partial\theta_s} = \left[\left(\left.\frac{\partial^2 W^*_k}{\partial v_k^2}\right|_{\tau=\hat{\tau}}\right)\frac{\partial\hat{v}_k}{\partial\theta_r}\right]\frac{\partial\hat{v}_k}{\partial\theta_s} + \left(\left.\frac{\partial W^*_k}{\partial v_k}\right|_{\tau=\hat{\tau}}\right)\frac{\partial^2\hat{v}_k}{\partial\theta_r\partial\theta_s},$$

where

$$\begin{aligned}
\frac{\partial^2\hat{v}}{\partial\theta_r\partial\theta_s} &= \left(\mathbf{Z}^T\hat{\mathbf{W}}*\mathbf{Z}+Q\right)^{-1}\left(\mathbf{Z}^T\hat{\mathbf{W}}'_s\mathbf{Z}+Q'_s\right)\left(\mathbf{Z}^T\hat{\mathbf{W}}*\mathbf{Z}+Q\right)^{-1}Q'_r\hat{v} \\
&\quad - \left(\mathbf{Z}^T\hat{\mathbf{W}}*\mathbf{Z}+Q\right)^{-1}\left[Q''_{rs}\hat{v} + Q'_r\frac{\partial\hat{v}}{\partial\theta_s}\right] \qquad (6.39) \\
&= -\left(\mathbf{Z}^T\hat{\mathbf{W}}*\mathbf{Z}+Q\right)^{-1}\left[\left(\mathbf{Z}^T\hat{\mathbf{W}}'_s\mathbf{Z}+Q'_s\right)\frac{\partial\hat{v}}{\partial\theta_r} + Q''_{rs}\hat{v} + Q'_r\frac{\partial\hat{v}}{\partial\theta_s}\right]
\end{aligned}$$

is a $2q\times 1$ vector and $\partial^2\hat{v}_k/\partial\theta_r\partial\theta_s$ is the first q elements of (6.39) if $k = 1$ and the second q elements if $k = 2$. The term $\partial^2 W^*_k/\partial v^2$ is found by twice differentiating $W^*_k(\beta_k, v_k) = \mu_k - N_k R_k C_k (R_k N_k)^T$ with respect to v_k. □

6.7.3 Proof of Estimating Equations in (6.23)

Given the frailty parameters θ, the MHLEs of $\tau = (\beta^T, v^T)^T$ are obtained by solving the joint estimating equations, $\partial h_w^*/\partial\tau = 0$. Here, the calculations in Ha and Lee (2003) and Ha et al. (2001) showed that

$$\partial h_{pw}/\partial\tau = \partial h_w/\partial\tau|_{\lambda_{01}^s=\widehat{\lambda}_{01}^w} = \{E^T(\delta - \mu) - F\tau\}|_{\lambda_{01}^s=\widehat{\lambda}_{01}^w} \tag{6.40}$$

since $\partial h_w/\partial\tau = (\partial\eta/\partial\tau)(\partial h_w/\partial\eta)$ with $\eta = X\beta + Zv = E\tau$. Here $E = (X, Z)$, $F = \text{BD}(0, U) = \text{BD}(0, \Sigma_1^{-1}, \ldots, \Sigma_q^{-1})$, and δ and μ are the $n \times 1$ vectors of δ_{ij}'s, and μ_{ij}'s, respectively. Note that the vector μ can be written as a simple form by using a weighted risk indicator matrix M which contains the weight w_{ij} as well as the risk set $R_{(r)}$. Let L be an $n \times 1$ vector of L_{ij}'s with $L_{ij} = \Lambda_{01}^s(y_{ij})w_{ij}$. Since $\Lambda_{01}^s(y_{ij}) = \sum_r \lambda_{01r}^s I\{(i, j) \in R_{(r)}\}$ and $w_{ij} = \widehat{G}(y_{(r)})/\widehat{G}(y_{ij} \wedge y_{(r)})$, we have $L = MAJ$, where M is an $n \times D$ matrix of weighted-risk indicators whose (ij, k)th element is $m_{ij,r}$, $A = \text{diag}(\lambda_{01r}^s)$ is a $D \times D$ diagonal matrix and J is a $D \times 1$ vector with ones. This gives $\mu = W_0(MAJ)$, where $W_0 = \text{diag}\{\exp(\eta_{ij})\}$. Note here that $m_{ij,r}$ are constructed by combining $R_{(r)}$ and w_{ij} as in Ruan and Gray (2008):

$$\begin{aligned} m_{ij,r} &= I\{y_{ij} \geq y_{(r)} \text{ or } (y_{ij} \leq y_{(r)} \text{ and } \xi_{ij} > 1)\}\{\widehat{G}(y_{(r)})/\widehat{G}(y_{ij} \wedge y_{(r)})\} \\ &= I\{y_{ij} \geq y_{(r)}\} + I\{y_{ij} \leq y_{(r)} \text{ and } \xi_{ij} > 1\}\{\widehat{G}(y_{(r)})/\widehat{G}(y_{ij})\}. \end{aligned} \tag{6.41}$$

This is also equivalent to the weights by Katsahian et al. (2006) and Katsahian and Boudreau (2011) because $m_{ij,r}$ are equal to one as long as individuals have not failed by time $y_{(r)}$ (i.e., $y_{ij} \geq y_{(r)}$), and below 1 and decreasing over time if they failed from another type (Type 2) before $y_{(r)}$ (i.e., $y_{ij} \leq y_{(r)}$ and $\xi_{ij} > 1$), and zero otherwise (e.g., they failed from Type 1 or have been right censored).

Furthermore, using the computation of Ha and Lee (2003), we have

$$-\partial^2 h_{pw}/\partial\tau^2 = E^T W^* E + F, \tag{6.42}$$

where $W^* = W_1 - W_2$, $W_1 = \text{diag}(\mu)$, $W_2 = (W_0M)C^{-1}(W_0M)^T$, $C = \text{diag}\{d_{(r)}/(\lambda_{01r}^s)^2\}$ is a $D \times D$ diagonal matrix, and $F = \text{BD}(0, U)$. Following Ha and Lee (2003) and (6.42), we can show that given θ, the MHL estimators of $\tau = (\beta^T, v^T)^T$ are obtained from the following score equations:

$$(E^T W^* E + F)\hat{\tau} = E^T w^*,$$

leading to (6.23). Here, $w^* = W^*\eta + (\delta - \mu)$. Note here that the λ_{01r}^s terms in W^* and w^* are evaluated at their estimates $\hat{\lambda}_{01r}^w = d_{(r)}/M_r^T\psi$, where M_r is the rth component vector of $M = (M_1, \ldots, M_D)$ and ψ is a vector of $\exp(\eta_{ij})$'s. This completes the proof. □

Chapter 7
Variable Selection for Frailty Models

7.1 Variable Selection

Including only the relevant variables in the model is crucial in statistical inference, improving the quality of estimation, prediction, and interpretation. If there exist many potential variables with equal status, i.e., no prior preference among them, then having as few variables as possible in the model would often facilitate clearer interpretation. When there are many potential variables, over-fitting can become a serious problem. However, missing relevant variables would be also undesirable.

There are many classical techniques for variable selection such as forward selection, backward elimination, stepwise selection, and best-subset selection. The stepwise selection is fast and convenient, but has inferior performance compared to the best-subset method. The latter is preferable, but very quickly becomes impractical because, with p variables, we need to compare 2^p models. Furthermore, the subset selection methods are often highly variable and they can not be used when the number of variables p is greater than the sample size n.

Recently, variable-selection methods using a penalized likelihood with various penalty functions have been widely studied in the linear models, GLMs and Cox's PH models. These methods select relevant variables and estimate the regression coefficients, simultaneously, i.e., they delete insignificant variables by estimating their coefficients as zero. However, in the semiparametric frailty models, variable-selection methods have been relatively less studied because the marginal likelihood of such models often involves analytically intractable integrals to eliminate the frailties. In this chapter, we investigate three variable-selection methods in survival analysis:

- least absolute shrinkage and selection operator (LASSO, Tibshirani 1996),
- smoothly clipped absolute deviation (SCAD, Fan and Li 2001),
- h-likelihood (HL) penalty (Lee and Oh 2014).

Specifically, we present the methods of variable selection of fixed effects in the various frailty models for clustered survival data. First, we show how to apply the

I.D. Ha et al., *Statistical Modelling of Survival Data with Random Effects*,
Statistics for Biology and Health, https://doi.org/10.1007/978-981-10-6557-6_7

variable-selection methods to survival data via the h-likelihood, and then illustrate the h-likelihood variable-selection methods with practical examples. We also compare three variable-selection procedures via the **frailtyHL** package.

7.2 Implied Penalty Functions from the Frailty Models

Following Lee and Oh (2014), we describe a random-effect model that generates a family of penalties, including the normal-type (bell-shaped L_2), LASSO-type (cusped L_1), and a new unbounded penalty at the origin to achieve selection consistency as we illustrate. Suppose that we have a linear predictor

$$\eta = \beta_0 + x_1\beta_1 + \cdots + x_{p-1}\beta_{p-1} + Zv,$$

where β_j are the unknown fixed effects, but we expect that many of them are zeros. Here Z is the model matrix of random effects v. It is well known that the random-effect estimators are shrinkage estimators toward zero and the shrinkage is beneficial to prediction. The ridge regression estimators shrink toward zero, but cannot be exactly zero, while the LASSO does. We want to find a class of frailty distributions which shrinks and allows the estimates of 0.

Suppose that conditional on a_j, we have

$$\beta_j|a_j \sim N(0, a_j\theta), \tag{7.1}$$

where θ is a fixed dispersion parameter, and a_j's are iid realizations from the gamma distribution with a parameter w such that

$$f_w(a_j) = (1/w)^{1/w}\frac{1}{\Gamma(1/w)}a_j^{1/w-1}e^{-a_j/w},$$

having $\mathrm{E}(a_j) = 1$ and $\mathrm{Var}(a_j) = w$. In this random-effect model, sparseness or selection can be achieved in a transparent way, since if $a_j = 0$ then $\beta_j = 0$.

Model (7.1) can be rewritten as $\beta_j = \sqrt{\tau_j}e_j$, with $e_j \sim N(0, 1)$ and

$$\log\tau_j = \log\theta + b_j,$$

and $b_j \equiv \log a_j$, which is a double HGLM (Lee et al. 2017b), having a random effect in the dispersion parameter. In this chapter, we consider the penalized partial h-likelihood for variable selection, defined by

$$h_v = h_p + h_2, \tag{7.2}$$

where h_p is the partial h-likelihood, and

$$h_2 = \sum_{j=1}^{p}\{\log f_\theta(\beta_j|a_j) + \log f_w(b_j)\},$$

with

$$\log f_\theta(\beta_j|a_j) = -\frac{1}{2}\left\{\log(2\pi\theta) + \log a_j + \beta_j^2/(\theta a_j)\right\},$$
$$\log f_w(b_j) = -\log(w)/w - \log\Gamma(1/w) + b_j/w - \exp(b_j)/w.$$

The outline of the estimation scheme using the ILS procedure is as follows:

- Given (β, w, θ), we estimate a_j's by solving

  $$\partial h_v/\partial a = 0,$$

  which gives the random-effect estimator

  $$\widehat{a}_j \equiv \widehat{a}_j(\beta) = [\{8w\beta_j^2/\theta + (2-w)^2\}^{1/2} + (2-w)]/4. \tag{7.3}$$

- Then, given $\widehat{a}$, we update β using the ILS procedure.

From model (7.1), it is clear that $\widehat{\beta}_j = 0$ when $\widehat{a}_j = 0$, which is how we achieve the sparseness. The estimator for β is obtained by maximizing the penalized partial h-likelihood via profiling, i.e.,

$$h_v^* = (h_p + h_2)|_{a=\hat{a}},$$

where $\hat{a}$ solves $dh_v/da = 0$. Then, the implied penalty of the penalized partial h-likelihood is

$$p_{\gamma,w}(\beta) = -h_2|_{a=\widehat{a}}, \tag{7.4}$$

where $\gamma = 1/\theta$ and $\widehat{a}_j$ is computed in the first step of the ILS above. Specifically, for fixed w, taking only the terms that involve β_j and $\widehat{a}_j$, the jth term of the penalty function is

$$p_{\gamma,w}(\beta_j) = \frac{\gamma\beta_j^2}{2\widehat{a}_j} + \frac{(w-2)}{2w}\log\widehat{a}_j + \frac{\widehat{a}_j}{w}.$$

Thus, the frailty model (7.1) leads to a family of the penalty functions $p_{\gamma,w}(\beta)$ that is indexed by w and $\gamma = 1/\theta$. This penalty stems from a statistical model with extra parameters γ and w, and includes the ridge and LASSO regressions as special cases:

- $w \to 0$: the ridge penalty
- $w = 2$: the LASSO penalty
- $w > 2$: the penalty with infinite value at 0.

This shows that ridge and LASSO estimators are viewed as ones from the normal and double exponential distributions of the random effects, respectively. The ridge, LASSO, and HL penalties all assume a proper density of the random effect with respect to β, i.e., $\int f_{\gamma,w}(\beta)d\beta = 1$. However, the SCAD estimator can be viewed as the random-effect estimator from an improper distribution with respect to β, i.e., $\int f_{\gamma,w}(\beta)d\beta = \infty$.

The penalty functions $p_{\gamma,w}(\cdot)$ at $w = 0$, 2 and 30 with $\gamma = 1/\theta = 1$ are shown in Fig. 7.1. As the convexity near the origin increases, the sparsity of the local solutions increases, and as the slope becomes flat, the amount of shrinkage lessens. From Fig. 7.1, we see that the HL controls the sparsity and shrinkage amount simultaneously. The form of the penalty changes from a quadratic shape ($w = 0$) for the ridge regression, to a cusped form ($w = 2$) for the LASSO, and then to an unbounded form ($w > 2$) at the origin. The quadratic penalties correspond to the ridge (shrinkage) estimates, which often lead to better prediction, while cusped ones lead to simultaneous variable selection and estimation of the LASSO and SCAD (Fan 1997). Given $w > 2$, the amount of shrinkage becomes larger as γ increases. In case of $w > 2$, it allows an infinite gain at zero, and the resulting penalty has a significant merit in variable selection as we shall discuss.

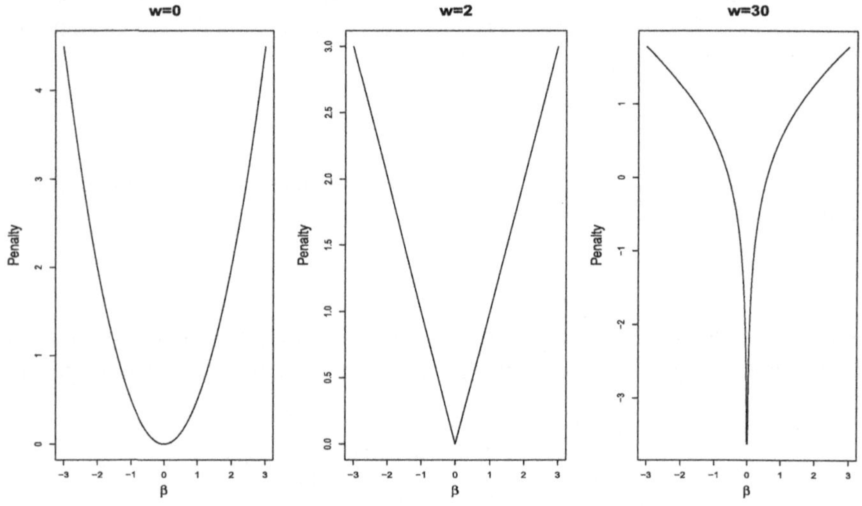

Fig. 7.1 HL penalties with $w = 0$, 2 and 30

7.3 Variable Selection via the H-Likelihood

7.3.1 Penalty Function for Variable Selection

From (7.2) and (7.4), we consider variable selection of the fixed effects β in various frailty models via maximization of a penalized partial h-likelihood $h_v(\beta, v, \alpha)$, given by

$$h_v(\beta, v, \alpha) = h_p - p_{\gamma,w}(\beta).$$

To implement h_v above, following Fan and Li (2001) and Ha et al. (2014b), we consider a penalty function $J(\cdot)$ such that $p(\cdot) = nJ(\cdot)$. In this book, we thus use the following penalized partial h-likelihood,

$$h_v(\beta, v, \alpha) = h_p - n \sum_{j=1}^{p} J_{\gamma,w}(|\beta_j|), \tag{7.5}$$

where $nJ_{\gamma,w}(\cdot) = p_{\gamma,w}(\cdot)$ is a penalty function that controls model complexity using the tuning parameters γ and w such that no penalty at $\gamma = 0$.

Here the true model is h_p with β fixed and from (7.4) $\exp\{-nJ_{\gamma,w}(|\beta_j|)\}$ is viewed as the frailty distribution of β_j. The maximization of h_v implies that the maximum penalized partial h-likelihood estimators (MPPHLEs) for β are used. Thus, even though β_j's are the fixed unknowns, the use of the penalized partial h-likelihood h_v implies that the MPPHLEs for β lead to variable selection. The use of the random-effect estimators for the fixed effects is beneficial not only to prediction but also to variable selection.

Typically, setting $\gamma = 0$ results in the standard frailty model with the partial h-likelihood h_p, whereas the regression coefficient estimates $\hat{\beta}$ tend to 0 as $\gamma \to \infty$. That is, a larger value of γ tends to choose a simple model, whereas a smaller value of γ inclines to a complex model. As w increases, the sparsity increases.

In variable selection, the true model is h_p with β fixed. Thus, the penalty term $\exp\{-nJ_{\gamma,w}(|\beta_j|)\}$ is not in the model, so that h_v is called the penalized partial h-likelihood and γ and w are the tuning parameters. In this chapter, we consider the following three penalty functions, LASSO, SCAD, and HL with appropriate tuning parameters.

- LASSO (Tibshirani 1996):

$$p_\gamma(|\beta|) = nJ_\gamma(|\beta|) = \gamma|\beta|, \tag{7.6}$$

- SCAD (Fan and Li 2001):

$$p'_{\gamma,w}(|\beta|) = nJ'_{\gamma,w}(|\beta|) = \gamma I(|\beta| \leq \gamma) + \frac{(w\gamma - |\beta|)_+}{w-1} I(|\beta| > \gamma), \quad (7.7)$$

where x_+ denotes the positive part of x; i.e., x_+ is x if $x > 0$, zero otherwise.

- HL (Lee and Oh 2014):

$$p_{\gamma,w}(|\beta|) = nJ_{\gamma,w}(|\beta|) = \frac{\gamma\beta^2}{2a(|\beta|)} + \frac{(w-2)\log a(|\beta|)}{2w} + \frac{a(|\beta|)}{w}, \quad (7.8)$$

where $a(|\beta|) = [\{8w\gamma\beta^2 + (2-w)^2\}^{1/2} + (2-w)]/4$.

The LASSO is the most common penalty as the L_1 penalty and has been known to give a good prediction. However, the LASSO has been criticized on the grounds that it typically ends up selecting a model with too many variables to prevent over-shrinkage of the regression coefficients (Radchenko and James 2008); otherwise, the regression coefficients for selected variables are often overshrunk. To improve the LASSO, Fan and Li (2001) proposed the SCAD, which also has two tuning parameters. For the SCAD, they proposed using $w = 3.7$. Fan and Li (2001) showed that the SCAD satisfies the oracle property that asymptotically selects the correct subset model and estimates the true non-zero coefficients in the linear models, simultaneously. The HL satisfies the oracle property and gives shrinkage estimators when $w > 2$ (Kwon et al. 2016). Lee and Oh (2014) proposed using $w = 30$ and showed by extensive simulation studies the results were not much sensitive to the choice of w. They also showed that the HL outperforms the LASSO and SCAD. For a better performance, in this book we choose both tuning parameters, (γ, w). For computational efficiency, we consider only a few values of w, e.g., $w = 2.1, 3, 10, 30, 50$ representing small, medium, and large values of w.

Recently, Ng et al. (2016) showed that in change point problems, the HL estimator provides consistent estimation of the number of change points, their locations and sizes of changes, while the LASSO and SCAD cannot. Ha et al. (2014b) have shown via simulation studies that in the frailty models, the HL has higher probability of choosing the true model than the LASSO and SCAD methods without losing the prediction accuracy. Lee et al. (2017b) showed that the HL gives the transparent estimation procedures when the regression coefficients have hierarchies with various constraints; for example, we can select an interaction term only when corresponding two main effects are present.

7.3.2 Penalized Partial H-Likelihood Procedure

For simplicity, we consider a simple frailty model (4.1) with a frailty parameter α. However, our results can be straightforwardly extended to general frailty models including multicomponents and competing-risks models. The variable-selection procedure based on the penalized partial h-likelihood h_v in (7.5) is as follows:

- **Estimation of** (β, v): Given the estimate of α, the MPPHLEs of (β, v) are obtained by solving the joint estimating equations of β and v, $\partial h_v/\partial(\beta, v) = 0$. In Appendix 7.7.1, we show that the ILS equations for $\tau = (\beta^T, v^T)^T$ can be explicitly expressed as an extended form of (4.12):

$$\begin{pmatrix} X^T W^* X + n\Sigma_{\gamma,w} & X^T W^* Z \\ Z^T W^* X & Z^T W^* Z + Q \end{pmatrix} \begin{pmatrix} \widehat{\beta} \\ \widehat{v} \end{pmatrix} = \begin{pmatrix} X^T w^* \\ Z^T w^* + R \end{pmatrix}, \qquad (7.9)$$

where $w^* = W^*\eta + (\delta - \mu)$ with $\mu = \exp(\log \Lambda_0 + \eta)$ and $R = Qv + (\partial \ell_2/\partial v)$ ($R = 0$ if the log-frailty $v \sim N(0, \alpha I_q)$), and $\Sigma_{\gamma,w} = \text{diag}\{J'_{\gamma,w}(|\beta_j|)/|\beta_j|\}$, i.e. $n\Sigma_{\gamma,w} = \text{diag}\{p'_{\gamma,w}(|\beta_j|)/|\beta_j|\}$. The equations above are also simply expressed as an extended form of (4.13):

$$(\mathbf{P}^T \mathbf{V} \mathbf{P} + n\Sigma_{\gamma,w})\hat{\tau} = \mathbf{P}^T \mathbf{y}_0^*.$$

- For the Cox PH model without frailty, they reduce to

$$(X^T W^* X + n\Sigma_{\gamma,w})\widehat{\beta} = X^T w^*. \qquad (7.10)$$

For the Cox model, Tibshirani (1997) developed the LASSO procedure for $w = 2$. We thus see that the ILS Eq. (7.9) extends the LASSO procedure under the Cox model to the frailty models.

- **Estimation of** α: For estimation of the frailty parameter α, we use a penalized partial restricted h-likelihood,

$$p_\tau(h_v) = \left[h_v - \frac{1}{2} \log \det \left\{ \frac{H(h_v; \tau)}{2\pi} \right\} \right] \Bigg|_{\tau=\hat{\tau}}, \qquad (7.11)$$

where $\hat{\tau} = \hat{\tau}(\alpha) = (\hat{\beta}^T(\alpha), v^T(\alpha))^T$. The estimate of α is obtained by solving the score equation $\partial p_\tau(h_v)/\partial\alpha = 0$.

- **Standard Error Formula:** An approximated standard error (SE) of $\hat{\beta}$ is obtained from a sandwich formula based on h_v:

$$\text{cov}(\hat{\beta}) = (H_{\beta\beta} + n\Sigma_{\gamma,w})^{-1} H_{\beta\beta} (H_{\beta\beta} + n\Sigma_{\gamma,w})^{-1}, \qquad (7.12)$$

where $H_{\beta\beta} = \{(X^T W^* X) - (X^T W^* Z)(Z^T W^* Z + Q)^{-1}(Z^T W^* X)\}|_{v=\hat{v}}$. The derivation of (7.12) is given in Appendix 7.7.2.

- **Tuning Parameter Selection:** For the choice of tuning parameters γ and w, Ha et al. (2014b) used a BIC-type criterion based on the penalized partial h-likelihood, defined by

$$\text{BIC}(\gamma, w) = -2p_v(h_p) + e(\gamma, w)\log(n), \tag{7.13}$$

where $p_v(h_p)$ is the first-order Laplace approximation to the marginal partial likelihood m_p in (4.8) and $e(\gamma, w) = \text{tr}[\{H_{\beta\beta} + n\Sigma_{\gamma,w}\}^{-1}H_{\beta\beta}]$ is the effective number of parameters (Lee and Nelder 1996; Ha et al. 2007a).

In summary, the variable-selection procedure above is easily implemented for survival data via a slight modification to the existing partial h-likelihood procedures because the penalty can be viewed as another frailty distribution for β.

An outline of variable-selection algorithm can be described as follows.

1. In the inner loop, we maximize h_v for $\tau = (\beta^T, v^T)$ (i.e., we solve (7.9)) and $p_\tau(h_v)$ in (7.11) for α, respectively.
2. In the outer loop, we find (γ, w) that minimizes BIC(γ, w) in (7.13).
3. At convergence, we compute the estimates of the standard errors for $\hat{\beta}$ using (7.12).

Remark 7.1 (i) To avoid a numerical difficulty when $\hat{\beta}_j = 0$ in solving (7.9), we employ $\Sigma_{\gamma,w,\epsilon} = \text{diag}\{J'_{\gamma,w}(|\beta_j|)/(|\beta_j| + \epsilon)\}$ for a small positive value of ϵ, say, $\epsilon = 10^{-8}$, instead of $\Sigma_{\gamma,w}$. Then $\Sigma_{\gamma,w,\epsilon}$ is always defined and with a small ϵ the diagonal elements of $\Sigma_{\gamma,w,\epsilon}$ are very close to those of $\Sigma_{\gamma,w}$. In fact, this algorithm extends Hunter and Li (2005) algorithm for improvement of the local quadratic approximation (LQA, Fan and Li 2001) to survival data. Here, we report $\hat{\beta} = 0$ if all five printed decimals are zero. We use a LASSO solution as the initial value for the SCAD and HL penalties.
(ii) To choose tuning parameters (γ, w), the generalized cross validation (GCV) statistic has been extensively used (Tibshirani 1997; Fan and Li 2001, 2002; Androulakis et al. 2012). However, Wang et al. (2007) showed that the GCV approach cannot select the tuning parameters satisfactorily, with a nonignorable overfitting effect in the resulting model (Fan and Lv 2010; Zhang 2010). In the spirit of Wang et al. (2007) and Ha et al. (2014b), in this book we use the BIC-type criterion in (7.13).
(iii) In the variable-selection procedure, we use the log-normal frailty distribution because it is useful for modelling correlated or multicomponent frailties in Chap. 5. For the gamma frailty distribution, we use the second-order approximation $s_v(h_v)$, where the marginal likelihood is also available: see Appendix 7.7.3.

7.4 Examples

Example 7.1 (*Kidney Infection Data: Univariate Frailty Models*) We first consider five covariates in the kidney infection data: Age, Sex (1=female, 0=male), and three indicator variables for GN, AN, and PKD which are different types of kidney diseases. Here, only Age is standardized as other covariates are binary. We fitted the univariate log-normal model (4.1) using the penalized partial h-likelihood procedures. The variable-selection procedures (LASSO, SCAD, and HL) were developed by creating a new R function, `frailty.vs()`, in the **frailtyHL** package. The R codes and results are presented below.

```
########################## LASSO #################################
> data(kidney, package="frailtyHL")
> attach(kidney)
> kidney$age<-(age-mean(age))/sd(age)
> kidney$GN<-as.numeric(disease=="GN")
> kidney$AN<-as.numeric(disease=="AN")
> kidney$PKD<-as.numeric(disease=="PKD")
> detach(kidney)
>
> la_result<-frailty.vs(Surv(time,status)~age+sex+GN+AN+PKD+(1|id),
+     model="lognorm", penalty="lasso",data=kidney,
+     B=c(0.074,-1.659, 0.173, 0.387, -1.161),tun1=seq(0,0.1,0.001))
[1] "Result of variable selection in frailty model"
[1]"Fitted model: log-normal"
[1]"penalty : lasso"
[1]"formula :"
Surv(time, status) ~ age + sex + GN + AN + PKD + (1 | id)
[1] "converge"
[1]"Fixed coefficients"
    Estimate Std. Error
age  0.00000    0.00000
sex -0.93163    0.28093
GN   0.00000    0.00000
AN   0.06912    0.06215
PKD -0.14137    0.08677
[1] "Dispersion parameter"
[1] 0.32288
[1] "Tuning parameter"
[1] 0.032
```

The LASSO procedure is implemented by specifying both initial values **`B`** and tuning parameter **`tun1`**$(=\gamma/n)$. Here the initial values are estimates from the log-normal frailty model without penalty, which are directly obtained (not shown) by specifying **`B=c(0,0,0,0,0)`** and **`tun1=0`** in the codes for the LASSO above.

The SCAD and HL procedures are similarly implemented using the LASSO solutions as initial values. In particular, the HL procedure requires specification of two tuning parameters, **`tun1`**$(=w)$ and **`tun2`**$(=\gamma)$.

```
########################## SCAD ################################
> sc_result<-frailty.vs(Surv(time,status)~age+sex+GN+AN+PKD+(1|id),
+     model="lognorm",penalty="scad",data=kidney,
+     B=c(0,-0.932,0,0.069,-0.141),tun1=seq(0,0.1,0.001))
[1] "Fixed coefficients"
    Estimate Std. Error
age  0.00000    0.00000
sex -1.55713    0.40993
GN   0.00000    0.00000
AN   0.00000    0.00000
PKD -1.30923    0.64930
[1] "Dispersion parameter"
[1] 0.25669
[1] "Tuning parameter"
[1] 0.075
```

```
############################ HL #################################
> hl_result<-frailty.vs(Surv(time,status)~age+sex+GN+AN+PKD+(1|id),
+     model="lognorm",penalty="hl",data=kidney,
+     B=c(0,-0.932,0,0.069,-0.141),tun1=c(2.1,3,10,30,50),
+     tun2=seq(0.001,0.25, 0.001))
[1] "Fixed coefficients"
    Estimate Std. Error
age  0.00000    0.00000
sex -1.00910    0.29822
GN   0.00000    0.00000
AN   0.00000    0.00000
PKD -0.29155    0.16719
[1] "Dispersion parameter"
[1] 0.31095
[1] "Tuning parameter"
[1] 2.100 0.224
```

The output shows that the selected values of the tuning parameters by the BIC in (7.13) were $\gamma = 0.032$, $\gamma = 0.078$, and $(\gamma, w) = (0.224, 2.1)$ for the LASSO, SCAD, and HL, respectively. The estimates of the frailty parameter σ_0^2 for no-penalty (standard frailty model), LASSO, SCAD, and HL are 0.418, 0.323, 0.257, and 0.311, respectively. The estimated regression coefficients and their standard errors are summarized under Case 1 in Table 7.1. It is known that the LASSO selects many covariates with excessive shrinkage in nonzero regression coefficients. The covariate Sex is significant in all of the four methods. The LASSO chooses three covariates (Sex, AN, and PKD) out of the five covariates, whereas the SCAD and HL choose two covariates (Sex and PKD). Note that the LASSO selects one more covariate, AN, which is not significant under no-penalty. The LASSO shrinks most, while the SCAD shrinks least. The nonzero estimates $(\hat{\beta}_2, \hat{\beta}_5)$ by the SCAD are similar to the corresponding estimates without penalty $(\gamma = 0)$. Thus, the SCAD is the least shrinkage estimator, while the HL gives the shrinkage (frailty) estimator, which is beneficial in prediction. Through extensive simulation studies, it has been found that the HL gives consistent variable selection without losing prediction accuracy in finite samples (Lee et al. 2017b).

Table 7.1 Variable selection for kidney infection data: estimated coefficients (SEs) for the univariate frailty model

Variable	No-penalty	LASSO	SCAD	HL
Case 1				
Age	0.074 (0.211)	0 (0)	0 (0)	0 (0)
Sex	−1.659 (0.447)	−0.932 (0.281)	−1.557 (0.410)	−1.009 (0.298)
GN	0.173 (0.520)	0 (0)	0 (0)	0 (0)
AN	0.387 (0.521)	0.069 (0.062)	0 (0)	0 (0)
PKD	−1.161 (0.793)	−0.141 (0.087)	−1.309 (0.649)	−0.292 (0.167)
Case 2				
Age	0.091 (0.209)	0.039 (0.111)	0 (0)	0 (0)
Sex	−2.689 (0.694)	−1.787 (0.360)	−1.913 (0.384)	−1.709 (0.345)
GN	−0.396 (0.868)	0.001 (0.004)	0 (0)	0 (0)
AN	−0.477 (0.948)	0.043 (0.086)	0 (0)	0 (0)
PKD	−3.433 (1.136)	−2.064 (0.570)	−2.840 (0.867)	−2.276 (0.641)
Sex*GN	0.675 (0.978)	0 (0)	0 (0)	0 (0)
Sex*AN	1.173 (1.000)	0.361 (0.288)	0 (0)	0 (0)
Sex*PKD	4.330 (1.361)	2.496 (0.774)	3.465 (1.110)	2.742 (0.863)

HL, h-likelihood penalty function

We also fitted the model with additional interaction terms between Sex and three kidney disease types. The estimated results are shown under Case 2 in Table 7.1. As expected, the LASSO still selects more variables while the SCAD and HL choose the three variables, Sex, PKD, and Sex*PKD. The HL shrinks more than the SCAD does.

Example 7.2 (*Bladder Cancer Data: Univariate Frailty Models*) Consider the bladder cancer data again, described in Sect. 6.5.2. We consider following covariates:

- main treatment; CHEMO (0="no", 1="yes")
- Age (0="$\leq$65 years", 1="$>$ 65 years")
- Sex (0="male", 1="female")
- prior recurrent rate; PRIORREC (0="primary", 1="$\leq$ 1/yr", 2="$>$ 1/yr")
- number of tumors; NOTUM (0="single", 1="2-7 tumors", 2="$\geq$ 8 tumors")
- tumor size; TUM3CM (0="$<$3cm", 1="$\geq$3cm")
- T category; TLOCC (0="Ta", 1="T1")
- carcinoma in situ; CIS (0="no", 1="yes")
- G grade; GLOCAL (0="G1", 1="G2", 2="G3")

For covariates with three categories (PRIORREC, NOTUM, and GLOCAL), we generated two indicator covariates. For example, with the variable PRIORREC, we

Table 7.2 Variable selection for bladder cancer data: estimated coefficients (SEs) for the univariate frailty model

Variable	No-penalty	LASSO	SCAD	HL
x_1: CHEMO	−0.879 (0.188)	−0.598 (0.142)	−0.875 (0.182)	−0.731 (0.162)
x_2: Age	−0.264 (0.147)	−0.128 (0.078)	0 (0)	0 (0)
x_3: Sex	0.005 (0.210)	0 (0)	0 (0)	0 (0)
x_4: PRIORREC1	0.311 (0.252)	0 (0)	0 (0)	0 (0)
x_5: PRIORREC2	0.549 (0.201)	0.346 (0.120)	0.440 (0.179)	0.355 (0.134)
x_6: NOTUM1	0.700 (0.168)	0.463 (0.118)	0.688 (0.164)	0.553 (0.137)
x_7: NOTUM2	1.230 (0.285)	0.700 (0.172)	1.230 (0.272)	0.944 (0.222)
x_8: TUM3CM	0.155 (0.176)	0.007 (0.007)	0 (0)	0 (0)
x_9: TLOCC	0.198 (0.175)	0.143 (0.082)	0 (0)	0 (0)
x_{10}: CIS	0.260 (0.280)	0 (0)	0 (0)	0 (0)
x_{11}: GLOCAL1	0.532 (0.166)	0.280 (0.104)	0.549 (0.159)	0.391 (0.126)
x_{12}: GLOCAL2	0.845 (0.275)	0.328 (0.122)	0.954 (0.251)	0.648 (0.196)

coded PRIORREC1 = I(PRIORREC=1) and PRIORREC2 = I(PRIORREC=2). Similarly, with the variables NOTUM and GLOCAL, we have used the respective indicators (NOTUM1, NOTUM2) and (GLOCAL1, GLOCAL2). Thus, total 12 covariates were included in the model. Here, patients with missing covariates were excluded, so that the remaining 396 patients from 21 centers were included; 196 patients (49.5%) among 396 patients were censored since for simplicity, we considered Type 1 event as the main event and Type 2 event as censoring.

We fitted the univariate log-normal model (4.1) as in Example 7.1. The estimates of the frailty parameter σ_0^2 for the no-penalty, LASSO, SCAD, and HL are 0.112, 0.070, 0.108, and 0.088, respectively. The estimated regression coefficients and their standard errors (SEs) are shown in Table 7.2. The main covariate, CHEMO ($\{x_1\}$) is significant by all of the four methods. The LASSO chooses nine covariates $\{x_1, x_2, x_5, x_6, x_7, x_8, x_9, x_{11}, x_{12}\}$ out of 12 covariates, whereas both the SCAD and HL choose six covariates $\{x_1, x_5, x_6, x_7, x_{11}, x_{12}\}$. Between the SCAD and HL, the SCAD shrinks less. Among nine LASSO covariates $\{x_1, x_2, x_5, x_6, x_7, x_8, x_9, x_{11}, x_{12}\}$, three of them $\{x_2, x_8$ and $x_9\}$, not selected by the SCAD and HL methods, are not significant under the standard frailty model (no-penalty).

Example 7.3 (*The CGD Data: Multicomponent Frailty Models*) In the CGD data, survival times from a given patient or those from a given hospital is likely to be correlated. Thus, we may use the multicomponent log-normal frailty models. For an illustration, we model the recurrent infection times T_{ijk}, with the eight covariates $x_{ijk} = (x_{ijk1}, \ldots, x_{ijk8})^T$, without hospital region x_{ijk9} and longitudinal covariate x_{ijk10}, because with those two covariates included, the hospital frailty variance estimate becomes zero. Here, three covariates (age x_{ijk3}, height x_{ijk4}, and weight x_{ijk5}) are standardized as other covariates are binary.

Table 7.3 Variable selection: estimated coefficients (SEs) for multilevel frailty model with the CGD data

Variable	No-penalty	LASSO	SCAD	HL
Gamma-IFN	−1.093 (0.357)	−0.760 (0.228)	−1.074 (0.335)	−0.898 (0.276)
Inheritance	−0.576 (0.409)	0 (0)	0 (0)	0 (0)
Age	−0.904 (0.446)	−0.201 (0.115)	0 (0)	0 (0)
Height	0.200 (0.462)	0 (0)	0 (0)	0 (0)
Weight	0.336 (0.480)	0 (0)	0 (0)	0 (0)
Corticosteroids	1.756 (0.941)	0 (0)	0 (0)	0 (0)
Prophylactic	−0.591 (0.480)	0 (0)	0 (0)	0 (0)
Sex	−0.630 (0.566)	0 (0)	0 (0)	0 (0)

We have $(\hat{\sigma}_h^2, \hat{\sigma}_p^2) = (0.027, 1.028)$ under the multicomponent (multilevel) frailty model (5.4) without a penalty. The two estimates $(\hat{\sigma}_h^2, \hat{\sigma}_p^2)$ under the LASSO, SCAD, and HL are (0.000, 0.929), (0.026, 0.982), and (0.020, 0.964), respectively. Table 7.3 shows that the LASSO chooses two covariates (Gamma-IFN, Age) out of eight covariates, whereas the SCAD and HL choose only one covariate (Gamma-IFN). Again the SCAD shrinks less than HL.

The R codes and outputs from the variable-selection procedures for the CGD data via the multilevel model are presented below.

```
> data(cgd, package="frailtyHL")
> cgd$age<-(cgd$age-mean(cgd$age))/sd(cgd$age)
> cgd$height<-(cgd$height-mean(cgd$height))/sd(cgd$height)
> cgd$weight<-(cgd$weight-mean(cgd$weight))/sd(cgd$weight)
############################  No penalty #########################
> no_penalty<-frailty.vs(Surv(tstop-tstart,status)~treat+inherit+age
+ +height+weight+steroids+propylac+sex+(1|center)+(1|id), model=
+"lognorm",penalty="lasso",data=cgd,B=c(0,0,0,0,0,0,0,0),tun1=c(0))
[1]"Result of variable selection in frailty model"
[1]"==Fitted model=="
[1]"model : lognorm"
[1]"penalty : lasso"
[1]"formula :"
Surv(tstop - tstart, status) ~ treat + inherit + age + height +
    weight + steroids + propylac + sex + (1 | center) + (1 |
    id)
[1]"converge"
[1]"==Fixed coefficients=="
                 Estimate Std. Error
treatrIFN-g      -1.09271    0.35662
inheritautosomal  0.57630    0.40907
age              -0.90405    0.44644
height            0.19966    0.46220
weight            0.33614    0.47960
steroids          1.75590    0.94130
propylac         -0.59125    0.47954
```

```
sexfemale        -0.63003    0.56606
[1]"==Dispersion parameter=="
[1] 0.02656 1.02808
[1]"==Tuning parameter=="
[1] 0
#################################  LASSO ##########################
> la_result<-frailty.vs(Surv(tstop-tstart,status)~treat+inherit+age
+  +height+weight+steroids+ propylac+sex +(1|center)+(1|id),model=
+"lognorm",penalty="lasso",data=cgd,B=c(-1.093,-0.576,-0.904,0.200,
+  0.336,1.756,-0.591,-0.630), tun1=seq(0,0.02,0.001))
[1]"Result of variable selection in frailty model"
[1]"==Fitted model=="
[1]"model : lognorm"
[1]"penalty : lasso"
[1]"formula :"
Surv(tstop - tstart, status) ~ treat + inherit + age + height
+    +weight + steroids + propylac + sex + (1 | center) + (1 |id)
[1]"converge"
[1]"==Fixed coefficients=="
                 Estimate Std. Error
treatrIFN-g      -0.75964    0.22759
inheritautosomal  0.00000    0.00000
age              -0.20053    0.11480
height            0.00000    0.00000
weight            0.00000    0.00000
steroids          0.00000    0.00000
propylac         -0.00002    0.00001
sexfemale         0.00000    0.00000
[1]"==Dispersion parameter=="
[1] 0.00000 0.92942
[1]"==Tuning parameter=="
[1] 0.015

#################################  SCAD ##########################
> sc_result<-frailty.vs(Surv(tstop-tstart,status)~treat+inherit+age
+  +height+weight+ steroids+ propylac+ sex + (1|center)+(1|id),model=
+ "lognorm",penalty="scad",data=cgd,B=c(-0.760,0,-0.201,0,0,0,0,0),
+  tun1=seq(0,0.1,0.001))
[1]"Result of variable selection in frailty model"
[1]"==Fitted model=="
[1]"model : lognorm"
[1]"penalty : scad"
[1]"formula :"
Surv(tstop - tstart, status) ~ treat + inherit + age + height
+    +weight + steroids + propylac + sex + (1 | center) + (1 |id)
[1]"converge"
[1]"==Fixed coefficients=="
                 Estimate Std. Error
treatrIFN-g      -1.07394    0.33532
inheritautosomal  0.00000    0.00000
age               0.00000    0.00000
height            0.00000    0.00000
weight            0.00000    0.00000
steroids          0.00000    0.00000
propylac          0.00000    0.00000
sexfemale         0.00000    0.00000
```

```
[1]"==Dispersion parameter=="
[1] 0.02620 0.98167
[1]"==Tuning parameter=="
[1] 0.066

################################ HL #############################
> hl_result<-frailty.vs(Surv(tstop-tstart,status)~treat+inherit+age
+ +height+weight+steroids+ propylac+sex +(1|center)+(1|id),model=
+"lognorm",penalty="hl",data=cgd,B=c(-0.760,0,-0.201,0,0,0,0,0),
+ tun1=c(2.1,3,10,30,50), tun2=seq(0.001,0.06,0.01))
[1]"Result of variable selection in frailty model"
[1]"==Fitted model=="
[1]"model : lognorm"
[1]"penalty : hl"
[1]"formula :"
Surv(tstop - tstart, status) ~ treat + inherit + age + height
+    +weight + steroids + propylac + sex + (1 | center) + (1 |id)
[1]"converge"
[1]"==Fixed coefficients=="
                  Estimate Std. Error
treatrIFN-g       -0.89992     0.2765
inheritautosomal   0.00000     0.0000
age                0.00000     0.0000
height             0.00000     0.0000
weight             0.00000     0.0000
steroids           0.00000     0.0000
propylac           0.00000     0.0000
sexfemale          0.00000     0.0000
[1]"==Dispersion parameter=="
[1] 0.01953 0.96393
[1]"==Tuning parameter=="
[1] 50.000  0.051
[1]"==BIC=="
[1] 697.7575
```

Example 7.4 (*Multicenter Lung Cancer Data: Correlated Frailty Models*) Consider the multicenter lung cancer data with the following five dichotomous covariates: treatment (x_{ij1}, Trt, is 0 for CAV and 1 for CAV-HEM), presence (1) or absence (0) of bone metastases (x_{ij2}, Bone), presence (1) or absence (0) of liver metastases (x_{ij3}, Liver), whether the subject was ambulatory ($x_{ij4} = 1$ if PS=1) or confined to bed or chair ($x_{ij4} = 0$ if PS=0), and whether there was weight loss prior to entry (x_{ij5}, WL).

Let v_{i1} and v_{i2} be random treatment and random bone metastases effects, respectively. We consider the following correlated frailty model (5.7) with v_{i1} and v_{i2}:

$$\eta_{ij} = (\beta_1 + v_{i1})x_{ij1} + (\beta_2 + v_{i2})x_{ij2} + \beta_3 x_{ij3} + \beta_4 x_{ij4} + \beta_5 x_{ij5},$$

where a correlation ρ is assumed between v_{i1} and v_{i2}.

Table 7.4 Variable selection: estimated coefficients (SEs) for the correlated frailty model with lung cancer data

Variable	No-penalty	LASSO	SCAD	HL
Case 1				
Trt	−0.233 (0.099)	−0.103 (0.048)	0 (0)	0 (0)
Bone	0.261 (0.127)	0.064 (0.032)	0 (0)	0 (0)
Liver	0.391 (0.093)	0.315 (0.071)	0.446 (0.091)	0.349 (0.077)
PS	−0.652 (0.108)	−0.527 (0.088)	−0.689 (0.108)	−0.584 (0.096)
WL	0.210 (0.090)	0.138 (0.055)	0 (0)	0.130 (0.053)
Case 2				
Trt	−0.116 (0.263)	0 (0)	0 (0)	0 (0)
Bone	0.372 (0.158)	0.083 (0.038)	0 (0)	0 (0)
Liver	1.005 (0.273)	0.299 (0.069)	0.432 (0.091)	0.325 (0.073)
PS	−0.100 (0.220)	−0.417 (0.083)	−0.552 (0.119)	−0.458 (0.090)
WL	0.687 (0.236)	0.130 (0.052)	0 (0)	0.112 (0.046)
Trt*Bone	−0.216 (0.196)	0 (0)	0 (0)	0 (0)
Trt*Liver	0.179 (0.187)	0 (0)	0 (0)	0 (0)
Trt*PS	−0.208 (0.225)	−0.231 (0.067)	−0.304 (0.112)	−0.216 (0.067)
Trt*WL	−0.028 (0.181)	0 (0)	0 (0)	0 (0)
Liver*PS	−0.595 (0.227)	0 (0)	0 (0)	0 (0)
Liver*WL	−0.377 (0.190)	0 (0)	0 (0)	0 (0)
PS*WL	−0.399 (0.237)	0 (0)	0 (0)	0 (0)

Trt, treatment; PS, ambulatory performance status; WL, weight loss

The frailty-parameter estimates $(\hat{\sigma_1^2}, \hat{\sigma_2^2}, \hat{\rho})$ for the LASSO, SCAD, and HL were (0.066, 0.181, −0.217), (0.093, 0.206, −0.401), and (0.095, 0.212, −0.389), respectively. The estimated coefficients and their SEs are reported for the main effects-only model, i.e., Case 1 in Table 7.4. The LASSO chooses all five covariates. Gray (1994) and Vaida and Xu (2000) have shown that there are substantial variations in the treatment (x_1) effect over centers, implying that x_1 may not be significant. The SCAD selects two covariates without x_1, while the HL selects three covariates without x_1.

The random-effect estimates and their 95% Wald intervals for each institution are plotted in Fig. 7.2 [(a) random treatment effects and (b) random bone effects] from the HL variable- selection procedure under the correlated model above. Figure 7.2b shows substantial institutional variation in the bone effects among institutions and the Wald intervals for the two institutions (22, 29) do not include zero.

Since the covariate of main interest in this study is Trt and the HL selects the three variables (Liver, PS, and WL), we considered all two-way interaction terms between these four caovariates and an additional interaction Trt∗Bone. The results are presented under Case 2 in Table 7.4. All three methods (LASSO, SCAD, and HL) select a new interaction Trt∗PS. The LASSO still selects more variables, and the HL selects an additional covariate WL compared with the SCAD. Here the interaction term Trt∗PS is selected even though the main effect Trt is not selected. It is often

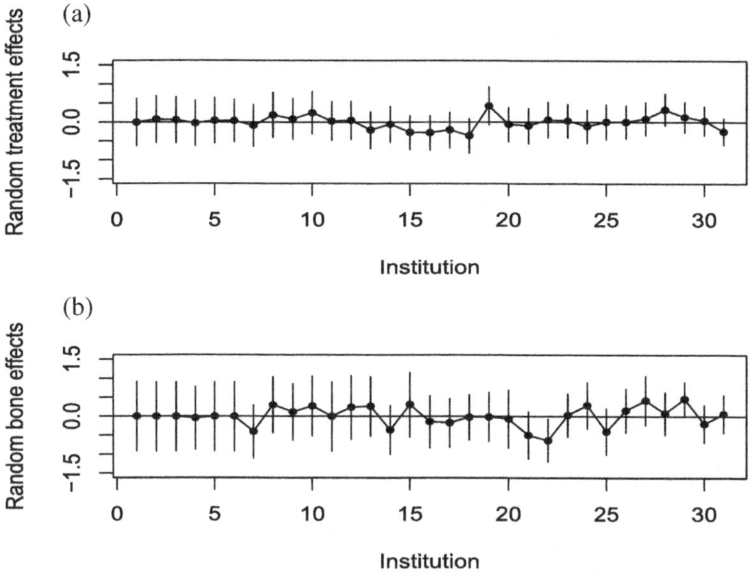

Fig. 7.2 Random effects and their 95% confidence intervals under the HL variable selection in the correlated frailty model allowing dependency between random treatment and random bone in the lung cancer data; **(a)** random treatment effects (v_1); **(b)** random bone effects (v_2); 31 institutions are sorted by the increasing order of number of patients

preferred to impose a strong hierarchy constraint that the presence of an interaction term requires including both of the corresponding main effects in the model. The HL can be easily modified to allow such a hierarchical constraint (Lee et al. 2017b).

Below are the R codes and outputs from the variable-selection procedure for the lung cancer data via the correlated model.

```
> lung.formula<-Surv(y,del)~treat + bone + liver + ps + wtlss
+   +(treat|center)+(bone|center)
################################# No-penalty #######################
> Nop_result<-frailty.vs(lung.formula,model="lognorm",penalty="lasso",
+ data=lung_d,B=c(-0.23,0.26,0.39,-0.65,0.21),alpha=c(0.03,0.03,0.01),
+ tun1=c(0))
[1]"Result of variable selection in frailty model"
[1]"==Fitted model=="
[1]"model : lognorm"
[1]"penalty : lasso"
[1]"formula :"
Surv(y, del) ~ treat +bone +liver +ps +wtlss +(treat|center) +
+ (bone | center)
[1]"converge"
[1]"==Fixed coefficients=="
      Estimate Std. Error
treat -0.23341    0.09937
bone   0.26074    0.12798
```

```
liver  0.39061    0.09284
ps    -0.65246    0.10755
wtlss  0.21044    0.08985
[1]"==Dispersion parameter=="
        [,1]
[1,] 0.05280
[2,] 0.13452
[3,] 0.01574
[1]"==Tuning parameter=="
[1] 0
[1]"==BIC=="
[1] 6118.688
################################# LASSO ###########################
> la_result<-frailty.vs(lung.formula,model="lognorm",penalty="lasso",
+data=lung_d,B=c(-0.23,0.26,0.39,-0.65,0.21),alpha=c(0.03,0.03,0.01),
+tun1=seq(0,0.03,0.001))
[1]"Result of variable selection in frailty model"
[1]"==Fitted model=="
[1]"model : lognorm"
[1]"penalty : lasso"
[1]"formula :"
Surv(y, del) ~ treat + bone + liver + ps + wtlss + (treat | center) +
    (bone | center)
[1]"converge"
[1]"==Fixed coefficients=="
      Estimate Std. Error
treat -0.10286    0.04822
bone   0.06414    0.03237
liver  0.31511    0.07079
ps    -0.52748    0.08799
wtlss  0.13847    0.05450
[1]"==Dispersion parameter=="
         [,1]
[1,]  0.06595
[2,]  0.18154
[3,] -0.02379
[1]"==Tuning parameter=="
[1] 0.019
[1]"==BIC=="
[1] 6112.012
################################# SCAD ############################
> sc_result<-frailty.vs(lung.formula,model="lognorm",penalty="scad",
+data=lung_d,B=c(-0.103,0.064,0.315,-0.527,0.138),alpha=c(0.03,0.03,
+ 0.01), tun1=seq(0,0.06,0.001))
[1]"Result of variable selection in frailty model"
[1]"==Fitted model=="
[1]"model : lognorm"
[1]"penalty : scad"
[1]"formula :"
Surv(y, del) ~ treat + bone + liver + ps + wtlss + (treat | center) +
    (bone | center)
[1]"converge"
[1]"==Fixed coefficients=="
      Estimate Std. Error
treat  0.00000    0.00000
bone   0.00000    0.00000
```

```
liver  0.44648    0.09146
ps    -0.68874    0.10788
wtlss  0.00000    0.00000
[1]"==Dispersion parameter=="
         [,1]
[1,]  0.09258
[2,]  0.20591
[3,] -0.05539
[1]"==Tuning parameter=="
[1] 0.049
[1]"==BIC=="
[1] 6112.563
################################ HL ################################
> hl_result<-frailty.vs(lung.formula,model="lognorm",penalty="hl",
+ data=lung_d,B=c(-0.23,0.26,0.39,-0.65,0.21),alpha=c(0.03,0.03,0.01),
+ tun1=c(2.1,3,10,30,50),tun2=seq(0.001,0.06,0.01))
[1]"Result of variable selection in frailty model"
[1]"==Fitted model=="
[1]"model : lognorm"
[1]"penalty : hl"
[1]"formula :"
Surv(y, del) ~ treat + bone + liver + ps + wtlss + (treat | center) +
    (bone | center)
[1]"converge"
[1]"==Fixed coefficients=="
      Estimate Std. Error
treat  0.00000    0.00000
bone   0.00000    0.00000
liver  0.34904    0.07730
ps    -0.58383    0.09558
wtlss  0.13024    0.05328
[1]"==Dispersion parameter=="
         [,1]
[1,]  0.09456
[2,]  0.21212
[3,] -0.05514
[1]"==Tuning parameter=="
[1] 5e+01 1e-03
[1]"==BIC=="
[1] 6111.287
```

7.5 Variable Selection for the Competing-Risks Frailty Models

The variable-selection procedures for the fixed effects β can be extended to the subhazard competing-risks frailty models using a penalized weighted partial h-likelihood, defined by

$$h_{vw}(\beta, v, \alpha) = h_{pw} - n \sum_{j=1}^{p} J_{\gamma,w}(|\beta_j|),$$

where h_{pw} is the weighted partial h-likelihood in (6.22). Below we present how to select relavent variables via h_{vw}, using two real-data examples from multicenter trials where patients within a center may have correlated outcomes.

Example 7.5 (*Bladder Cancer Data: Subhazard Univariate Frailty Models*) Consider the multicenter bladder cancer data with 396 patients again. In this section we focus on the following two competing events:
(i) time to first bladder recurrence (an event of interest; Type 1 event),
(ii) time to death prior to recurrence (competing event; Type 2 event).
Of 396 patients, 200 (50.51%) had recurrence of bladder cancer and 81 (20.45%) died prior to recurrence. 115 patients (29.04%) who were still alive without recurrence were censored at the date of the last available follow-up. The numbers of patients per center varied from 3 to 78, with the mean of 18.9 and the median of 14.

We fitted the subhazard univariate frailty models with 12 covariates ($x_1, x_2, \ldots, x_{12}$) in Example 7.2. The selected values of the tuning parameters γ by the BIC were $\gamma = 0.012$, 0.084, and $(\gamma, w) = (0.011, 50)$ for the LASSO, SCAD, and HL, respectively. The estimates of the frailty parameter σ_0^2 for no-penalty, LASSO, SCAD, and HL are 0.106, 0.072, 0.107, and 0.088, respectively. The estimated coefficients and their standard errors for bladder cancer recurrence (i.e., Type 1 event) are given in Table 7.5. The main covariate, CHEMO ($\{x_1\}$), is very significant in all of the four methods. The LASSO chooses nine covariates $\{x_1, x_2, x_5, x_6, x_7, x_8, x_9, x_{11}, x_{12}\}$ out of the twelve covariates, whereas the SCAD and HL choose six $\{x_1, x_5, x_6, x_7, x_{11}, x_{12}\}$ and seven $\{x_1, x_2, x_5, x_6, x_7, x_{11}, x_{12}\}$ covariates, respectively. We prefer the HL because extensive simulation studies show

Table 7.5 Variable selection: estimated coefficients (SEs) from the subhazard univariate frailty model for bladder cancer data

Variable	No-penalty	LASSO	SCAD	HL
x_1: CHEMO	−0.933 (0.187)	−0.666 (0.166)	−0.929 (0.182)	−0.785 (0.174)
x_2: Age	−0.343 (0.147)	−0.214 (0.120)	0 (0)	−0.218 (0.119)
x_3: Sex	0.058 (0.208)	0 (0)	0 (0)	0 (0)
x_4: PRIORREC1	0.276 (0.249)	0 (0)	0 (0)	0 (0)
x_5: PRIORREC2	0.514 (0.200)	0.327 (0.149)	0.395 (0.180)	0.294 (0.150)
x_6: NOTUM1	0.713 (0.168)	0.494 (0.139)	0.688 (0.164)	0.593 (0.150)
x_7: NOTUM2	1.307 (0.283)	0.816 (0.229)	1.293 (0.272)	1.051 (0.249)
x_8: TUM3CM	0.213 (0.175)	0.060 (0.094)	0 (0)	0 (0)
x_9: TLOCC	0.171 (0.173)	0.127 (0.115)	0 (0)	0 (0)
x_{10}: CIS	0.266 (0.278)	0 (0)	0 (0)	0 (0)
x_{11}: GLOCAL1	0.474 (0.165)	0.250 (0.126)	0.491 (0.159)	0.384 (0.137)
x_{12}: GLOCAL2	0.808 (0.274)	0.347 (0.189)	0.910 (0.250)	0.610 (0.222)

that it achieves selection consistency without losing prediction accuracy in the finite samples.

Example 7.6 (*Breast Cancer Data: Subhazard Correlated Frailty Models*) We again consider the multicenter breast cancer data, analyzed in Sect. 6.4.2. For simplicity, we consider only the subset of older 1763 patients (i.e., age $\geq$ 50) from the original data set. The number of patients per center varied from 1 to 114, with the mean of 11.8 and the median of 6. Type 1 is an event of interest (465 patients; 26.38%), Type 2 is a competing event (469 patients; 26.60%), and patients without events were censored at the last follow-up (1200 patients; 47.02%). We studied the following ten covariates on time to local or regional recurrence (Type 1 event): treatment group (GROUP; placebo, and tamoxifen), race (RACE; white, black, and other), menopausal status (MENSE; premenopausal, perimenopausal, and postmenopausal), number of nodes removed (RNOD), tumor size (TSIZE), estrogen receptor level (ER), progesterone receptor level (PR), and surgery type (SURGTYPE; lumpectomy and mastectomy). We created two indicator covariates for variables RACE and MENSE (Table 7.6). Four continuous covariates (RNOD, TSIZE, ER, and PR) are standardized while other six covariates are binary, a total of 10 covariates being included in the model.

We fitted the subhazard correlated frailty model (6.17) where the random center effect v_{i0} and random treatment effect v_{i1} are correlated. The fitted results are as follows. The selected values of the tuning parameters are γ = 0.004, 0.026, and

Table 7.6 Variable selection: estimated coefficients (SEs) in the correlated subhazard frailty model (Type 1 event) with breast cancer data

Variable	No-penalty	LASSO	SCAD	HL
x_1: GROUP	−0.617 (0.107)	−0.528 (0.097)	−0.610 (0.106)	−0.521 (0.097)
x_2: RACE1	−0.202 (0.267)	0 (0)	0 (0)	0 (0)
x_3: RACE2	−0.165 (0.340)	0 (0)	0 (0)	0 (0)
x_4: MENSE1	0.112 (0.222)	0 (0)	0 (0)	0 (0)
x_5: MENSE2	−0.158 (0.265)	0 (0)	0 (0)	0 (0)
x_6: RNOD	−0.139 (0.051)	−0.124 (0.046)	−0.139 (0.050)	−0.109 (0.044)
x_7: TSIZE	0.272 (0.041)	0.254 (0.039)	0.266 (0.040)	0.253 (0.039)
x_8: ER	0.077 (0.037)	0.069 (0.035)	0.022 (0.019)	0.068 (0.032)
x_9: PR	0.058 (0.045)	0.052 (0.040)	0 (0)	0 (0)
x_{10}: SURGTYPE	−0.089 (0.101)	0 (0)	0 (0)	0 (0)

GROUP, treatment group; Group=I(tamoxifen);
RACE, race (white, black, other); RACE1=I(RACE=white), RACE2=I(RACE=black);
MENSE, menopausal status (premenopausal, perimenopausal, postmenopausal);
MENSE1=I(MENSE=premenopausal), MENSE2=I(MENSE=perimenopausal);
RNOD, number of nodes removed; TSIZE, tumor size (mm);
ER, estrogen receptor level; PR, progesterone receptor level;
SURGTYPE, surgery type; SURGTYPE=I(mastectomy)

$(\gamma, w) = (0.001, 50)$ for the LASSO, SCAD, and HL, respectively. The frailty-parameter estimates ($\hat{\sigma}_0^2$, $\hat{\sigma}_1^2$, and $\hat{\rho}$) for the no-penalty, LASSO, SCAD, and HL are (0.297, 0.116, and − 0.988), (0.290, 0.101, and − 0.996), (0.289, 0.115, and − 0.992), and (0.294, 0.101, and −0.996), respectively. The estimated coefficients and their SEs for Type 1 events are reported in Table 7.6. Ha et al. (2014b) and Christian et al. (2016) have shown that the main treatment effect (GROUP; x_1) is significant, which is also confirmed by the three methods (LASSO, SCAD, and HL). Here the LASSO chooses five covariates $\{x_1, x_6, x_7, x_8, x_9\}$, but the SCAD and HL select four covariates $\{x_1, x_6, x_7, x_8\}$. Again the HL shrinks more than the SCAD.

7.6 Discussion

Using the penalized partial h-likelihood, we have shown how to implement the variable-selection procedures in the frailty models and the subhazard competing-risks models. Ha et al. (2014a, b) have demonstrated via numerical studies and data analyses that the HL method is preferable to the SCAD method because it identifies zero and nonzero coefficients better without losing prediction accuracy. In the h-likelihood approach, the variable selection is equivalent to assuming some frailty distribution for the fixed effects, so that it can be straightforwardly extended to the cause-specific frailty models and high dimensional cases (i.e. $p > n$). For advantages of the HL penalty in structured variable selection and multivariate analysis, see Lee et al. (2017b).

In this chapter, we have considered selecting individual variables only. In many regression problems, the covariates often possess a natural group structure. For example, categorical variables are often represented by a group of indicator variables. In these situations, the problem of selecting relevant variables is that of selecting groups rather than selecting individual variables (Yuan and Lin 2006; Huang et al. 2012). Extension of the group penalization methods such as group LASSO (Yuan and Lin 2006) or group HL (Lee et al. 2015a, 2017b) to the general frailty models would be an interesting future topic.

7.7 Appendix

7.7.1 Derivation of Score Equations (7.9) for Variable Selection

Given the frailty parameter α, the MPPHLEs of (β, v) are obtained by solving the estimating equations,

$$\frac{\partial h_v}{\partial \beta_j} = \frac{\partial h_p}{\partial \beta_j} - n \sum_{j=1}^{p} [J_{\gamma,w}(|\beta_j|)]' = 0 \tag{7.14}$$

and

$$\frac{\partial h_v}{\partial v} = \frac{\partial h_p}{\partial v} = 0. \tag{7.15}$$

Here, (7.14) is an adjusted estimating equation induced by adding the penalty term, whereas (7.15) is the same as the standard estimating equation without penalty. However, with the three penalty functions considered in (7.6)–(7.8), $J_{\gamma,w}()$ in (7.14) becomes non-differentiable at the origin and it does not have continuous second-order derivatives. These lead to a difficulty in solving (7.14). Thus, we use a LQA to such penalty functions. That is, given an initial value $\beta^{(0)}$ that is close to the true value of β, the penalty function $J_{\gamma,w}()$ can be locally approximated by a quadratic function as

$$[J_{\gamma,w}(|\beta_j|)]' = J'_{\gamma,w}(|\beta_j|)\text{sgn}(|\beta_j|) \approx \{J'_{\gamma,w}(|\beta_j^{(0)}|)/|\beta_j^{(0)}|\}\beta_j \text{ for } \beta_j \approx \beta_j^{(0)}.$$

Then, from (4.35), the negative Hessian matrix of β and v based on h_v can be explicitly written as a simple matrix form:

$$H(h_v; \beta, v) = -\frac{\partial^2 h_v}{\partial(\beta, v)^2} = \begin{pmatrix} X^T W^* X + n\Sigma_{\gamma,w} & X^T W^* Z \\ Z^T W^* X & Z^T W^* Z + Q \end{pmatrix}, \tag{7.16}$$

Following (4.37) and (7.16), we can show that given α, the MPPHL equations of (β, v) become (7.9). □

7.7.2 *Derivation of the Standard Error Formula (7.12)*

Now, we are interested in computing an approximated covariance estimate of $\hat{\beta}$. For this we consider a further penalized profile h-likelihood after eliminating v from h_v in (7.5), defined by

$$\hat{h}_v(\beta, \alpha) \equiv h_v|_{v=\hat{v}} = \hat{h}^* - n \sum_{j=1}^{p} J_{\gamma,w}(|\beta_j|), \tag{7.17}$$

where $\hat{h}^* = \hat{h}^*(\beta, \alpha) = h_p(\beta, \alpha, v)|_{v=\hat{v}}$. In the frailty models, the regression parameters β and frailty parameter α are asymptotically orthogonal (Lee and Nelder 1996; Ha and Lee 2003; Ha et al. 2011). Therefore, to estimate the covariance matrix of $\hat{\beta}$ only, we do not have to consider the information loss caused by estimating α. Thus, the SEs of $\hat{\beta}$ are obtained from the following sandwich formula (Fan and Li 2002; Cai et al. 2005) based on $\hat{h}_v$ in (7.17):

$$\text{cov}(\hat{\beta}) = H(\hat{h}_v; \beta)^{-1}\text{cov}(\partial \hat{h}_v/\partial \beta) H(\hat{h}_v; \beta)^{-1}, \quad (7.18)$$

where $H(\hat{h}_v; \beta) \equiv -\partial^2 \hat{h}_v/\partial \beta^2 = H_{\beta\beta} + n\Sigma_{\gamma,w}$. Here, $H_{\beta\beta} \equiv H(\hat{h}; \beta) \equiv -\partial^2 \hat{h}/\partial \beta^2$ is explicitly computed as follows:

$$\begin{aligned} H_{\beta\beta} &= \{(-\partial^2 h_p/\partial \beta^2) - (-\partial^2 h_p/\partial \beta \partial v)(-\partial^2 h_p/\partial v^2)(-\partial^2 h_p/\partial v \partial \beta)\}|_{v=\hat{v}} \\ &= \{(X^T W^* X) - (X^T W^* Z)(Z^T W^* Z + Q)^{-1}(Z^T W^* X)\}|_{v=\hat{v}}, \end{aligned}$$

since $\partial \hat{h}/\partial \beta = \{(\partial h_p/\partial \beta) + (\partial h_p/\partial v)(\partial \hat{v}/\partial \beta)\}|_{v=\hat{v}}$ (Ha and Lee 2003; Ha et al. 2010). Here, we use $H_{\beta\beta}$ to estimate $\text{cov}(\partial \hat{h}_v/\partial \beta)$. □

7.7.3 Variable Selection via the Penalized Marginal Likelihood

For the gamma frailty, the variable-selection procedure is straightforward as in the log-normal frailty. To estimate the frailty parameter α, which is the variance of the gamma frailty, Ha et al. (2014b) used the second-order approximation, $s_v(h_v) = p_v(h_v) - F(h)$, where $F(h) = \sum_{i=1}^{q}\{-2(\alpha^{-1} + \sum_j \delta_{ij})^{-1}\}$. The corresponding marginal likelihood procedure is also available because it has an explicit form. Ha et al. (2014b) found out via simulations that in the gamma frailty model, the h-likelihood and the marginal likelihood approaches based on the SCAD and HL provide similar results and perform well.

Now, we outline a penalized marginal likelihood method. Consider maximizing a penalized profile marginal likelihood m_v (Fan and Li 2002), defined by

$$m_v(\beta, \alpha) = \hat{m}(\beta, \alpha) - n \sum_{j=1}^{p} J_{\gamma,w}(|\beta_j|), \quad (7.19)$$

where $\hat{m}(\beta, \alpha) = m(\beta, \lambda_0, \alpha)|_{\lambda_0=\hat{\lambda}_0}$ and $\hat{\lambda}_0$ solves $\partial m/\partial \lambda_0 = 0$ (Andersen et al. 1997; Ha et al. 2001). The likelihood m_v in (7.19) requires m which often involves an intractable integration, except for the gamma frailty model. Notice that in the gamma frailty, the formula (3.10) of Fan and Li (2002) omitted an extra term $\log \Gamma(\theta^{-1} + \delta_{i+})$ in m_v, which was also pointed out by Androulakis et al. (2012). The standard errors of $\hat{\beta}$ corresponding to (7.19) are obtained from a sandwich formula:

$$\text{cov}(\hat{\beta}) = \{M_{\beta\beta} + n\Sigma_\gamma\}^{-1}\text{cov}(\partial \hat{m}/\partial \beta)\{M_{\beta\beta} + n\Sigma_\gamma\}^{-1},$$

where $M_{\beta\beta} = -\partial^2 \hat{m}/\partial \beta^2$. We use $M_{\beta\beta}$ to estimate $\text{cov}(\partial \hat{m}/\partial \beta)$. To compute $M_{\beta\beta}$ in gamma frailty model, we also use $H_{\beta\beta}$ in (7.18) since, given α, $M_{\beta\beta}$ is the same as $H_{\beta\beta}$ (Ha et al. 2001, 2010). To choose (γ, w), we use the BIC corresponding to (7.13), defined by

$$\text{BIC}(\gamma, w) = -2\hat{m}(\hat{\beta}, \hat{\alpha}) + e^*(\gamma, w)\log(n),$$

where $e^*(\gamma, w) = \text{tr}\{(M_{\beta\beta}+n\Sigma_{\gamma,w})^{-1}M_{\beta\beta}\}$ with $M_{\beta\beta} = -\partial^2\hat{m}/\partial\beta^2$ is the effective number of parameters (Fan and Li 2002; Ha et al. 2007a). Note here that Fan and Li (2002) and Androulakis et al. (2012) have used the GCV method for the optimal choice of γ, which can not select the tuning parameters satisfactorily (Wang et al. 2007; Fan and Lv 2010; Zhang et al. 2010). □

Chapter 8
Mixed-Effects Survival Models

The frailty model accounts for dependence between survival times, by including a random effect acting multiplicatively on the individual hazard rate. The linear mixed model (LMM) has been introduced as an alternative in which the random effect acts linearly on each individual survival time. Thus, the fixed effect describes the mean survival time. The accelerated failure-time (AFT) random-effect model is the LMM under the log-transformation of survival time, an extension of the AFT model in Sect. 2.4. Various methods to obtain the MLE have been developed, but they tend to be computationally intensive, e.g., the EM (Pettitt 1986), Monte Carlo EM (Hughes 1999) and Newton–Raphson method (Klein et al. 1999). In this chapter, we show that the h-likelihood approach provides a conceptually simple, numerically efficient, and reliable inferential procedure for the LMM. Here, we present the h-likelihood methods for the LMM with censoring, mainly under the AFT models.

8.1 Linear Mixed Model with Censoring

Under Assumptions 3 and 4 given in Chap. 4, consider the AFT models with random effects: for $i = 1, \ldots, q$ and $j = 1, \ldots, n_i$,

$$\log T_{ij} = x_{ij}^T \beta + U_i + \epsilon_{ij}, \tag{8.1}$$

where $x_{ij} = (x_{ij1}, \ldots, x_{ijp})^T$ is a vector of the fixed covariates, β is a $p \times 1$ vector of the fixed effects, and $U_i \sim N(0, \sigma_u^2)$ and $\epsilon_{ij} \sim N(0, \sigma_\epsilon^2)$ are independent. Here, the dispersion or variance components σ_u^2 and σ_ϵ^2 stand for variability between and within individuals, respectively. Note that $E(\log T_{ij}) \neq \log E(T_{ij})$, so that care is needed in transforming conclusions based on a log-scale $E(\log T_{ij})$ back onto the original scale $E(T_{ij})$.

I.D. Ha et al., *Statistical Modelling of Survival Data with Random Effects*,
Statistics for Biology and Health, https://doi.org/10.1007/978-981-10-6557-6_8

8.1.1 Estimation Procedure

Inferential procedures are well developed for the LMM as shown in Chap. 3. However, those procedures cannot be directly applied to the AFT model because of censoring. The observable random variables for the AFT model are

$$Y_{ij} = \min(\log T_{ij}, \log C_{ij}) \text{ and } \delta_{ij} = I(T_{ij} \leq C_{ij}).$$

- **Pseudo-Response Variable**

Because of censoring,

$$E(Y_{ij}|U_i = u_i) \neq \mu_{ij},$$

where $\mu_{ij} = E(\log T_{ij}|U_i = u_i) = x_{ij}^T\beta + u_i$.

Buckley and James (1979) proposed using the pseudo-response variable Y_{ij}^* for the linear model under censoring. Here, we want to extend their method to the AFT model. Define

$$\begin{aligned} Y_{ij}^* &= E(\log T_{ij}|Y_{ij}, \delta_{ij}, U_i = u_i) \\ &= Y_{ij}\delta_{ij} + E(\log T_{ij}|\log T_{ij} > Y_{ij}, U_i = u_i)(1 - \delta_{ij}). \end{aligned}$$

In Appendix 8.6.1, we show that

$$\begin{aligned} E(Y_{ij}^*|U_i = u_i) &= E(\log T_{ij}|U_i = u_i) \\ &= \mu_{ij}. \end{aligned} \tag{8.2}$$

Let y_{ij} be the observed value for Y_{ij} and let $y_{ij}^* = Y_{ij}^*|_{Y_{ij}=y_{ij}}$ be the pseudo-response variables, computed based upon the observed data $Y_{ij} = y_{ij}$. Explicit formulae can be obtained under certain models. Suppose that

$$\log T_{ij}|U_i = u_i \sim N(\mu_{ij}, \sigma_\epsilon^2).$$

Let $V(\cdot) = \phi(\cdot)/\bar{\Phi}(\cdot)$ be the hazard function of $N(0, 1)$, where $\phi(\cdot)$ and $\Phi(\cdot)(= 1 - \bar{\Phi}(\cdot))$ are the density and cumulative distribution functions of $N(0, 1)$, respectively, and

$$m_{ij} = \frac{(y_{ij} - \mu_{ij})}{\sigma_\epsilon}.$$

Then,

$$\begin{aligned} E(\log T_{ij}|\log T_{ij} > y_{ij}, U_i = u_i) &= \int_{y_{ij}}^{\infty} \{tf(t|U_i)\}/S(y_{ij})dt \\ &= \int_{m_{ij}}^{\infty} \{(\mu_{ij} + \sigma_\epsilon z)\phi(z)\}/\bar{\Phi}(m_{ij})dz \end{aligned}$$

$$= \mu_{ij} + \{\sigma_\epsilon/\bar{\Phi}(m_{ij})\} \int_{m_{ij}}^{\infty} z\phi(z)dz$$

$$= \mu_{ij} + \sigma_\epsilon V(m_{ij}),$$

where we use $\phi^{'}(z) = -z\phi(z)$ at the last step. Thus, we have the pseudo-responses

$$y_{ij}^* = y_{ij}\delta_{ij} + \{\mu_{ij} + \sigma_\epsilon V(m_{ij})\}(1 - \delta_{ij}). \tag{8.3}$$

• H-likelihood Procedure

Under Assumptions 3 and 4, the h-likelihood for the AFT model, denoted by h, is defined by

$$h = h(\beta, \sigma_\epsilon^2, \sigma_u^2) = \sum_{ij} \ell_{1ij} + \sum_i \ell_{2i}, \tag{8.4}$$

where

$$\ell_{1ij} = \ell_{1ij}(\beta, \sigma_\epsilon^2; y_{ij}, \delta_{ij}|u_i) = -\delta_{ij}\{\log(2\pi\sigma_\epsilon^2) + (m_{ij})^2\}/2 + (1-\delta_{ij})\log\{\bar{\Phi}(m_{ij})\}$$

is the logarithm of the conditional density function for Y_{ij} and δ_{ij} given $U_i = u_i$, and $\ell_{2i} = \ell_{2i}(\sigma_u^2; u_i) = -\{\log(2\pi\sigma_u^2) + (u_i^2/\sigma_u^2)\}/2$ is the logarithm of the density function for U_i. Then, the estimation procedures are as follows.

(1) Estimation of the fixed and random effects

Given the dispersion components $\theta = (\sigma_\epsilon^2, \sigma_u^2)$, the MHLEs of $\tau = (\beta^T, u^T)^T$ with $u = (u_1, \ldots, u_q)^T$ are obtained by solving

$$\frac{\partial h}{\partial \beta_k} = \frac{1}{\sigma_\epsilon} \sum_{ij} \Big\{ \delta_{ij} m_{ij} + (1 - \delta_{ij}) V(m_{ij}) \Big\} x_{ijk} = 0 \ (k = 1, \ldots, p) \tag{8.5}$$

and

$$\frac{\partial h}{\partial u_i} = \frac{1}{\sigma_\epsilon} \sum_{j} \Big\{ \delta_{ij} m_{ij} + (1 - \delta_{ij}) V(m_{ij}) \Big\} - \frac{1}{\sigma_u^2} u_i = 0 \ (i = 1, \ldots, q). \tag{8.6}$$

In Appendix 8.6.2, we show that the h-likelihood yields the IWLS estimating equations. Note here that in the AFT model, we can use the IWLS equations because W^* is the non-singular diagonal matrix:

$$\begin{pmatrix} X^T W^* X & X^T W^* Z \\ Z^T W^* X & Z^T W^* Z + Q \end{pmatrix} \begin{pmatrix} \widehat{\beta} \\ \widehat{u} \end{pmatrix} = \begin{pmatrix} X^T W^* w^* \\ Z^T W^* w^* \end{pmatrix}, \tag{8.7}$$

where $W^* = W/\sigma_\epsilon^2$, $W = \text{diag}(w_{ij})$ is the $n \times n$ diagonal matrix with the ijth element

$$w_{ij} = \delta_{ij} + (1 - \delta_{ij})\xi(m_{ij})$$

with $\xi(m_{ij}) = V(m_{ij})\{V(m_{ij}) - m_{ij}\}$, $Q = -\partial^2 \ell_2 / \partial u^2 = I_q / \sigma_u^2$, and w^* is the n dimensional vector with

$$w_{ij}^* = \mu_{ij} + w_{ij}^{-1}(y_{ij}^* - \mu_{ij}).$$

Note that the asymptotic covariance matrix for $\widehat{\tau} - \tau$ is obtained from H^{-1}, where

$$H = -\frac{\partial^2 h}{\partial(\beta, u)^2} = \begin{pmatrix} X^T W^* X & X^T W^* Z \\ Z^T W^* X & Z^T W^* Z + Q \end{pmatrix}. \tag{8.8}$$

Thus, the upper left-hand corner of H^{-1} gives the variance matrix of $\widehat{\beta}$;

$$\text{var}(\widehat{\beta}) = \sigma_\epsilon^2 (X^T \Sigma^{-1} X)^{-1}, \tag{8.9}$$

where $\Sigma = W^{-1} + \lambda^{-1} Z Z^T$ with $\lambda = \sigma_\epsilon^2 / \sigma_u^2$.

Let

$$\mathbf{P} = \begin{pmatrix} X & Z \\ \mathbf{0} & I_q \end{pmatrix} \quad \text{and} \quad \mathbf{V} = \begin{pmatrix} W^* & \mathbf{0} \\ \mathbf{0} & Q \end{pmatrix}.$$

Then, as in Chap. 3, the IWLS Eq. (8.7) again reduce to a new simple matrix form

$$(\mathbf{P}^T \mathbf{V} \mathbf{P})\hat{\tau} = \mathbf{P}^T \mathbf{V} \mathbf{y_0}, \tag{8.10}$$

where $\mathbf{y_0} = (w^{*T}, \mathbf{0}^T)^T$. In AFT models, the IWLS equations are numerically efficient, so that the ILS equations are not necessary. Note here that $H = \mathbf{P}^T \mathbf{V} \mathbf{P}$. The estimating Eq. (8.10) can be also viewed as the IWLS equations from an augmented weighted linear model:

$$\mathbf{y_0} = \mathbf{P}\tau + \boldsymbol{\epsilon}^*,$$

where the error term $\boldsymbol{\epsilon}^* \sim N(\mathbf{0}, \mathbf{V}^{-1})$.

Note that when there is no censoring ($W = I_n$), the IWLS Eq. (8.7) become the usual Henderson's (1975) LMM equations using data y_{ij}, and that both y^* and W in (8.7) depend on the censoring patterns.

(2) Estimation of dispersion parameters

To estimate the dispersion parameters $\theta = (\sigma_\epsilon^2, \sigma_u^2)^T$, we use the restricted h-likelihood $p_\tau(h)$ with $\tau = (\beta^T, u^T)^T$ from the Eq. 3.9. The estimating equations for θ,

$$\frac{\partial p_\tau(h)}{\partial \theta} = 0,$$

yield the REMLEs for σ_ϵ^2 and σ_u^2 (Appendix 8.6.3), given by

$$\widehat{\sigma_\epsilon^2} = \frac{\sum_{ij}(y_{ij}^* - \widehat{\mu}_{ij})^2}{n_0 - (p + q - \gamma_0)} \quad \text{and} \quad \widehat{\sigma_u^2} = \frac{\sum_i \widehat{u}_i^2}{q - \gamma_1}, \tag{8.11}$$

where $\widehat{\mu}_{ij} = x_{ij}^T \widehat{\beta} + \widehat{u}_i$, and n_0, γ_0, and γ_1 were defined in Appendix 8.6.3. Note that when there is no censoring ($W = I_n$), the two REML estimators (8.11) become those for the LMM in Chap. 3.

(3) Fitting algorithm

Step 0: Obtain the initial estimates $\widehat{\tau}$ and $\widehat{\theta}$ of τ and θ by treating censored observations as uncensored, i.e., by taking $y^* = y$.
Step 1: Given $y^* = \widehat{y^*}$ and $\theta = \widehat{\theta}$, new estimates $\widehat{\tau}$ are obtained by (8.7), and then given these $\widehat{y^*}$ and $\widehat{\tau}$, new estimates $\widehat{\theta}$ are obtained by (8.11).
Step 2: Repeat Steps 0 and 1 until the maximum absolute difference of the previous and current estimates for τ and θ is less than 10^{-5}. After convergence, we compute the estimates of var$(\widehat{\beta})$ in (8.9).

Remark 8.1 (i) Since we cannot observe all the y_{ij}^*'s, the unobserved y_{ij}^*'s are imputed; for example in (8.11), using the estimates of other quantities, we have

$$\widehat{y_{ij}^*} = y_{ij}\delta_{ij} + \{\widehat{\mu}_{ij} + \widehat{\sigma}_\epsilon V(\widehat{m}_{ij})\}(1 - \delta_{ij}),$$

where $\widehat{\mu}_{ij} = x_{ij}^T \widehat{\beta} + \widehat{u}_i$ and $\widehat{m}_{ij} = (y_{ij} - \widehat{\mu}_{ij})/\widehat{\sigma}_\epsilon$. Replacing y_{ij}^* by $\widehat{y_{ij}^*}$ increases the variance of $\widehat{\beta}$. In our algorithm, this variance increase, caused by censoring, is reflected in the REML estimation of θ via n_0, γ_0, and γ_1, defined in Appendix 8.6.3, so that our variance estimator, $\widehat{\text{var}}(\widehat{\beta})$ in (8.9), works reasonably well (Ha et al. 2002).

(ii) This algorithm is fast and almost always converges except for very few cases with heavy censoring in a small sample (e.g., 80% censoring in $q = 20$ pairs). We have also observed by simulation studies (Ha et al. 2002, 2007b) that our procedure for the LMM is robust against violations of the normal assumption if the censoring rate is not too high.

8.1.2 Comparison with Other Methods

From the expectation identity (8.2) for Y_{ij}^*, we see that the h-likelihood method implicitly implements the EM-type algorithm by imputing Y_{ij}^*. Pettitt (1986) developed an EM algorithm using the pseudo-responses without conditioning $U_i = u_i$, i.e.,

$$E(\log T_{ij} | Y_{ij} = y_{ij}, \delta_{ij}) = y_{ij}\delta_{ij} + E(\log T_{ij} | \log T_{ij} > y_{ij})(1 - \delta_{ij}).$$

However, due to some difficulty in evaluating $E(\log T_{ij} | \log T_{ij} > y_{ij})$ without conditioning on $U_i = u_i$, the method was limited to handle the univariate random-effect model only. Hughes (1999) avoided the integration in evaluating $E(\log T_{ij} | \log T_{ij} > y_{ij})$ by using the Monte Carlo method, which, however, requires heavy computation and extensive derivations in the E-step. Moreover, these marginal methods require a numerical integration which becomes intractable as the number of random components increases. Thus, the advantage of using the conditional form, derived from the h-likelihood, is immediate.

The random-effect model has been able to be fitted using the SAS procedure `PROC NLMIXED`; for the `model` statement, it is only necessary to specify a general likelihood function, the ℓ_1 term in the h-likelihood (8.4), with the response variable y. However, this marginal likelihood method is based upon Gauss–Hermite quadrature (GHQ), which may not handle nested or crossed random effects in general when a high dimensional integration is necessary (Wolfinger 1999).

Lee and Nelder (2001a) showed that in the LMM, without censoring, the h-likelihood method provides the MLEs for the fixed effects (using Henderson's (1975) equations) and the REMLEs for the dispersion parameters. Now we see that for the LMM with censoring, it implicitly implements an EM-type algorithm by imputing unobserved responses $\log T_{ij}$ with $E(\log T_{ij} | Y_{ij} = y_{ij}, \delta_{ij}, U_i = u_i)$ in the estimating equations: see Eqs. (8.7) and (8.11). Thus, this h-likelihood method is straightforwardly extended to the models with many random components. By using the h-likelihood, the numerically challenging E-step or numerical integration can be avoided by automatically imputing the censored responses to y_{ij}^*.

Example 8.1 (*Skin grafts data*) Batchelor and Hackett (1970) presented a small data set of 16 severely burned patients treated with skin allografts. They received skin allografts from one to four donors who may match closely or poorly with the patient's human lymphocyte antigen (HLA) tissue type. The survival outcome was time in days to rejection of the allograft because of an immune response by the patient. Survival times of some allografts were censored by the death of the patient. Survival times from the same patient are correlated because they depend on the degree of HLA matching between the patient and the donor and on the unobserved strength of the patient's immune response.

Now we analyze the closely and poorly matched skin graft data on 11 burned patients, presented in Table 8.1, by using the AFT model ($i = 1, \ldots, 11$; $j = 1, 2$):

$$\log T_{ij} = \beta_0 + \beta_1 x_{ij} + U_i + \epsilon_{ij},$$

Table 8.1 Survival times (days) of closely and poorly matched skin grafts on the same patient, adopted from Batchelor and Hackett (1970)

Case	4	5	7	8	9	10	11	12	13	15	16
Close	37	19	57[a]	93	16	22	20	18	63	29	60[a]
Poor	29	13	15	26	11	16.5	26	21	43	15	40

[a]Right censoring observation

Table 8.2 Results from fitting the univariate random-effect LMM to the skin graft data

Method	$\widehat{\beta}_0$	SE	$\widehat{\beta}_1$	SE	$\widehat{\sigma^2_\epsilon}$	$\widehat{\sigma^2_u}$
EM	3.305	0.150	0.253	0.082	0.148	0.167
GHQ	3.298	0.146	0.247	0.080	0.136	0.165
HL	3.298	0.154	0.247	0.085	0.153	0.182

EM, Pettitt (1986) results using the marginal EM method;
GHQ, SAS PROC NLMIXED using the marginal GHQ method;
HL, the h-likelihood method

where x_{ij} is a fixed covariate indicating 1 for close match or -1 for poor match. Table 8.2 shows that the three methods (EM, GHQ, and HL) give virtually identical fixed-effect estimates. Here, the results for $\hat{\beta}_1$ including the SE indicate that the close match significantly prolongs the time to rejection than the poor match does. For the dispersion estimation, the EM and GHQ methods give the marginal ML estimates, while the HL method gives the REML estimates for σ^2_u and σ^2_ϵ. As expected, the REMLEs in Table 8.2 are slightly larger than the two MLEs, resulting in larger standard error estimates for the fixed-effect estimators. However, for the estimated correlation $\widehat{\rho} = \widehat{\sigma^2_u}/(\widehat{\sigma^2_u} + \widehat{\sigma^2_\epsilon})$, the three methods give similar results; 0.53 for EM, 0.55 for GHQ, and 0.54 for HL. Thus, the h-likelihood approach is preferred because it is efficient and can be easily extended to the multicomponent models. In addition, the HL results are presented in Sect. 8.4, together with the R codes. □

8.2 Multicomponent Mixed Models with Censoring

8.2.1 Model and Estimation Procedure

• Model

For the purpose of illustration, we use the CGD data in Sect. 1.2.3. Let T_{ijk} be the infection time for the kth observation from the jth patient in the ith hospital. Let U_i be the unobservable random effect for the ith hospital and let U_{ij} be that for the jth patient in the ith hospital. On a log-scale of the infection times, we consider a two-component LMM

$$\log T_{ijk} = x^T_{ijk}\beta + U_i + U_{ij} + \epsilon_{ijk}, \quad (8.12)$$

where $x_{ijk} = (x_{ijk1}, \ldots, x_{ijkp})^T$ are the covariates, $\beta = (\beta_1, \ldots, \beta_p)^T$ are the fixed effects, and $U_i \sim N(0, \sigma^2_1)$, $U_{ij} \sim N(0, \sigma^2_2)$, and $\epsilon_{ijk} \sim N(0, \sigma^2_\epsilon)$ are mutually independent error components. This model allows an explicit expression of correlations between recurrent infection times;

$$\text{cov}(\log T_{ijk}, \log T_{i'j'k'}) = \begin{cases} 0 & \text{if } i \neq i', \\ \sigma_1^2 & \text{if } i = i', j \neq j', \\ \sigma_1^2 + \sigma_2^2 & \text{if } i = i', j = j', k \neq k', \\ \sigma_1^2 + \sigma_2^2 + \sigma_\epsilon^2 & \text{if } i = i', j = j', k = k'. \end{cases}$$

Thus, the intra-hospital (ρ_1) and intra-patient (ρ_2) correlations are defined as

$$\rho_1 = \sigma_1^2/(\sigma_1^2 + \sigma_2^2 + \sigma_\epsilon^2) \text{ and } \rho_2 = (\sigma_1^2 + \sigma_2^2)/(\sigma_1^2 + \sigma_2^2 + \sigma_\epsilon^2). \quad (8.13)$$

Here, the observed responses are

$$Y_{ijk} = \min(\log T_{ijk}, \log C_{ijk}) \text{ and } \delta_{ijk} = I(T_{ijk} \leq C_{ijk}).$$

Let $i = 1, \ldots, q_1$, $j = 1, \ldots, r_i$, and $k = 1, \ldots, n_{ij}$, where q_1 is the number of hospitals, r_i is the number of patients in the ith hospital, and n_{ij} is the number of recurrent infection times for the jth patient in the ith hospital. Let $q_2 = \sum_i r_i$ be the number of patients, $q = q_1 + q_2$ be the total number of hospitals and patients, and $n = \sum_{ij} n_{ij}$ be the total number of observations.

- **Extension of the h-likelihood**

Here, the h-likelihood is defined by

$$h = h(\beta, \sigma_1^2, \sigma_2^2, \sigma_\epsilon^2) = \sum_{ijk} \ell_{0ijk} + \sum_i \ell_{1i} + \sum_{ij} \ell_{2ij}, \quad (8.14)$$

where

$$\begin{aligned}\ell_{0ijk} = \ell_{0ijk}(\beta, \sigma_\epsilon^2; y_{ijk}, \delta_{ijk}|u_i, u_{ij}) &= -\ \delta_{ijk}\{\log(2\pi\sigma_\epsilon^2) + (m_{ijk})^2\}/2 \\ &+ (1 - \delta_{ijk}) \log\{1 - \Phi(m_{ijk})\}\end{aligned}$$

is the logarithm of the conditional density function for Y_{ijk} and δ_{ijk} given $U_i = u_i$ and $U_{ij} = u_{ij}$,

$$\ell_{1i} = \ell_{1i}(\sigma_1^2; u_i) = -\ \{\log(2\pi\sigma_1^2) + (u_i^2/\sigma_1^2)\}/2$$

is the logarithm of the density function for U_i and

$$\ell_{2ij} = \ell_{2ij}(\sigma_2^2; u_{ij}) = -\ \{\log(2\pi\sigma_2^2) + (u_{ij}^2/\sigma_2^2)\}/2$$

is that for U_{ij}. Here, $m_{ijk} = (y_{ijk} - \mu_{ijk})/\sigma_\epsilon$ and

$$\mu_{ijk} = E(\log T_{ijk}|U_i = u_i, U_{ij} = u_{ij}) = x_{ijk}^T\beta + u_i + u_{ij}.$$

Because

$$E(Y_{ijk}|U_i = u_i, U_{ij} = u_{ij}) \neq \mu_{ijk},$$

the h-likelihood method implicitly uses the pseudo-responses given by

$$\begin{aligned} y^*_{ijk} &= E(\log T_{ijk}|Y_{ijk} = y_{ijk}, \delta_{ijk}, U_i = u_i, U_{ij} = u_{ij}) \\ &= y_{ijk}\delta_{ijk} + A_{ijk}(1 - \delta_{ijk}), \end{aligned}$$

where

$$A_{ijk} = E(\log T_{ijk}| \log T_{ijk} > y_{ijk}, U_i = u_i, U_{ij} = u_{ij}) = \mu_{ijk} + \sigma_\epsilon V(m_{ijk}).$$

Note that

$$E(y^*_{ijk}|U_i = u_i, U_{ij} = u_{ij}) = \mu_{ijk}.$$

Thus, the h-likelihood procedure in Sect. 8.1 is straightforwardly extended to the multicomponent LMM, as shown in Appendix 8.6.4.

8.2.2 Application to the CGD Data

In the CGD study, the recurrent infection times for a given patient are likely to be correlated. However, since each patient belongs to one of the 13 hospitals, the correlation may also be due to a hospital effect. This data set was previously analyzed in Chap. 5 using the multicomponent log-normal frailty models with a single covariate x_{ijk} (= 0 for placebo and = 1 for gamma interferon).

We consider the following four models:

M1: ($\sigma_1^2 = 0, \sigma_2^2 = 0$), regression model without random effects,
M2: ($\sigma_2^2 = 0, \sigma_1^2 > 0$), one-component model without patient effects,
M3: ($\sigma_1^2 = 0, \sigma_2^2 > 0$), one-component model without hospital effects, and
M4: ($\sigma_1^2 > 0, \sigma_2^2 > 0$), two-component model requiring both patient and hospital effects.

The results from these LMMs are given in Table 8.3. Ignoring important random components may render invalid many of the traditional statistical analysis techniques (Goldstein 1995). For testing the need for a random component, we use the LRT (difference in $-2p_\tau(h)$ in Table 8.3) based upon the restricted likelihood $p_\tau(h)$. Because such a hypothesis is on the boundary of the parameter space, as shown in Sect. 4.3.2, the critical value is $\chi^2_{2\alpha}$ for a size α test.

Here, M4 ⊃ M3 ⊃ M1 and M4 ⊃ M2 ⊃ M1. The difference in the restricted likelihood between M3 and M4 is 0.45, which is not significant at a 5% level ($\chi^2_{1,0.10} = 2.71$), indicating the absence of the random hospital effects, i.e., accepting the null hypothesis of $\sigma_1^2 = 0$. The difference between M1 and M3 is 8.92., indicating that the random patient effects are indeed necessary (i.e., $\sigma_2^2 > 0$). The difference between

Table 8.3 Analysis results from the multilevel LMM for the CGD data

Model	$\widehat{\beta}_0$ (SE)	$\widehat{\beta}_1$ (SE)	$\widehat{\sigma_1^2}$	$\widehat{\sigma_2^2}$	$\widehat{\sigma_\epsilon^2}$	$-2p_\tau(h)$	rAIC
M1: LM ($\sigma_1^2 = \sigma_2^2 = 0$)	5.428 (0.185)	1.494 (0.322)	—	—	3.160	426.52	428.52
M2: One-component ($\sigma_1^2 > 0, \sigma_2^2 = 0$)	5.594 (0.249)	1.470 (0.313)	0.294	—	2.872	422.00	426.00
M3: One-component ($\sigma_1^2 = 0, \sigma_2^2 > 0$)	5.661 (0.202)	1.237 (0.331)	—	0.722	2.163	417.60	421.60
M4: Two-component ($\sigma_1^2 > 0, \sigma_2^2 > 0$)	5.698 (0.220)	1.255 (0.334)	0.067	0.710	2.185	417.15	423.15

M2 and M4 is 4.85(> 2.71), also indicating $\sigma_2^2 > 0$. The rAIC($= -2p_\tau(h) + 2df_r$) also chooses M3 as the final model. With the model M3, the estimated intra-patient correlation in (8.13) is $\widehat{\rho}_2 = \widehat{\sigma_2^2}/(\widehat{\sigma_2^2} + \widehat{\sigma_\epsilon^2}) = 0.250$ and $\widehat{\beta}_1 = 1.237$ (SE = 0.331), implying that gamma interferon significantly prolongs the recurrent infection times.

8.3 The AFT Models with LTRC

In this section, we investigate a genetic LMM for twin data with LTRC (Left Truncated Right Censored). Twin studies are the most widely used methods for quantifying the contribution of genetic and environmental factors on traits such as behavior or disease susceptibility. To do this, data on monozygotic (MZ) and dizygotic (DZ) twins are required (Neale and Cardon 1992) and they are analyzed using the random-effects models that allow for separating the genetic effect from the environment effect. Here, we are interested in the genetic analysis of age-at-onset traits using the LMM, with a specific application to the analysis of life span using the correlated survival data from twins.

The genetic LMM often encounters the LTRC problem, occurring during the data collection for the twin studies. That is, some twins may only be observed at a certain time (not the time origin of interest) and they may be still alive at the end of follow-up. We also show that the h-likelihood procedure results in a simple and fast computation in the analysis of large survival data sets with LTRC, such as one from Swedish Twin Register.

8.3.1 The Swedish Twin Survival Data with LTRC

The Swedish Twin Registry is the largest population-based twin registry in the world and includes various information about life span and disease status, etc., for the twins born in Sweden since 1886. Table 8.4 briefly presents the structure of survival data on life spans of a few twins in the Twin Registry. For the purpose of analysis, the

Table 8.4 Survival data in the Swedish Twin Registry, born since 1886

Number	Pairid	Zygalg	Birthday	Dead	Death.date	Eff.date	Sex
1	11	2	06JAN1900	1	03JAN1987	.	2
2	11	2	06JAN1900	1	15DEC1990	.	2
3	12	2	07JAN1900	1	23DEC1982	.	2
4	12	2	07JAN1900	1	20FEB1994	.	2
⋮	⋮	⋮	⋮	⋮	⋮	⋮	⋮
17	21	1	01JAN1926	0	.	30AUG1997	1
18	21	1	01JAN1926	0	.	30AUG1997	1
19	22	4	01JAN1926	0	.	21JUN2000	1
20	22	4	01JAN1926	1	15MAY1991	.	2
21	23	2	01JAN1926	0	.	03JAN2002	1
22	23	2	01JAN1926	1	13MAR1993	.	1
⋮	⋮	⋮	⋮	⋮	⋮	⋮	⋮
78205	244782	2	30DEC1958	0	.	13JUL2002	1
78206	244782	2	30DEC1958	0	.	02SEP1999	1
78207	244783	2	31DEC1958	0	.	06JUN2000	1
78208	244783	2	31DEC1958	0	.	02FEB2001	1
78209	244784	4	31DEC1958	0	.	21MAY1999	1
78210	244784	4	31DEC1958	0	.	30AUG1997	2

Number, number of twins; Pairid, ID of twin pairs
Zygalg, 1 = MZ and 2[4] = DZ same[opposite] sex
Dead, censoring indicator (1 = dead, 0 = alive)
Death.date, the date of death; Eff.date, the latest follow-up date
Sex (1 = male, 2 = female)

twins data are often divided into three different age cohorts: old, middle, and young cohorts. To make use of sufficient number of endpoints, we consider the old cohort only (Yashin et al. 1999). This cohort consists of all same-gender pairs born between 1886 and 1925, where both members of the pair were alive and living in Sweden in 1959. The survival outcome is defined as time to death from any cause (in years). The data are left truncated because *both* members of a twin pair had to survive until 1959, the beginning of follow-up; the left truncation time is thus calculated as (Jan 01, 1959 – birthday)/365. If an individual is still alive at the end of follow-up, the survival time is right censored; in this case, the date of death is replaced by the latest follow-up date.

The data used in the analyses are summarized in Table 8.5 and categorized according to the censoring status. The subgroups have the censoring rates of about 20 to 30%. For example, the censoring rate in male MZ twins is 19%, calculated by $(313 + 2 \times 159)/(2 \times 1646)$. The table shows that there are more female than male twins, which may be explained by the longer female life span. The ratio of MZ to DZ twin pairs is about a half: MZ = 3653 versus DZ = 6796, as described in other studies (Sham 1998, pp. 189).

Table 8.5 Composition of the old cohort of the Swedish twin survival data by sex, zygosity, and censoring status

Data	One censored	Both censored	None censored	Total (pairs)
Males				
MZ	313	159	1174	1646
DZ	620	258	2074	2952
Females				
MZ	450	396	1161	2007
DZ	931	686	2227	3844
Total	2314	1499	6636	10449

8.3.2 *The Model*

Consider modeling a direct relationship between survival time and covariates including observed or unobserved factors. Let T_{ij} be the survival time (e.g., age at death) for the jth member of the ith twin pair, which is subject to only partial observation due to LTRC variables (L_{ij}, R_{ij}), assumed to be independent of T_{ij}'s (Lai and Ying 1994). Here, L_{ij} and R_{ij} are left truncation and right censoring variables, respectively. Let $Y_{ij} = \min(\log T_{ij}, \log R_{ij})$ be the observed survival time, $\delta_{ij} = I(T_{ij} \leq R_{ij})$ be the event indicator, and $B_{ij} = \log L_{ij}$ be the truncation time. For the LTRC data, one observes $(Y_{ij}, \delta_{ij}, B_{ij})$ only when $Y_{ij} \geq B_{ij}$. Thus, the corresponding observed data consist of n observations $(y_{ij}, \delta_{ij}, b_{ij})$ with $y_{ij} \geq b_{ij}$ $(i = 1, \ldots, q,\ j = 1, 2,\ n = 2q)$.

Let g_{ij} and c_{ij} be the random genetic and shared environment effects for the jth individual in the ith twin pair, respectively. Because T_{ij}'s are positive-valued and likely to be skewed, we use $\log T_{ij}$ as the response. Now, we consider the LMM with two random effects; for $i = 1, 2, \ldots, q$ and $j = 1, 2$,

$$\log T_{ij} = x_{ij}^T\beta + g_{ij} + c_{ij} + \epsilon_{ij}, \tag{8.15}$$

where $g_{ij} \sim N(0, \sigma_g^2)$, $c_{ij} \sim N(0, \sigma_c^2)$, and $\epsilon_{ij} \sim N(0, \sigma_\epsilon^2)$ are mutually independent error components. The last error component can be interpreted as an unshared environmental component. Now we want that between-pair genetic and environment effects are independent, but within-pair effects are not. If the ith twin pair is MZ (denoted by MZ_i), we may have (Sham 1998, p. 189)

$$\text{corr}(g_{i1}, g_{i2}) = 1 \text{ and } \text{corr}(c_{i1}, c_{i2}) = 1,$$

and if it is DZ (denoted by DZ_i), we have

$$\text{corr}(g_{i1}, g_{i2}) = 0.5 \text{ and } \text{corr}(c_{i1}, c_{i2}) = 1.$$

It is this discrepancy in the genetic correlation between MZ and DZ twins that allows us to separate the genetic factor from the common environmental factor. It is also useful to reexpress the parameters accordingly; let $v_{ij} = g_{ij} + c_{i0}$ for $j = 1, 2$, where $c_{i0}(= c_{i1} = c_{i2})$ denotes the common environmental effect for the ith twin pair. Then, we have

$$\rho \equiv \text{corr}(v_{i1}, v_{i2}) = \frac{\sigma_{12} + \sigma_c^2}{\sigma_g^2 + \sigma_c^2},$$

where $\sigma_{12} = \text{cov}(g_{i1}, g_{i2})$. Note that $\rho = 1$ for MZ_i and $\rho = (0.5\sigma_g^2 + \sigma_c^2)/(\sigma_g^2 + \sigma_c^2) \in [0.5, 1.0]$ for DZ_i.

For the sake of interpretation, it is convenient to define the quantity

$$h_g^2 = \frac{\sigma_g^2}{\sigma_g^2 + \sigma_c^2 + \sigma_\epsilon^2},$$

known as *heritability*, which measures the importance of genetics relative to other factors in explaining the variability of a trait in a population (Sham 1998, p. 212).

As shown in Appendix 8.6.5, the two random component in the LMM (8.15) can be reexpressed as follows:

$$\log T_{ij} = x_{ij}^T \beta + z_{ij}^{*T} u_i + \epsilon_{ij}, \tag{8.16}$$

where z_{ij}^* is the jth component of $Z_i^*(\rho)$ in (8.37), $u_i \sim N(0, \sigma_v^2 I_k)$ with $\sigma_v^2 = \sigma_g^2 + \sigma_c^2$. Here, I_k is the k-dimensional identity matrix such that $k = 1$ for MZ_i and $k = 2$ for DZ_i. Now, the model (8.16) can be used for inference on the parameters in model (8.15) through the h-likelihood method described in Sect. 8.1.

8.3.3 Estimation Procedure Under LTRC

Here, the h-likelihood for model (8.16) under LTRC is defined by

$$h = h(\beta, \sigma_v^2, \sigma_\epsilon^2, \rho) = \sum_{ij} \ell_{1ij} + \sum_i \ell_{2i},$$

where

$$\begin{aligned}\ell_{1ij} &= \ell_{1ij}(\beta, \sigma_\epsilon^2, \rho; y_{ij}, \delta_{ij}, b_{ij}|u_i, y_{ij} \geq b_{ij}) \\ &= -\delta_{ij}\{\log(2\pi\sigma_\epsilon^2) + (m_{ij})^2\}/2 + (1-\delta_{ij})\log\{1-\Phi(m_{ij})\} \\ &\quad -\log\{1-\Phi(m_{ij}^*)\}\end{aligned}$$

is the log-conditional density of $(Y_{ij}, \delta_{ij}, B_{ij})$ given u_i and $Y_{ij} \geq B_{ij}$, and

$$\ell_{2i} = \ell_{2i}(\sigma_v^2; u_i) = -\{\log\det(2\pi\sigma_v^2 I_k) + (u_i^T u_i/\sigma_v^2)\}/2$$

is the log-density of u_i. Here, $m_{ij} = (y_{ij} - \mu_{ij})/\sigma_\epsilon$, $m_{ij}^* = (b_{ij} - \mu_{ij})/\sigma_\epsilon$, and

$$\mu_{ij} = E(\log T_{ij}|u_i, Y_{ij} \geq B_{ij}) = x_{ij}^T\beta + z_{ij}^{*T}u_i.$$

Given the dispersion parameters $\theta = (\sigma_\epsilon^2, \sigma_v^2)$ and ρ, the MHLEs of $\tau = (\beta^T, u^T)^T$ with $u = (u_1, \ldots, u_q)^T$ are obtained by solving the score equations

$$\frac{\partial h}{\partial \beta} = \frac{1}{\sigma_\epsilon}\sum_{ij} x_{ij}\left\{\delta_{ij}m_{ij} + (1-\delta_{ij})V(m_{ij}) - V(m_{ij}^*)\right\} = 0 \qquad (8.17)$$

and

$$\frac{\partial h}{\partial u_i} = \frac{1}{\sigma_\epsilon}\sum_j z_{ij}^*\left\{\delta_{ij}m_{ij} + (1-\delta_{ij})V(m_{ij}) - V(m_{ij}^*)\right\} - \frac{1}{\sigma_v^2}u_i = 0 \ (i = 1, \ldots, q). \qquad (8.18)$$

The two MHL Eqs. (8.17) and (8.18) can be simply written as

$$\frac{1}{\sigma_\epsilon^2}\sum_{ij} x_{ij}\left(y_{ij}^* - \mu_{ij}\right) = 0, \qquad (8.19)$$

and

$$\frac{1}{\sigma_\epsilon^2}\sum_j z_{ij}^*\left(y_{ij}^* - \mu_{ij}\right) - \frac{1}{\sigma_v^2}u_i = 0 \ (i = 1, \ldots, q), \qquad (8.20)$$

by using

$$y_{ij}^* \equiv y_{ij}\delta_{ij} + \{\mu_{ij} + \sigma_\epsilon V(m_{ij})\}(1-\delta_{ij}) - \sigma_\epsilon V(m_{ij}^*),$$

which is an extension of the pseudo-responses under right censoring in the previous section to the LTRC case.

- **A fast computational method for large data**

The score Eqs. (8.19) and (8.20) have the same form as those for the LMM in (8.25) and (8.26) in Appendix 8.6.1, which is for the right censoring. Therefore, with a moderate sample size, it is straightforward to fit model (8.16) by using the standard h-likelihood methods described in the previous two sections. However, for large data sets such as the Swedish twin data used in Sect. 8.3.1, the dimension of the model matrix Z^* of random effects u increases with q. In this case, it would be difficult to apply the standard h-likelihood procedure directly.

A simple and fast computational method using a partition matrix approach has been developed (Ha et al. 2007b). First, assume that ρ is known. Given $\theta = (\sigma_\epsilon^2, \sigma_v^2)^T$ and $y_i^* = (y_{i1}^*, y_{i2}^*)^T$, the MHLEs of $\tau = (\beta^T, u^T)^T$ are obtained by solving the following two score equations iteratively:

$$\left(\sum_i X_i^T X_i\right)\widehat{\beta} = \sum_i X_i^T y_i^* - \sum_i X_i^T Z_i^* \widehat{u}_i, \tag{8.21}$$

$$(Z_i^{*T} Z_i^* + \lambda I_k)\widehat{u}_i = Z_i^{*T} y_i^* - Z_i^{*T} X_i \widehat{\beta} \;\; (i = 1, \ldots, q). \tag{8.22}$$

The derivations of (8.21) and (8.22) are given in Appendix 8.6.6. A fast computational procedure for the asymptotic variance of $\widehat{\beta}$ is also given in (8.43).

To estimate the dispersion parameters $\theta = (\sigma_\epsilon^2, \sigma_v^2)^T$, $p_\tau(h)$ can be used. However, the inverse of $D(h, \tau) = -\partial^2 h/\partial \tau^2$ could be computationally intensive for large samples because the REMLEs from $p_\tau(h)$ are asymptotically equivalent to the MLEs from $p_u(h)$ (Noh and Lee 2007). Furthermore, the inversion of $D(h, u) = -\partial^2 h/\partial u^2$ in $p_u(h)$ is very simple because $D(h, u) = H_{22}/\sigma_\epsilon^2$, where $H_{22} = Z^{*T} W Z^* + \Lambda$ in (8.42), is a diagonal matrix.

So, the resulting MLEs for σ_ϵ^2 and σ_v^2 by using $p_u(h)$ are given by

$$\widehat{\sigma_\epsilon^2} = \frac{\sum_{ij}(y_{ij}^* - \widehat{\mu}_{ij})^2}{n_0 - (q^* - \gamma_0)} \quad \text{and} \quad \widehat{\sigma_v^2} = \frac{\sum_i \widehat{u}_i^T \widehat{u}_i}{q^* - \gamma_1}, \tag{8.23}$$

where

$$n_0 = \sum_{ij}[w_{ij} - 2\{(y_{ij}^* - \mu_{ij})/\sigma_\epsilon\}V(m_{ij}^*)],$$

$$\gamma_0 = \sigma_\epsilon^2 \text{trace}\{{H_{22}}^{-1}(\partial H_{22}/\partial \sigma_\epsilon^2)\},$$
$$\gamma_1 = -\sigma_v^2 \text{trace}\{{H_{22}}^{-1}(\partial H_{22}/\partial \sigma_v^2)\}.$$

Appendix 8.6.3 gives the formulas for $\partial H_{22}/\partial \sigma_\epsilon^2$ and $\partial H_{22}/\partial \sigma_v^2$, and the trace terms in γ_0 and γ_1 are easily calculated using the partition matrix. Note that since we cannot observe all of the y_{ij}^*'s due to the LTRC, we update them by using $\widehat{y_{ij}^*}$, in each iteration.

The fitting algorithm is summarized as follows:

(Step 1) Given ρ and hence $L_i(\rho)$, estimate τ and $\theta = (\sigma_\epsilon^2, \sigma_v^2)^T$ using (8.21), (8.22), and (8.23).

(Step 2) Given τ and θ, estimate ρ by maximizing $p_u(h)$.

(Step 3) Iterate (Step 1) and (Step 2) until convergence is achieved.

After convergence, we compute the estimates of σ_g^2 and σ_c^2 from (8.38) and those of $\text{var}(\widehat{\beta})$ from (8.43), respectively.

8.3.4 Application

In this section, we use the twin data presented in Sect. 8.3.1 as an example.

- **Estimation of probability of left truncation**

Table 8.6 shows the extent of left truncation of the old cohort. First, the probability p of left truncation is calculated as follows. When the ith ($i = 1, \ldots, 10449$) twin pair was born in year L (for $L = 1886, \ldots, 1925$), we have that, for $j = 1, 2$, $p = P(T_{ij} \leq B_i)$, where $B_i = 1959 - L_i$. However, for the ith twin pair, the survival times T_{i1} and T_{i2} may be correlated. To simplify the computation, we randomly select one of the survival times T_{i1} and T_{i2}, and obtain a sample of independent survival times—say T_i—and calculate $p = P(T_i \leq B_i)$.

Assuming $\log T_i = \beta_0 + \epsilon_i \sim N(\beta_0, \sigma_\epsilon^2)$, we can then obtain the estimated values of p; for example, for males $\widehat{p} = 0.22603$ at $L = 1890$, $\widehat{p} = 0.03604$ at $L = 1900$, and $\widehat{p} \approx 0$ at $L = 1910$ and at $L = 1920$. These values are close to zero except at early birth years. Overall, the estimated ratios of population sizes are somewhat

Table 8.6 Extent of left truncation in the old cohort (M: Male, F: Female)

Data	Birth range	Birth year	$\widehat{p}$	Sample size (%)	Estimated pop size (%)
M	1886–1895	1890	0.22603	407 (8.9)	526 (11.1)
	1896–1905	1900	0.03604	901 (19.6)	935 (19.7)
	1906–1915	1910	0.00118	1519 (33.0)	1521 (32.0)
	1916–1925	1920	0.00000	1771 (38.5)	1771 (37.3)
Total				4598	4753
F	1886–1895	1890	0.12402	598 (10.2)	683 (11.5)
	1896–1905	1900	0.01252	1210 (20.7)	1225 (20.6)
	1906–1915	1910	0.00021	1863 (31.8)	1863 (31.3)
	1916–1925	1920	0.00000	2180 (37.3)	2180 (36.6)
Total				5851	5951
M + F	1886–1895	1890	0.16920	1005 (9.6)	1210 (11.3)
	1896–1905	1900	0.02293	2111 (20.2)	2161 (20.2)
	1906–1915	1910	0.00062	3382 (32.4)	3384 (31.6)
	1916–1925	1920	0.00000	3951 (37.8)	3951 (36.9)
Total				10449	10706

$\widehat{p}$, the estimated probability of left truncation
Estimated pop (population) size x is calculated using x = sample size/$(1 - \widehat{p})$

close to the ratios of sample sizes. Here, the sample size is obtained from the birth range in the old cohort data of Table 8.5, and the population size x is calculated using x = sample size/$(1 - \widehat{p})$.

• Separate analysis

First, we analyze separately the MZ and DZ twins among males and females. Note, however, that within each zygosity group, the shared environment effect c_{ij} is not distinguishable from the genetic effect g_{ij}. Thus, in these analyses, from (8.15), we consider the LMM with only one random component (i.e., one-way random-effect model): for $i = 1, 2, \ldots, q$ and $j = 1, 2$,

$$\log T_{ij} = \beta_0 + u_i + \epsilon_{ij},$$

where β_0 is the intercept, and $u_i \sim N(0, \sigma_b^2)$ and $\epsilon_{ij} \sim N(0, \sigma_\epsilon^2)$ are mutually independent. Here, the random effect u_i stands for the pair effect. In particular, this model provides the within-pair or intraclass correlation of the log survival times, given by $\kappa = \sigma_u^2/(\sigma_u^2 + \sigma_\epsilon^2)$. The results are given in Table 8.7. As expected, in both genders, the estimated correlation $\widehat{\kappa}$ is higher for the MZ twins than for the DZ twins. The correlations are generally smaller than those ($\widehat{\kappa}^*$ in Table 8.7) from Henderson's mixed model equation for the LMM without the LTRC information.

In both genders, the MZ twins have a larger estimated between-pair variance $\widehat{\sigma_u^2}$ than the DZ twins do. To test the necessity of a random component (i.e., $\sigma_u^2 = 0$), we use the LRT based upon the change in the adjusted profile likelihood $p_u(h)$, i.e., difference in $-2p_u(h)$ in Table 8.7. We consider the two models,
(i) E: $\log T = \beta_0 + \epsilon$ with $\sigma_b^2 = 0$ and

Table 8.7 Separate analyses using the between-pair LMM for the old cohort

Data	Model	$\widehat{\beta_0}$ (SE)	$\widehat{\sigma_b^2}$	$\widehat{\sigma_\epsilon^2}$	$\widehat{\kappa}$ ($\widehat{\kappa^*}$)	$-2p_u(h)$
Males						
MZ	E	4.349 (0.0028)	–	0.0221	–	−1618.9
	BE	4.351 (0.0030)	0.0066	0.0150	0.31 (0.39)	−1922.2
DZ	E	4.339 (0.0021)	–	0.0229	–	−2748.3
	BE	4.340 (0.0022)	0.0029	0.0199	0.13 (0.20)	−2890.3
Females						
MZ	E	4.405 (0.0025)	–	0.0207	–	−1326.0
	BE	4.405 (0.0027)	0.0052	0.0154	0.25 (0.34)	−1505.2
DZ	E	4.401 (0.0018)	–	0.0214	–	−2539.8
	BE	4.401 (0.0019)	0.0024	0.0190	0.11 (0.21)	−2651.5

β_0, intercept; SE, the corresponding standard error
E, $\log T = \beta_0 + \epsilon$ with $\sigma_b^2 = 0$
BE, $\log T = \beta_0 + b + \epsilon$ with $\sigma_b^2 > 0$
$\kappa = \sigma_b^2/(\sigma_b^2 + \sigma_\epsilon^2)$, within-pair correlation
κ^*, correlation from Henderson's mixed model equation for the LMM

(ii) BE: $\log T = \beta_0 + b + \epsilon$ with $\sigma_b^2 > 0$.
We see that since the hypothesis H_0: $\sigma_b^2 = 0$ is on the boundary of the parameter space, the critical value is $\chi^2_{2\alpha}$ for a size α test. For example, for the male MZ twins, the likelihood difference between the E and BE models (B for between) is 303.3, indicating that the pair effects are highly significant, i.e., $\sigma_u^2 > 0$. In fact, the pair effects are highly significant in each subgroup in Table 8.7.

The fixed-effect estimates provide a useful information about life span. As expected, for both MZ and DZ twins, from the value of $\widehat{\beta}_0$ in the BE model, we observe that males tend to have shorter life span than females. For example, the estimated median life span is $\exp(4.351) = 77.6$ years for male MZ and $\exp(4.405) = 81.9$ years for female MZ.

• Combined Analysis

Table 8.8 shows the results from the model (8.15) when MZ and DZ twins are combined within each gender group. We consider the four models,
E, LMM with $\sigma_g^2 = \sigma_c^2 = 0$,
CE, LMM with $\sigma_g^2 = 0$ and $\sigma_c^2 > 0$,
GE, LMM with $\sigma_g^2 > 0$ and $\sigma_c^2 = 0$,
GCE, LMM with $\sigma_g^2 > 0$ and $\sigma_c^2 > 0$.
To test the necessity of a random component (i.e., $\sigma_g^2 = 0$ or $\sigma_c^2 = 0$), we again use the LRT based on $-2p_u(h)$ as in Table 8.7. For males, the difference in $-2p_u(h)$ between GE and GCE is 0.00, indicating no evidence of the shared environmental effects (i.e., $\sigma_c^2 = 0$). The difference between CE and GCE is 96.2, indicating that the genetic effects are highly significant, i.e., $\sigma_g^2 > 0$. In addition, the difference between E and GE is 481.9, indicating that the genetic effects are indeed highly significant with or without random environmental effects. The results for females are similar to those for males.

For model selection among nested models, we use the testing procedure described in the above. However, for model selection among non-nested models such as CE and GE, the following mAIC can be considered:

$$\text{mAIC} = -2p_u(h) + 2df_m, \tag{8.24}$$

where df_m is the number of fixed and dispersion parameters, not the number of random effects. From Table 8.8, for males, the mAIC chooses GE as the best model, with estimated heritability $\widehat{h_g^2} = 26\%$. For females, the GE model is again best, with $\widehat{h_g^2} = 21\%$.

We then fitted model (8.15) with an additional fixed covariate x_{ij} representing zygosity ($= 1$ for MZ$_i$ and $= 0$ for DZ$_i$). The results are in the second block in Table 8.8; GE models are the best models according to the mAIC. From the estimates of β_1, we observe the following interesting findings:
(i) For both male and female MZ twins, the estimated life expectancy is longer than that for respective DZ twins;
(ii) For male twins, the MZ tends to have significantly longer life span than for the

Table 8.8 Combined (MZ & DZ) analyses using the genetic LMM for the old cohort (M, Male; F, Female)

Data	Model	$\widehat{\beta}_0$ (SE)	$\widehat{\beta}_1$ (SE)	$\widehat{\sigma_g^2}$	$\widehat{\sigma_c^2}$	$\widehat{\sigma_\epsilon^2}$	$\widehat{h_g^2}$	$-2p_u(h)$	mAIC	cAIC
M	E	4.342(0.0017)	–	–	–	0.0226	–	−4357.2	479.9	4549.4
	CE	4.344(0.0018)	–	–	0.0042	0.0182	–	−4742.9	96.2	1407.4
	GE	4.344(0.0018)	–	0.0058	–	0.0166	0.26	−4839.1	0	0
	GCE	4.344(0.0018)	–	0.0058	0.0000	0.0166	0.26	−4839.1	2.0	0.0
F	E	4.402(0.0015)	–	–	–	0.0212	–	−3872.5	313.4	4412.9
	CE	4.402(0.0016)	–	–	0.0033	0.0178	–	−4131.6	56.3	1270.8
	GE	4.402(0.0015)	–	0.0044	–	0.0166	0.21	−4187.9	0	0
	GCE	4.402(0.0015)	–	0.0044	0.0000	0.0166	0.21	−4187.9	2.0	0.0
			$\widehat{\beta}_1^{MZ}$(SE)							
M	E	4.339(0.0021)	0.010(0.0035)	–	–	0.0226	–	−4364.9	485.5	4536.1
	CE	4.340(0.0022)	0.011(0.0037)	–	0.0042	0.0182	–	−4757.1	95.3	1400.2
	GE	4.341(0.0022)	0.011(0.0037)	0.0057	–	0.0166	0.26	−4852.4	0	0
	GCE	4.341(0.0022)	0.011(0.0037)	0.0057	0.0000	0.0166	0.26	−4852.4	2.0	0.0
F	E	4.401(0.0018)	0.005(0.0031)	–	–	0.0211	–	−3865.1	322.9	4405.0
	CE	4.401(0.0019)	0.005(0.0033)	–	0.0033	0.0178	–	−4133.9	56.1	1265.1
	GE	4.401(0.0019)	0.005(0.0033)	0.0044	–	0.0166	0.21	−4190.0	0	0
	GCE	4.401(0.0019)	0.005(0.0033)	0.0044	0.0000	0.0166	0.21	−4190.0	2.0	0.0
			$\widehat{\beta}_1^{Male}$(SE)							
M+F	E	4.403(0.0015)	−0.059(0.0022)	–	–	0.0218	–	−8229.6	783.5	8992.2
	CE	4.403(0.0016)	−0.060(0.0024)	–	0.0037	0.0180	–	−8867.2	147.9	2682.2
	GE	4.403(0.0016)	−0.060(0.0024)	0.0050	–	0.0166	0.23	−9015.1	0	0
	GCE	4.403(0.0016)	−0.060(0.0024)	0.0050	0.0000	0.0166	0.23	−9015.1	2.0	0.0

E, LMM with $\sigma_g^2 = \sigma_c^2 = 0$; CE, LMM with $\sigma_g^2 = 0$ and $\sigma_c^2 > 0$
GE, LMM with $\sigma_g^2 > 0$ and $\sigma_c^2 = 0$; GCE, LMM with $\sigma_g^2 > 0$ and $\sigma_c^2 > 0$
$\hat{h}_g^2 = \sigma_g^2/(\sigma_g^2 + \sigma_c^2 + \sigma_\epsilon^2)$
AIC, AIC differences with the minimum set to zero

DZ ($\widehat{\beta_1} = 0.011$ with SE $= 0.0037$), but for female twins, the difference is no longer significant ($\widehat{\beta_1} = 0.005$ with SE $= 0.0033$).

Finally, we combined male and female twins data to investigate the pattern of the expected life span between both genders. For this, we also fitted model (8.15) allowing a different intercept for males and females (i.e., allowing a fixed covariate x_{ij} representing gender, coded as 1 if the ith twin pair is male and as 0 if it is female), but forcing common variance parameters. The results are given in Table 8.8, where GE model was again chosen as the best model. From the estimates of $\widehat{\beta_1}$, we also found that males tend to have significantly shorter life span than females ($\widehat{\beta_1} = -0.060$ with SE $= 0.0024$). The estimated variance and heritability parameters are in between the corresponding values from the separate analyses.

In addition, we conducted model selection using the cAIC in Table 8.8. We define the cAIC corresponding to the mAIC in (8.24), given by

$$\text{cAIC} = -2\ell_1 + 2df_c,$$

where $df_c = p$ is the number of regression parameters. The cAIC selects either the GE or GCE model as the best model for combined data cases in Table 8.8. However, care is still necessary for the cAIC because the estimate of random-effect variance is near zero ($\widehat{\sigma_c^2} \approx 0$) as in Sect. 5.3.

8.4 Software and Examples Using R

8.4.1 Skin Grafts Data: LMM with Censoring

Below are the R codes and results from fitting the LMM with censoring using the HL method with the skin grafts data in Table 8.1. The R outputs are summarized in Table 8.2.

```
######################## LMM with censoring ###################
> case<-c(4,5,7,8,9,10,11,12,13,15,16,4,5,7,8,9,10,11,12,13,15,16)
> time<-c(37,19,57,93,16,22,20,18,63,29,60,29,13,15,26,11,16.5,26,
+   21,43,15,40)
> status<-c(1,1,0,1,1,1,1,1,1,1,0, 1,1,1,1,1,1,1,1,1,1,1)
> group<-c(rep(1,11),rep(-1,11))
> data_surv<-data.frame(case,time,status,group)
> mlmc1<-jointmodeling(Model="mean",RespDist="AFT",Link="log",
+  LinPred=Surv(time,status)~group+(1|case),RandDist="gaussian")
> ress<-mlmfit(mlmc1,data_surv,Maxiter=200)
######################### Output #################################
[1] "iterations :"
[1] 46
[1] "convergence :"
[1] 7.759592e-07
    beta_h      se_beh    t_value      p_value
```

```
 3.2981484 0.15392021 21.427650 0.000000000
 0.2472033 0.08463752  2.920729 0.003492136
        alpha_h     phi_h
[1,] 0.1818075 0.1531119
> ress$V.Est
 [1]  0.1343752 -0.3824222  0.1283264  0.4202422 -0.5016637 -0.2469572
 [7] -0.1204953 -0.2327105  0.4602254 -0.1832934  0.5243730
```

8.4.2 CGD Data: Multilevel LMM with Censoring

In this section, the R codes and results from fitting the multilevel LMM with the CGD data are provided, specifically the results from the model "M4" in Table 8.3.

```
############### Multilevel LMM with censoring ###########
> data(cgd, package="frailtyHL")
> data_surv <- cgd
> mlmc1<-jointmodeling(Model="mean",RespDist="AFT",Link="log",
+  LinPred=Surv(tstop-tstart,status)~treat+(1|center)+(1|id),
+  RandDist="gaussian")
> res_cgd<-mlmfit(mlmc1,data_surv, Maxiter=300)
######################### OUTPUT #########################
[1]"iterations :"
[1]91
[1]"convergence :"
[1]9.516475e-07
   beta_h     se_beh    t_value       p_value
 5.696051 0.2215316 25.712141 0.0000000000
 1.271542 0.3372314  3.770533 0.0001628992
       alpha1_h  alpha2_h     phi_h
[1,] 0.06614283 0.7146174 2.251489
```

8.5 Discussion

The LMM can be useful in analyzing survival data with random effects. Even for the single random-effect LMM, the marginal likelihood method leads to a very complicated Newton–Raphson procedure (Klein et al. 1999). The h-likelihood method provides the marginal MLEs for the fixed effects and the REMLEs for the dispersion parameters, based upon an analytic Laplace approximation. Alternatively, the marginal MLEs can be obtained via numerical methods such as GHQ, as used in the `NLMIXED` procedure in SAS or in the `gllamm` function in Stata (Rabe-Hesketh et al. 2001, 2002). Recently, R packages such as `nlme` or `lme4` also become available. However, such numerical methods cannot be applied when the number of random components, and hence the dimension of the integral, increases.

By using the h-likelihood, troublesome integration can be avoided, giving a computationally fast and statistically efficient inferential procedure for the LMM with any complex random-effect structure. This procedure can be easily extended to crossed structure by taking $U_{ij} = U_j$ in (8.12). In the LMM without censoring, the h-likelihood estimators for the fixed effects are the same as the MLEs, which are well known to be asymptotically consistent under violation of the normality assumption and misspecification of the covariance matrix; the REMLEs for dispersions, which are derived under the normality assumption, are known to be consistent, even if the normality assumption fails (Jiang 1997). The simulation studies by Ha et al. (2002, 2007b) show that the h-likelihood estimators under censoring are still robust against various violations of the model assumptions: see also Butler and Louis (1992) and Verbeke and Lesaffre (1997). The specification of other distributions for U_i and ϵ_{ij} in the LMM (8.1) is also possible under the h-likelihood framework; for example, a log-gamma distribution for U_i and an extreme value distribution for ϵ_{ij}.

In the CGD data, the multiple infections are recorded per patient from the same hospital, so there may be temporal correlations rather than the compound symmetry correlation within a patient, as in the frailty models used in Chap. 5. It would be interesting to investigate whether an autoregressive structure is necessary. The LMM approach for analysis of correlated survival data was also reviewed, with an application to genetic analysis of life span using data from the Swedish Twin Register. The results under LTRC in Table 8.8 suggest that the h-likelihood method is very useful in practice. Furthermore, the h-likelihood approach using y^* can also be easily extended to left censoring (Ha 2008).

Finally, the h-likelihood methods are somewhat robust but fully parametric. An extension of the h-likelihood method to semiparametric models (e.g., Pan and Louis 2000) would merit future research. It would be also interesting to extend a robust method (Lai and Ying 1994) for the linear models under LTRC to the LMMs via the h-likelihood approach.

8.6 Appendix

8.6.1 Proof of the Expectation Identity in (8.2)

From the pseudo-response variable

$$Y_{ij}^* = Y_{ij}\delta_{ij} + E(T_{ij}|T_{ij} > Y_{ij}, U_i = u_i)(1 - \delta_{ij}),$$

we obtain the following equation:

$$\begin{aligned} E(Y_{ij}^*|U_i = u_i) = & \; E\{T_{ij}\, I(T_{ij} \le C_{ij})|U_i = u_i\} \\ & + \; E\{E(T_{ij}|T_{ij} > C_{ij}, U_i = u_i)\, I(T_{ij} > C_{ij})|U_i = u_i\}. \end{aligned}$$

Now by the conditional independence of T_{ij} and C_{ij} in Assumption 1, the first term on the right-hand side (RHS) of the above equation is

$$E[T_{ij}\, I(T_{ij} \le C_{ij})|U_i = u_i] = E[E\{T_{ij}\, I(T_{ij} \le C_{ij})|T_{ij}, U_i\}|U_i = u_i]$$
$$= E(T_{ij}|U_i = u_i) - \int_0^\infty t\, G_{ij}(t|u)\, dF_{ij}(t|u),$$

and the second term on the RHS is also given by

$$E\{E(T_{ij}|T_{ij} > C_{ij}, u_i)\, I(T_{ij} > C_{ij})|U_i = u_i\} = \int_0^\infty t\, G_{ij}(t|u)\, dF_{ij}(t|u),$$

where $G_{ij}(\cdot|u)$ and $F_{ij}(\cdot|u)$ are arbitrary continuous conditional distribution functions of $C_{ij}|U_i = u_i$ and $T_{ij}|U_i = u_i$, respectively. Thus, by combining the two equations, we obtain the expectation identity

$$E(Y_{ij}^*|U_i = u_i) = E(T_{ij}|U_i = u_i)$$
$$= \mu_{ij}.$$
□

8.6.2 Proofs of the IWLS Equations (8.7)

For this proof, we use matrix manipulations given in Appendix 4.7.4. First, substituting (8.3) into the two MHL Eqs. (8.5) and (8.6) reduces them, respectively, to

$$\sigma_\epsilon^{-2} \sum_{ij} \left(y_{ij}^* - \mu_{ij} \right) x_{ijk} = 0 \; (k = 1, \ldots, p), \tag{8.25}$$

and

$$\sigma_\epsilon^{-2} \sum_{j} \left(y_{ij}^* - \mu_{ij} \right) - \sigma_u^{-2} u_i = 0 \; (i = 1, \ldots, q). \tag{8.26}$$

Let $E = (X, Z)$ and let μ be an $n \times 1$ vector with the ijth element μ_{ij}. Here, X and Z are $n \times p$ and $n \times q$ model matrices for β and v whose ijth row vectors are x_{ij}^T and z_{ij}^T, respectively. Then, μ_{ij} can be expressed as

$$\mu = X\beta + Zu = E\tau.$$

The two score equations of (8.25) and (8.26) can also be expressed as

$$\frac{\partial h}{\partial \tau} = \sigma_\epsilon^{-2} E^T (y^* - \mu) + b, \tag{8.27}$$

where y^* be an $n \times 1$ vector with the ijth element y_{ij}^* and $b = (0^T, -\sigma_u^{-2} u^T)^T$.

Next, from (8.5) and (8.6), we have the negative second partial derivatives with respect to β_k and u_i:

$$\frac{-\partial h^2}{\partial \beta_k \partial \beta_l} = \sigma_\epsilon^{-2} \Sigma_{ij} x_{ijk} w_{ij} x_{ijl},$$
$$\frac{-\partial h^2}{\partial \beta_k \partial u_r} = \sigma_\epsilon^{-2} \Sigma_{ij} x_{ijk} w_{ij} z_{ijl},$$
$$\frac{-\partial h^2}{\partial u_k \partial u_l} = \sigma_\epsilon^{-2} \Sigma_{ij} z_{ijk} w_{ij} z_{ijl} + \sigma_u^{-2} I(k = l),$$

where $w_{ij} = \delta_{ij} + (1 - \delta_{ij})\xi(m_{ij})$ with $\xi(x) = \partial V(x)/\partial x = V(x)\{V(x) - x\}$. Then, the three derivatives above are expressed as a simple matrix form, H, in (8.8) and it can also be written as

$$H = E^T W^* E + F, \tag{8.28}$$

where $F = BD(0, Q)$ with $Q = \sigma_u^{-2} I_q$.

From $\widehat{\tau} = \tau + H^{-1}(\partial h/\partial \tau)$, (8.27), and (8.28), we obtain

$$\begin{aligned}(E^T W^* E + F)\widehat{\tau} &= (E^T W^* E + F)\tau + \sigma_\epsilon^{-2} E^T (y^* - \mu) + b \\ &= (E^T W^* E)\tau + \sigma_\epsilon^{-2} E^T (y^* - \mu) + g \\ &= E^T W^* w^*\end{aligned}$$

since $g = F\tau + b = 0$ and $w^* = \mu + W^{-1}(y^* - \mu)$. This completes the proof of (8.7). □

8.6.3 *Proofs of the Two Dispersion Estimators in (8.11)*

From (8.8), the Hessian matrix H can be written as

$$H = H(h; \tau) = -\frac{\partial^2 h}{\partial \tau^2} = \frac{H_0}{\sigma_\epsilon^2}$$

with

$$H_0 = \begin{pmatrix} X^T W X & X^T W Z \\ Z^T W X & Z^T W Z + \sigma_\epsilon^2 Q \end{pmatrix},$$

where $W = \mathrm{diag}(w_{ij})$. Thus, from (3.9) and (8.4), we have

$$p_\tau(h) = \hat{h} - \frac{1}{2}\log\{\det(\hat{H}_0)\} + \frac{p+q}{2}\log(2\pi\sigma_\epsilon^2), \tag{8.29}$$

where $\hat{h} = h|_{\tau=\hat{\tau}(\theta)} = h(\hat{\tau}(\theta), \theta)$ and $\hat{H}_0 = H_0|_{\tau=\hat{\tau}(\theta)} = H_0(\hat{\tau}(\theta), \theta)$ with $\theta = (\sigma_\epsilon^2, \sigma_u^2)^T$. Note that in solving $\partial p_\tau(h)/\partial\theta_k = 0$ $(k = 1, 2)$, we allow the term $\partial\hat{u}/\partial\theta_k$, not $\partial\hat{\beta}/\partial\theta_k$. From (8.29), we have

$$\frac{\partial p_\tau(h)}{\partial\sigma_\epsilon^2} = \frac{\partial\ell_1}{\partial\sigma_\epsilon^2}|_{u=\hat{u}} + \frac{1}{2\sigma_\epsilon^2}\{(p+q) - \gamma_0\}, \tag{8.30}$$

where $\ell_1 = \sum_{ij}\ell_{1ij}$,

$$\partial\ell_1/\partial\sigma_\epsilon^2 = [\ -r + \sum_{ij\in D_0}(m_{ij})^2 + \sum_{ij\in C_0}\{m_{ij}V(m_{ij})\}\ ]/(2\sigma_\epsilon^2),$$

$r = \sum_{ij}\delta_{ij}$, $m_{ij} = (y_{ij} - \mu_{ij})/\sigma_\epsilon$, D_0 and C_0 are the index sets for uncensored and censored observations, respectively, and $\gamma_0 = \sigma_\epsilon^2\text{trace}\{\hat{H}_0^{-1}(\partial\hat{H}_0/\partial\sigma_\epsilon^2)\}$. Now, substituting Eq. (8.3) into $\partial\ell_1/\partial\sigma_\epsilon^2$ in Eq. (8.30) gives

$$\frac{\partial\ell_1}{\partial\sigma_\epsilon^2} = -\frac{r}{2\sigma_\epsilon^2} + \frac{1}{2\sigma_\epsilon^2}\sum_{ij}\frac{(y_{ij}^* - \mu_{ij})^2}{\sigma_\epsilon^2} - \frac{1}{2\sigma_\epsilon^2}\sum_{ij\in C_0}\xi(m_{ij}).$$

Thus, Eq. (8.30) reduces to

$$\frac{\partial p_\tau(h)}{\partial\sigma_\epsilon^2} = -\frac{1}{2\sigma_\epsilon^2}\{n_0 - (p + q - \gamma_0)\} + \frac{1}{2\sigma_\epsilon^2}\sum_{ij}\frac{1}{\sigma_\epsilon^2}(y_{ij}^* - \hat{\mu}_{ij})^2, \tag{8.31}$$

where $n_0 = r + \sum_{ij\in C}\xi(\hat{m}_{ij}) = \sum_{ij} w_{ij}$ with $w_{ij} = \delta_{ij} + (1 - \delta_{ij})\xi(\hat{m}_{ij})$. Here, the term $\partial H_0/\partial\sigma_\epsilon^2$ in γ_0 is calculated as follows:

$$\frac{\partial\hat{H}_0}{\partial\sigma_\epsilon^2} = \begin{pmatrix} X^T W' X & X^T W' Z \\ Z^T W' X & Z^T W' Z + \sigma_u^{-2} I_q \end{pmatrix},$$

where $W' = \text{diag}(w'_{ij})$ is an $n \times n$ diagonal matrix with the ijth element

$$w'_{ij} = \partial w_{ij}/\partial\sigma_\epsilon^2 = (1 - \delta_{ij})d_{ij}$$

and $d_{ij} = \partial\xi(\hat{m}_{ij})/\partial\sigma_\epsilon^2 = \{m_{ij}/2\sigma_\epsilon^2 + \sigma_\epsilon^{-1}\partial(\hat{u}_i/\partial\sigma_\epsilon^2)\}V(m_{ij})\{3m_{ij}V(m_{ij}) - 2V^2(m_{ij}) - m_{ij}^2 + 1\}$. Similarly, from (8.29)

$$\frac{\partial p_\tau(h)}{\partial \sigma_u^2} = -\frac{1}{2\sigma_u^2}(q - \gamma_1) + \frac{1}{2\sigma_u^2}\sum_i \frac{\hat{u}_i^2}{\sigma_u^2}, \tag{8.32}$$

where $\gamma_1 = -\sigma_u^2 \,\text{trace}\{\hat{H}_0^{-1}(\partial \hat{H}_0/\partial \sigma_u^2)\}$.

Note that the computation of the terms $\partial \hat{H}_0/\partial \theta_k$ in γ_0 and γ_1 requires evaluating $\partial \hat{u}/\partial \theta_k$. Now we show how to implement those terms. Let $u = (u_1, \ldots, u_q)^T$. Then, we obtain

$$\frac{\partial h}{\partial u} = \sigma_\epsilon^{-1} Z^T a - D^{-1}u,$$

where a is an $n \times 1$ vector with the ijth element $a_{ij} = \delta_{ij} m_{ij} + (1 - \delta_{ij})V(m_{ij})$ and $D = \text{diag}(\sigma_u^2 I_q)$ is a $q \times q$ diagonal matrix. From the h-likelihood h, given $\theta = (\sigma_u^2, \sigma_\epsilon^2)^T$, let $\widehat{u}$ be the solution of $f(\theta) = \partial h/\partial u|_{\tau=\hat{\tau}} = \sigma_\epsilon^{-1} Z^T \widehat{a} - D^{-1}\widehat{u} = 0$. First, from $\partial f(\theta)/\partial \sigma_\epsilon^2 = -\frac{1}{2}\sigma_\epsilon^{-3}(Z^T\widehat{a} + Z^T W\widehat{m}) - \sigma_\epsilon^2(Z^T W Z + \Lambda)(\partial \widehat{u}/\partial \sigma_\epsilon^2) = 0$, where m is an $n \times 1$ vector with the ijth element m_{ij} and $\widehat{m} = \sigma_\epsilon^{-1}(y - X\widehat{\beta} - Z\widehat{u})$, and $\Lambda = \sigma_\epsilon^2 Q$, we have

$$\frac{\partial \widehat{u}}{\partial \sigma_\epsilon^2} = -\frac{1}{2}\sigma_\epsilon^{-1}(Z^T W Z + \Lambda)^{-1}\{Z^T(\widehat{a} + W\widehat{m})\}.$$

Next, from $\partial f(\theta)/\partial \sigma_u^2 = -D'\widehat{u} - \sigma_\epsilon^2(Z^T W Z + \Lambda)(\partial \widehat{u}/\partial \sigma_u^2) = 0$, where $D' = \partial D^{-1}/\partial \sigma_u^2$, we have

$$\frac{\partial \widehat{u}}{\partial \sigma_u^2} = -\sigma_\epsilon^2(Z^T W Z + \Lambda)^{-1}D'\widehat{u}.$$

These two results can also be obtained using (4.41). Accordingly, application of the two Eqs. (8.31) and (8.32) to the estimating equations $\partial p_\tau(h)/\partial \theta = 0$ completes the proof of (8.11). □

8.6.4 *H-Likelihood Procedure for Fitting the Multicomponent LMM*

We shall demonstrate that the h-likelihood procedure can be extended to arbitrarily structured multicomponent (nested and/or crossed) models. Without loss of generality, we consider model (8.12). Let μ be an $n \times 1$ vector with the ijkth element μ_{ijk}, i.e.,

$$\mu = X\beta + Zu,$$

where X is an $n \times p$ model matrix for the $p \times 1$ fixed effects β, $Z = (Z_1, Z_2)$ is an $n \times q$ model matrix for the $q \times 1$ random effects $u = (u^{(1)T}, u^{(2)T})^T$ and $Zu = Z_1 u^{(1)} + Z_2 u^{(2)}$. Here, Z_i $(i = 1, 2)$ are an $n \times q_i$ model matrix for an

$q_i \times 1$ random effects $u^{(i)}$, $u^{(1)} = (u_1, \ldots, u_{q_1})^T$ and $u^{(2)} = (u_1^{(2)T}, \ldots, u_{q_1}^{(2)T})^T$ with $u_i^{(2)} = (u_{i1}, \ldots, u_{ir_i})^T$ for $i = 1, \ldots, q_1$. Let y^* be an $n \times 1$ vector with the ijkth element y^*_{ijk}.

Given the dispersion parameters $\theta = (\sigma_1^2, \sigma_2^2, \sigma_\epsilon^2)^T$ and y^*, the score equations for the MHLEs of $\tau = (\beta^T, u^T)^T$ are directly extended. That is, the IWLS equations in (8.7) are straightforwardly extended to the multicomponent LMM with $Z = (Z_1, Z_2)$, $Q = \text{BD}(Q_1, Q_2)$, and $W = \text{diag}(w_{ijk})$, where $w_{ijk} = \delta_{ijk} + (1 - \delta_{ijk})\xi(m_{ijk})$.

The REML estimators for θ are also easily obtained using the corresponding restricted h-likelihood $p_\tau(h)$. That is, they are given by

$$\widehat{\sigma_\epsilon^2} = \frac{(y^* - \widehat{\mu})^T(y^* - \widehat{\mu})}{n_0 - (p + q - \gamma_0)} \quad \text{and} \quad \widehat{\sigma_i^2} = \frac{\widehat{u}^{(i)T}\widehat{u}^{(i)}}{q_i - \gamma_i} \; (i = 1, 2), \tag{8.33}$$

where $\widehat{\mu} = X\widehat{\beta} + Z\widehat{u}$, $n_0 = \sum_{ijk} w_{ijk}$, $\gamma_0 = \sigma_\epsilon^2 \text{trace}\{H_0{}^{-1}(\partial H_0/\partial\sigma_\epsilon^2)\}$ and $\gamma_i = -\sigma_i^2 \text{trace}\{H_0{}^{-1}(\partial H_0/\partial\sigma_i^2)\}$. Appendix 8.6.3 gives the formulae for the terms $\partial H_0/\partial\sigma_\epsilon^2$ and $\partial H_0/\partial\sigma_i^2$. Note that since we cannot observe all the y^*_{ijk}'s due to censoring, we substitute their estimates, say $\widehat{y^*_{ijk}}$, in each iteration. □

8.6.5 Derivation of Model (8.16)

From $v_{ij} = g_{ij} + c_{i0}$ for $j = 1, 2$, the model (8.15) can be expressed as a simple matrix form:

$$\log T_i = X_i\beta + Z_i v_i + \epsilon_i, \tag{8.34}$$

where $T_i = (T_{i1}, T_{i2})^T$, $X_i = (x_{i1}, x_{i2})^T$ is a $2 \times p$ model matrix for β, Z_i is a model matrix for v_i, $\epsilon_i = (\epsilon_{i1}, \epsilon_{i2})^T \sim N(0, \sigma_\epsilon^2 I_2)$, and I_2 is a 2×2 identity matrix. For MZ$_i$, $Z_i = (1, 1)^T$ and $v_i (= v_{i1} = v_{i2}) \sim N(0, \sigma_v^2)$, but for DZ$_i$, $Z_i = I_2$ and $v_i = (v_{i1}, v_{i2})^T \sim N(0, \sigma_v^2 \Sigma_i)$ with a compound symmetric correlation structure

$$\Sigma_i = \begin{pmatrix} 1 & \rho \\ \rho & 1 \end{pmatrix}.$$

Here,

$$\sigma_v^2 = \sigma_g^2 + \sigma_c^2, \tag{8.35}$$

and

$$\rho = \text{corr}(v_{i1}, v_{i2}) = \frac{0.5\sigma_g^2 + \sigma_c^2}{\sigma_g^2 + \sigma_c^2}, \tag{8.36}$$

where $\rho \in [0.5, 1.0]$. The use of ρ leads to some useful results. From (8.36), we see that σ_g^2 is much larger than σ_c^2 (i.e., $\sigma_g^2 \gg \sigma_c^2$) if ρ goes to 0.5, but $\sigma_g^2 \ll \sigma_c^2$ if ρ goes to 1.0. In particular, the model (8.34) reduces to model (8.15) without the random environmental effects c_{ij} if $\rho = 0.5$ (i.e., $\sigma_c^2 = 0$), while it becomes model (8.15) without the random genetic effects g_{ij} if $\rho = 1.0$ (i.e., $\sigma_g^2 = 0$).

Following Lee and Nelder (2001b), the random effects v_i for DZ$_i$ are assumed to have the form $L_i(\rho)u_i$, where $u_i \sim N(0, \sigma_v^2 I_2)$. For DZ_i, using the Cholesky decomposition, we have a lower triangular matrix L_i such that $\Sigma_i = L_i L_i^T$. Here, we choose

$$L_i(\rho) = \begin{pmatrix} 1 & 0 \\ \rho & \sqrt{1-\rho^2} \end{pmatrix},$$

and so the random effects $v_i = L_i u_i \sim N(0, \sigma_v^2 L_i L_i^T)$.

Thus, model (8.34) can be written as

$$\log T_i = X_i\beta + Z_i^* u_i + \epsilon_i, \tag{8.37}$$

where $u_i \sim N(0, \sigma_v^2 I_k)$, and $Z_i^* = (1, 1)^T$ and $I_k = 1$ for MZ_i, and $Z_i^* = L_i(\rho)$ and $I_k = I_2$ for DZ$_i$. Note that from (8.35) and (8.36), we obtain σ_g^2 and σ_c^2 as follows:

$$\sigma_g^2 = \sigma_v^2 - \sigma_c^2 \text{ and } \sigma_c^2 = 2(\rho - 0.5)\sigma_v^2. \tag{8.38}$$

Then, the jth element of model (8.37) becomes the model (8.16). □

8.6.6 Derivations of the Score Equations in (8.21) and (8.22), and Computation of Variance of $\widehat{\beta}$

Let μ be an $n \times 1$ vector with the ijth element μ_{ij},

$$\mu = X\beta + Z^* u,$$

where $X = (X_1^T, \ldots, X_q^T)^T$ is an $n \times p$ model matrix for the $p \times 1$ fixed effects β and $Z^* = \text{BD}(Z_1^*, \ldots, Z_q^*)$ is an $n \times q^*$ block diagonal matrix for the $q^* \times 1$ random effects $u = (u_1, \ldots, u_q)^T$. Here, $q^* = q_1 + 2q_2$, q_1 being the number of MZ twin pairs and q_2 that of DZ twin pairs. Note that $q = q_1 + q_2$. Let $y^* = (y_1^{*T} \ldots, y_q^{*T})^T$ be an $n \times 1$ vector with the ith vector $y_i^* = (y_{i1}^*, y_{i2}^*)^T$. Assume that ρ is known. Given $\theta = (\sigma_\epsilon^2, \sigma_v^2)^T$ and y^*, from (8.19) and (8.20), the score equations for the MHLEs of $\tau = (\beta^T, u^T)^T$ become Henderson's (1975) mixed model equations with pseudo-response variables y^*:

$$\begin{pmatrix} X^T X & X^T Z^* \\ Z^{*T} X & Z^{*T} Z^* + \Lambda \end{pmatrix} \begin{pmatrix} \widehat{\beta} \\ \widehat{u} \end{pmatrix} = \begin{pmatrix} X^T y^* \\ Z^{*T} y^* \end{pmatrix}, \tag{8.39}$$

where $\Lambda = \lambda I_{q^*}$, $\lambda = \sigma_\epsilon^2/\sigma_v^2$ and I_{q^*} is a $q^* \times q^*$ identity matrix. The Eq. (8.39) can be expressed as the two equations:

$$(X^T X)\widehat{\beta} + (X^T Z^*)\widehat{u} = X^T y^* , \tag{8.40}$$

$$(Z^{*T} X)\widehat{\beta} + (Z^{*T} Z^* + \lambda I_{q^*})\widehat{u} = Z^{*T} y^* . \tag{8.41}$$

Substituting

$$\begin{aligned} X &= (X_1^T, \ldots, X_q^T)^T, \\ Z^* &= \text{BD}(Z_1^*, \ldots, Z_q^*) \text{ and} \\ y^* &= (y_1^{*T}, \ldots, y_q^{*T})^T \end{aligned}$$

into (8.40) and (8.41) reduces them to Eqs. (8.21) and (8.22).

The asymptotic covariance matrix for $\widehat{\tau} - \tau$ is given by H^{-1} with

$$H = -\frac{\partial^2 h}{\partial \tau^2} = \frac{1}{\sigma_\epsilon^2} H_0, \tag{8.42}$$

where

$$H_0 = \begin{pmatrix} X^T W X & X^T W Z^* \\ Z^{*T} W X & Z^{*T} W Z^* + \Lambda \end{pmatrix}.$$

Here, $W = \text{diag}(w_{ij})$ is an $n \times n$ diagonal matrix with the ijth element $w_{ij} = \delta_{ij} + (1 - \delta_{ij})\xi(m_{ij}) - \xi(m_{ij}^*)$ and $\xi(x) = V(x)\{V(x) - x\}$. So, the upper left-hand corner of H^{-1} in (8.42) gives the variance matrix of $\widehat{\beta}$, which is also easily computed for large samples as follows. Let H_0^{11} be the upper left-hand corner of $H_0{}^{-1}$ in (8.42). Then we have

$$\text{var}(\widehat{\beta}) = \sigma_\epsilon^2 H_0^{11}, \tag{8.43}$$

where

$$\begin{aligned} H_0^{11} &= \{(X^T W X) - (X^T W Z^*)(Z^{*T} W Z^* + \lambda I_{q^*})^{-1}(Z^{*T} W X)\}^{-1} \\ &= \{\sum_i X_i^T W_i X_i - \sum_i (X_i^T W_i Z_i^*)(Z_i^{*T} W_i Z_i^* + \lambda I_k)^{-1}(Z_i^{*T} W_i X_i)\}^{-1} . \end{aligned}$$

Here, W_i is the ith component matrix of W. □

Chapter 9
Joint Model for Repeated Measures and Survival Data

In this chapter, through the h-likelihood approach we study the joint model, for which the response variables of interest would involve repeated measurements over time on the same subject as well as time to an event of interest with or without competing risks. The analyses presented here will further extend the multivariate analyses performed by Lee et al. (2017b) for the HGLM to a joint model where at least one outcome is time-to-event.

9.1 Introduction

Consider a case where a subject has two outcomes, y_1 and y_2. Then, y_1 and y_2 may be correlated due to a shared individual effect. Modeling jointly these outcomes would be more informative because a separated analysis ignoring the association can result in biases (Guo and Carlin 2004). For this joint modeling, an unobserved random effect can be used to describe an association between the two outcomes.

Let v be a common random effect for the same subject and θ be an unknown parameter. To analyze this type of data, we may use the marginal likelihood, defined by

$$L(\theta) = f_\theta(y_1, y_2) = \int f_\theta(y_1, y_2|v) f_\theta(v) dv, \tag{9.1}$$

where $f_\theta(y_1, y_2)$ is the joint density function of (y_1, y_2), and $f_\theta(\cdot|v)$ is the conditional density of (y_1, y_2) given the random effect v, and $f_\theta(v)$ is the density of v. Here, the h-likelihood is defined by

$$h = \log f_\theta(y_1, y_2|v) + \log f_\theta(v). \tag{9.2}$$

Up to now, we have considered a single response variable y. For an extension to the multiple response variables $y_1, \cdots, y_K$, $f_\theta(y|v)$ is simply replaced by the joint density function $f_\theta(y_1, \cdots, y_K|v)$ in the h-likelihood h.

I.D. Ha et al., *Statistical Modelling of Survival Data with Random Effects*, Statistics for Biology and Health, https://doi.org/10.1007/978-981-10-6557-6_9

9.2 Joint Model for Repeated Measures and a Single Event-Time Data

• **A motivating example**: Consider an example of renal transplant data (Ha et al. 2003). Data were available from 87 male and 25 female renal transplanted patients who survived more than 4 years after transplant. The aim of this study was to investigate the chronic renal allograft dysfunction in renal transplants. For each patient, both repeated-measure outcomes (y_1: serum creatinine levels) at several time points and a terminating event time (y_2: graft-loss time) were observed. Here, the renal function was evaluated from the serum creatinine level. Since these two types of observations (y_1, y_2) were collected from the same patient, they are correlated.

• **A joint model with frailty model**: It is of interest to investigate the covariate effects on these two types of responses. For the analysis of renal transplant data, we consider a joint model with the LMM for y $(= y_1)$ and the frailty model for T $(= y_2)$. Let y_{ij} be the jth repeated response of the ith patient at time point t $(i = 1, \ldots, q;\ j = 1, \ldots, n_i)$, and let T_i be a single event time of the ith patient and let C_i be the corresponding censoring time. Denote by v_i a shared random effect of the ith patient. In this section, we assume both $y_i = (y_{i1}, \ldots y_{in_i})^T$ and T_i given v_i are conditionally independent, and that T_i and C_i given v_i are also conditionally independent. Then, conditional on v_i, y_i and T_i are assumed to have the following joint model:

$$\text{(i) } y_{ij} = x_{ij1}^T\beta_1 + v_i + \epsilon_{ij},$$

where $\epsilon_{ij} \sim N(0, \phi)$, and

$$\text{(ii) } \lambda_i(t|v_i) = \lambda_0(t)\exp(x_{i2}^T\beta_2 + \gamma v_i),$$

where $v_i \sim N(0, \alpha)$ and $\lambda_0(\cdot)$ is a completely unspecified baseline hazard function, and v_i and ϵ_{ij} are independent. Here, β_1 and β_2 are $p_1 \times 1$ and $p_2 \times 1$ regression parameter vectors corresponding to the vectors of covariates x_{ij1} and x_{i2}, respectively; x_{i2} may be a subset of x_{ij1}. This is a shared random-effect model. Note that γ is a real-valued association parameter that allows the magnitude of the association to be different between two outcomes, y_{ij} and T_i; if $\gamma > 0$ $(\gamma < 0)$, then y_{ij} and the hazard rate tend to be positively (negatively) correlated, and if $\gamma = 0$ they are not associated.

• **Construction of the h-likelihood**:

All observable random variables are repeated-measure responses y_{ij} and time-to-event data with

$$t_i^* = \min(T_i, C_i) \text{ and } \delta_i = I(T_i \le C_i).$$

Construction of the h-likelihood for the joint model above is immediate. We take $y_1 = y$ and $y_2 = (t^*, \delta)$ in (9.2). By the assumptions of conditional independence between

$y_i = (y_{i1}, \ldots y_{in_i})^T$ and T_i and the noninformative censoring, the h-likelihood is defined by

$$h = \sum_{ij} \ell_{1ij} + \sum_i \ell_{2i} + \sum_i \ell_{3i}, \tag{9.3}$$

where

$$\begin{aligned}
\ell_{1ij} &= \ell_{1ij}(\beta_1, \phi; y_{ij}|v_i) = \log f_{\beta_1,\phi}(y_{ij}|v_i) \\
&= -\log(2\pi\phi)/2 - (y_{ij} - \eta_{1ij})^2/(2\phi), \\
\ell_{2i} &= \ell_{2i}(\beta_2, \lambda_0; t_i^*, \delta_i|v_i) = \log f_{\beta_2,\lambda_0}(t_i^*, \delta_i|v_i) \\
&= \delta_i(\log \lambda_0(t_i^*) + \eta_{2i}) - \Lambda_0(t_i^*)\exp(\eta_{2i}), \\
\ell_{3i} &= \ell_{3i}(\alpha; v_i) = \log f_\alpha(v_i) = -\log(2\pi\alpha)/2 - v_i^2/(2\alpha).
\end{aligned}$$

where ℓ_{1ij} is the conditional log-likelihood for y_{ij} given v_i, ℓ_{2i} is that for (t_i^*, δ_i) given v_i, and ℓ_{3i} is the log-likelihood for v_i. Here, we have two linear predictors

$$\eta_{1ij} = x_{ij1}^T \beta_1 + v_i$$

and

$$\eta_{2i} = x_{i2}^T \beta_2 + \gamma v_i.$$

9.2.1 Estimation Procedure

Because the functional form of $\lambda_0(\cdot)$ from ℓ_{2i} in (9.3) is unknown, we again define the baseline cumulative hazard function $\Lambda_0(t) = \int_0^t \lambda_0(u)du$ to be a step function with jumps λ_{0r} at the observed event times $t_{(r)}$:

$$\Lambda_0(t) = \sum_{r:t_{(r)} \leq t} \lambda_{0r}, \tag{9.4}$$

where $t_{(r)}$ is the rth $(r = 1, \ldots, D)$ smallest distinct event time among the t_i^*'s and $\lambda_{0r} = \lambda_0(t_{(r)})$. By substituting (9.4) into (9.3), the second term $\sum_i \ell_{2i}$ in (9.3) becomes

$$\sum_i \ell_{2i} = \sum_r d_{(r)} \log \lambda_{0r} + \sum_i \delta_i \eta_{2i} - \sum_r \lambda_{0r} \left\{ \sum_{i \in R_{(r)}} \exp(\eta_{2i}) \right\},$$

where $d_{(r)}$ is the number of events at $t_{(r)}$ and $R_{(r)} = \{i : t_i^* \geq t_{(r)}\}$ is the risk set at $t_{(r)}$. As the number of λ_{0r}'s in $\sum_i \ell_{2i}$ increases with the number of events, the function $\lambda_0(t)$ is potentially of high dimension. Here, the profile h-likelihood is given by

$$h^* = h|_{\lambda_0=\widehat{\lambda}_0} = \sum_{ij} \ell_{1ij} + \sum_i \ell_{2i}^* + \sum_i \ell_{3i}, \tag{9.5}$$

where

$$\sum_i \ell_{2i}^* = \sum_i \ell_{2i}|_{\lambda_0=\widehat{\lambda}_0} = \sum_r d_{(r)} \log \widehat{\lambda}_{0r} + \sum_i \delta_i \eta_{2i} - \sum_r d_{(r)}.$$

Here,

$$\widehat{\lambda}_{0r} = \widehat{\lambda}_{0r}(\beta_2, v) = \frac{d_{(r)}}{\sum_{i \in R_{(r)}} \exp(\eta_{2i})}$$

are the solutions of the estimating equations, $\partial h/\partial \lambda_{0r} = 0$, for $r = 1, \ldots, D$. Thus, h^* does not depend on λ_{0r} and it is proportional to the partial h-likelihood h_p, given by

$$h_p = \sum_{ij} \ell_{1ij} + \sum_i \delta_i \eta_{2i} - \sum_r d_{(r)} \log\left\{ \sum_{i \in R_{(r)}} \exp(\eta_{2i}) \right\} + \sum_i \ell_{3i}. \tag{9.6}$$

Accordingly, the h-likelihood procedure can be derived via h_p. Let X_1, X_2, and Z be model matrices for vectors β_1, β_2 and $v = (v_1, \ldots, v_q)^T$, respectively. The score equations for fixed and random effects (β_1, β_2, v) given dispersion parameters $\psi = (\phi, \alpha, \gamma)^T$ are

$$\begin{aligned}
\partial h_p/\partial \beta_1 &= X_1^T (y - \mu_1)/\phi, \\
\partial h_p/\partial \beta_2 &= X_2^T (\delta - \hat{\mu}_2), \\
\partial h_p/\partial v &= {Z_1}^T (y - \mu_1)/\phi + \gamma {Z_2}^T (\delta - \hat{\mu}_2) - v/\alpha,
\end{aligned}$$

where $\mu_1 = X_1\beta_1 + Z_1 v = \eta_1$, and $\hat{\mu}_2 = \exp(\log \hat{\Lambda}_0(t^*) + \eta_2)$ with $\eta_2 = X_2\beta_2 + \gamma Z_2 v$, and Z_1 is an $n \times q$ group indicator matrix whose ijkth element z_{ijk} is $\partial \eta_{1ij}/\partial v_k$ and $Z_2 = I_q$. Here, $\hat{\Lambda}_0(t) = \sum_{r: t_{(r)} \leq t} \hat{\lambda}_{0r}$ is the Breslow-type estimator of the cumulative baseline hazard. Thus, the ILS equations (4.12) in Chap. 4 are extended as

$$\begin{pmatrix} {X_1}^T W_1 X_1 & 0 & X_1^T W_1 Z_1 \\ 0 & X_2^T W_2^* X_2 & X_2^T (\gamma W_2^*) Z_2 \\ Z_1^T W_1 X_1 & {Z_2}^T (\gamma W_2^*) X_2 & \mathbf{Z}^T \mathbf{W} \mathbf{Z} + Q \end{pmatrix} \begin{pmatrix} \hat{\beta}_1 \\ \hat{\beta}_2 \\ \hat{v} \end{pmatrix} = \begin{pmatrix} {X_1}^T W_1 w_1 \\ {X_2}^T w_2^* \\ \mathbf{Z}^T \mathbf{w}^* \end{pmatrix},$$

where $W_1 = -\partial^2 h_p/\partial \eta_1 \partial \eta_1^T = \phi^{-1} I_{p_1}$, and $W_2^* = -\partial^2 h_p/\partial \eta_2 \partial \eta_2^T$, $Q = -\partial^2 \ell_3/\partial v \partial v^T = \alpha^{-1} I_q$, $w_1 = y$, $w_2^* = W_2^* \eta_2 + (\delta - \hat{\mu}_2)$, and

$$\mathbf{Z} = \begin{pmatrix} Z_1 \\ \gamma Z_2 \end{pmatrix}, \quad \mathbf{W} = \begin{pmatrix} W_1 & \mathbf{0} \\ \mathbf{0} & W_2^* \end{pmatrix} \quad \text{and} \quad \mathbf{w}^* = \begin{pmatrix} W_1 w_1 \\ w_2^* \end{pmatrix}.$$

Note here that $\mathbf{Z}^T\mathbf{W}\mathbf{Z} = Z_1^T W_1 Z_1 + Z_2^T(\gamma^2 W_2^*)Z_2$ and that $\mathbf{Z}^T\mathbf{w}^* = Z_1^T W_1 w_1 + \gamma {Z_2}^T w_2^*$.

Again the ILS equations above for $\tau = (\beta_1^T, \beta_2^T, v^T)^T$ lead to a simple matrix form:

$$(\mathbf{P}^T\mathbf{V}\mathbf{P})\hat{\tau} = \mathbf{P}^T\mathbf{y}_\mathbf{0}^*, \tag{9.7}$$

where

$$\mathbf{P} = \begin{pmatrix} \mathbf{X} & \mathbf{Z} \\ \mathbf{0} & I_q \end{pmatrix}, \quad \mathbf{X} = \begin{pmatrix} X_1 & \mathbf{0} \\ \mathbf{0} & X_2 \end{pmatrix}, \quad \mathbf{V} = \begin{pmatrix} \mathbf{W} & \mathbf{0} \\ \mathbf{0} & Q \end{pmatrix},$$

and $\mathbf{y}_\mathbf{0}^* = (w_1^{*T}, w_2^{*T}, \mathbf{0}^T)^T$ with $w_1^* = W_1 w_1$. Note that $H_p = -\partial^2 h_p/\partial\tau\partial\tau^T = \mathbf{P}^T\mathbf{V}\mathbf{P}$.

To estimate $\psi = (\phi, \alpha, \gamma)^T$, we again use the partial restricted likelihood $p_\tau(h_p)$. The PREMLEs of ψ are obtained by solving

$$\partial p_\tau(h_p)/\partial\psi = 0,$$

leading to the PREMLEs for ϕ and α

$$\hat{\phi} = \frac{(y-\hat{\mu}_1)^T(y-\hat{\mu}_1)}{n-\kappa_0} \quad \text{and} \quad \hat{\alpha} = \frac{\hat{v}^T\hat{v}}{q-\kappa_1},$$

where $\kappa_0 = -\phi\text{tr}\{\hat{H}_p^{-1}(\partial\hat{H}_p/\partial\phi)\}$, $\kappa_1 = -\alpha\text{tr}\{\hat{H}_p^{-1}(\partial\hat{H}_p/\partial\alpha)\}$, $\hat{H}_p = \mathbf{P}^T\mathbf{V}\mathbf{P}|_{\tau=\hat{\tau}(\psi)}$, and $\hat{\mu}_1 = X_1\hat{\beta}_1 + Z_1\hat{v}$. The estimator of γ is also easily implemented via the Newton–Raphson method using the first and second derivatives, $\partial p_\tau(h_p)/\partial\gamma$ and $\partial^2 p_\tau(h_p)/\partial\gamma^2$. Thus, we see the h-likelihood procedure in Chap. 4 can be straightforwardly extended to the joint models.

9.2.2 *Numerical Study*

A simulation study, based on 500 replications, is presented to evaluate the performance of the proposed joint modeling approach. For simplicity, we consider a joint model for repeated measures and a single event time. The simulation scheme is as follows. First, we generate the random effects $v_i \sim N(0, \alpha = 0.5)$ for $i = 1, \ldots, 50$. Given v_i, the repeated-measure responses y_{ij} for $j = 1, 2, 3, 4$ are generated from the LMM with two covariates (time_{ij}, Trt_i):

$$y_{ij} \sim N(\beta_{10} + \beta_{11}\text{time}_{ij} + \beta_{12}\text{Trt}_i + v_i, \phi = 1),$$

where we set $\beta_{10} = -0.5$, $\beta_{11} = 0.5$, $\beta_{12} = 1$, $\text{time}_{ij} = 0, 2, 4, 8$ (weeks), and Trt_i are generated from Bernoulli distribution with the equal probability of 0.5. Here, "Trt" denotes a treatment group with a new drug (coded as 1) or placebo (coded as 0), which are assigned to subjects by randomization without replacement. Then, given v_i, survival times T_i's, for $i = 1, \ldots, 50$, are also generated from the frailty model with one covariate (Trt_i):

$$\lambda_i(t|v_i) = \lambda_0(t) \exp(\beta_{21}\text{Trt}_i + \gamma v_i),$$

where we set $\lambda_0(t) = 1$, $\beta_{21} = -1$, and $\gamma = -1$ or 1. Finally, the corresponding censoring times C_i's are generated from an exponential distribution to induce about 30% censoring rate. We set the maximum follow-up time to be 8. With 500 replications, we computed the mean, the standard deviation (SD), and the mean of the estimated standard errors (SE) for $\hat{\beta} = (\hat{\beta}_{10}, \hat{\beta}_{11}, \hat{\beta}_{12}, \hat{\beta}_{21})^T$ and $(\hat{\phi}, \hat{\alpha}, \hat{\gamma})$. In addition, we calculated the empirical coverage probability (CP) for a nominal 95% confidence interval for β based on the SE.

The simulation results are summarized in Table 9.1. First, under $\gamma = -1$, we find the following results. Overall, $\hat{\beta}$ estimate the true values well. The standard-error estimators of $\hat{\beta}$ also work well as judged by the good agreement between SE and SD. The CPs for β are also reasonable with a range of 94–96%. For the dispersion parameters, $(\hat{\phi}, \hat{\alpha})$ perform well, but $\hat{\gamma}$ seems to introduce a slight bias. A reason of the bias may be the relative scarcity of the outcomes in the frailty model, i.e., the repeated-measure data have four outcomes for each subject, whereas the corresponding survival data have one single outcome for the same subject, leading to a larger variation (i.e., SD) for $\hat{\gamma}$ in the frailty model in Table 9.1. Table 9.1 shows that the simulation results for $\gamma = 1$ are similar to those for $\gamma = -1$.

Table 9.1 Simulation results with 500 replications under the joint model

Parameter	True	Mean	SD	SE (CP)	Mean	SD	SE (CP)
		$\gamma = -1$			$\gamma = 1$		
LMM for y							
β_{10}	−0.5	−0.501	0.195	0.195 (0.940)	−0.496	0.191	0.193 (0.956)
β_{11}	0.5	0.500	0.023	0.024 (0.954)	0.501	0.023	0.024 (0.946)
β_{12}	1	0.993	0.251	0.249 (0.956)	0.998	0.245	0.246 (0.952)
ϕ	1	0.989	0.117	–	0.989	0.117	–
α	0.5	0.506	0.153	–	0.503	0.158	–
Frailty model for T							
β_{21}	−1	−1.051	0.461	0.450 (0.954)	−1.050	0.457	0.451 (0.944)
γ	±1	−1.115	0.515	–	1.117	0.512	–

CP, empirical coverage probability of a nominal 95% confidence interval for β

9.3 Joint Model for Repeated Measures and Competing-Risks Data

In this section, we extend the h-likelihood approach in Sect. 9.2 to the joint models for competing-risks data. Let T_{ik} be event time from cause k for the ith subject ($i = 1, \ldots, q$; $k = 1, \ldots, K$). For simplicity, we consider two types of events ($k = 1, 2$). Suppose that the ith subject has the repeated-measure responses y_{ij} ($j = 1, \ldots, n_i$) and also two types of event times, T_{i1} and T_{i2}.

• **A joint model with competing risks**:

Consider a joint model conditional on v_i,

$$\begin{aligned}
\text{(i) } y_{ij} &= x_{ij1}^T\beta_1 + v_i + \epsilon_{ij}, \\
\text{(ii) } \lambda_{1i}(t|v_i) &= \lambda_{01}(t)\exp(x_{i2}^T\beta_2 + \gamma_1 v_i), \\
\text{(iii) } \lambda_{2i}(t|v_i) &= \lambda_{02}(t)\exp(x_{i2}^T\beta_3 + \gamma_2 v_i),
\end{aligned}$$

where $v_i \sim N(0, \alpha)$ and $\epsilon_{ij} \sim N(0, \phi)$ are independent, and $\lambda_{01}(\cdot)$ and $\lambda_{02}(\cdot)$ are the unknown baseline hazard functions for cause $k = 1, 2$, respectively. Here, β_1, β_2, and β_3 are $p_1 \times 1$, $p_2 \times 1$, and $p_2 \times 1$ vectors of regression parameters, respectively. Note here that γ_1 and γ_2 are the dispersion parameters to represent associations among submodels via v_i. That is, γ_1 [γ_2] represents an association between submodels (i) and (ii) [(i) and (iii)], respectively.

The arguments from Remark 6.1 can be applied here to establish identifiability of the model parameters in the above joint model with competing risks.

• **H-likelihood construction**:

Let C_i denote independent censoring time. In addition, we assume that given v_i, C_i is independent of (T_{ik}, δ_{ik}) for $k = 1, 2$. We observe the event time and event indicator, which are, respectively, given by

$$t_i^* = \min(T_{i1}, T_{i2}, C_i) \text{ and } \delta_{ik} = I(t_i^* = T_{ik}).$$

Thus, all observable random variables are $(y_{ij}, t_i^*, \delta_{ik})$ ($i = 1, \ldots, q$; $j = 1, \ldots, n_i$; $k = 1, 2$). Here, the h-likelihood is

$$h = \sum_{ij} \ell_{1ij} + \sum_{ik} \ell_{2ik} + \sum_{i} \ell_{3i},$$

where ℓ_{1ij} and ℓ_{3i} are given in (9.3), and for $k = 1, 2$

$$\ell_{2ik} = \ell_{2ik}(\beta_{k+1}, \lambda_{0k}; (t_i^*, \delta_{ik})|v_i) = \delta_{ik}\{\log \lambda_{0k}(t_i^*) + \eta_{2ik}\} - \Lambda_{0k}(t_i^*)\exp(\eta_{2ik}),$$

where $\eta_{2ik} = x_{i2}^T\beta_{k+1} + \gamma_k v_i$.

For an unknown $\lambda_{0k}(t)$, we again use the profile h-likelihood h^* with λ_{0k} eliminated, given by

$$h^* = h|_{\lambda_{0k}=\widehat{\lambda}_{0k}} = \sum_{ij} \ell_{1ij} + \sum_{ik} \ell^*_{2ik} + \sum_{i} \ell_{3i},$$

where

$$\sum_{ik} \ell^*_{2ik} = \sum_{ik} \ell_{2ik}|_{\lambda_{0k}=\widehat{\lambda}_{0k}} = \sum_{kr} d_{(kr)} \log \widehat{\lambda}_{0kr} + \sum_{ik} \delta_{ik}\eta_{2ik} - \sum_{kr} d_{(kr)},$$

where $d_{(kr)}$ is the number of events at time $t_{(kr)}$ and $R_{(kr)} = \{i : t_i^* \geq t_{(kr)}\}$ is the risk set at $t_{(kr)}$ which is the rth ($r = 1, \ldots, D_k$) smallest distinct event time for Type k event among t_i^*'s. Here,

$$\widehat{\lambda}_{0kr}(\beta_{k+1}, v) = \frac{d_{(kr)}}{\sum_{i \in R_{(kr)}} \exp(\eta_{2ik})}$$

are the solutions of the estimating equations, $\partial h/\partial \lambda_{0kr} = 0$, for $r = 1, \ldots, D_k$. This is again equivalent to using the partial h-likelihood

$$h_p = \sum_{ij} \ell_{1ij} + \sum_{ik} \delta_{ik}\eta_{2ik} - \sum_{kr} d_{(kr)} \log\left\{ \sum_{i \in R_{(kr)}} \exp(\eta_{2ik}) \right\} + \sum_{i} \ell_{3i}. \tag{9.8}$$

- **Estimation procedure**:

From (9.8), the score equations for the fixed and random effects $(\beta_1, \beta_{k+1}, v)$ ($k = 1, 2$), given the dispersion parameters $\psi = (\phi, \alpha, \gamma_k)^T$, are

$$\begin{aligned}
\partial h_p/\partial \beta_1 &= X_1^T (y - \mu_1)/\phi, \\
\partial h_p/\partial \beta_{k+1} &= X_2^T (\delta_k - \hat{\mu}_{2k}), \\
\partial h_p/\partial v &= {Z_1}^T (y - \mu_1)/\phi + \gamma_1 {Z_2}^T (\delta_1 - \hat{\mu}_{21}) + \gamma_2 {Z_2}^T (\delta_2 - \hat{\mu}_{22}) - v/\alpha,
\end{aligned}$$

where $\mu_1 = X_1\beta_1 + Z_1 v = \eta_1$, and $\hat{\mu}_{2k} = \exp(\log \hat{\Lambda}_{0k} + \eta_{2k})$ with $\eta_{2k} = X_2\beta_{k+1} + \gamma_k Z_2 v$, Z_1 and Z_2 are defined in the previous section, and δ_k is a vector of δ_{ik}'s for each k. Here, $\hat{\Lambda}_{0k}(t) = \sum_{r:t_{(kr)} \leq t} \hat{\lambda}_{0kr}$ is the Breslow-type estimator for the cumulative baseline hazard Λ_{0k} for event type k as in Chap. 6. The corresponding ILS equations for $\tau = (\beta_1^T, \beta_2^T, \beta_3^T, v^T)^T$ are given by

$$\begin{pmatrix}
{X_1}^T W_1 X_1 & 0 & 0 & X_1^T W_1 Z_1 \\
0 & X_2^T W_2^* X_2 & 0 & X_2^T (\gamma_1 W_2^*) Z_2 \\
0 & 0 & X_2^T W_3^* X_2 & X_2^T (\gamma_2 W_3^*) Z_2 \\
Z_1^T W_1 X_1 & {Z_2}^T (\gamma_1 W_2^*) X_2 & {Z_2}^T (\gamma_2 W_3^*) X_2 & \mathbf{Z}^T \mathbf{W} \mathbf{Z} + Q
\end{pmatrix}
\begin{pmatrix} \hat{\beta}_1 \\ \hat{\beta}_2 \\ \hat{\beta}_3 \\ \hat{v} \end{pmatrix}
=
\begin{pmatrix} {X_1}^T W_1 w_1 \\ {X_2}^T w_2^* \\ {X_2}^T w_3^* \\ \mathbf{Z}^T \mathbf{w}^* \end{pmatrix},$$

where $W_1 = -\partial^2 h_p/\partial\eta_1\partial\eta_1^T = \phi^{-1}I_{p_1}$, $W_2^* = -\partial^2 h_p/\partial\eta_{21}\partial\eta_{21}^T$, $W_3^* = -\partial^2 h_p/\partial\eta_{22}\partial\eta_{22}^T$, $Q = -\partial^2 \ell_3/\partial v \partial v^T = \alpha^{-1}I_q$, $w_1 = y$, $w_2^* = W_2^*\eta_{21} + (\delta_1 - \hat{\mu}_{21})$, $w_3^* = W_3^*\eta_{22} + (\delta_2 - \hat{\mu}_{22})$, and

$$\mathbf{Z} = \begin{pmatrix} Z_1 \\ \gamma_1 Z_2 \\ \gamma_2 Z_2 \end{pmatrix}, \quad \mathbf{W} = \begin{pmatrix} W_1 & \mathbf{0} & \mathbf{0} \\ \mathbf{0} & W_2^* & \mathbf{0} \\ \mathbf{0} & \mathbf{0} & W_3^* \end{pmatrix}, \quad \text{and} \quad \mathbf{w}^* = \begin{pmatrix} W_1 w_1 \\ w_2^* \\ w_3^* \end{pmatrix}.$$

Note here that $\mathbf{Z}^T\mathbf{W}\mathbf{Z} = Z_1^T W_1 Z_1 + Z_2^T(\gamma_1^2 W_2^*)Z_2 + Z_2^T(\gamma_2^2 W_3^*)Z_2$ and that $\mathbf{Z}^T\mathbf{w}^* = Z_1^T W_1 w_1 + \gamma_1 {Z_2}^T w_2^* + \gamma_2 {Z_2}^T w_3^*$.

As before, the ILS equations above leads to a simple form

$$(\mathbf{P}^T\mathbf{V}\mathbf{P})\hat{\tau} = \mathbf{P}^T\mathbf{y}_0^*, \tag{9.9}$$

where

$$\mathbf{P} = \begin{pmatrix} \mathbf{X} & \mathbf{Z} \\ \mathbf{0} & I_q \end{pmatrix} \text{ and } \mathbf{V} = \begin{pmatrix} \mathbf{W} & \mathbf{0} \\ \mathbf{0} & Q \end{pmatrix},$$

with

$$\mathbf{X} = \begin{pmatrix} X_1 & \mathbf{0} & \mathbf{0} \\ \mathbf{0} & X_2 & \mathbf{0} \\ \mathbf{0} & \mathbf{0} & X_2 \end{pmatrix}.$$

Here, $\mathbf{y}_0^* = (w_1^{*T}, w_2^{*T}, w_3^{*T}, \mathbf{0}^T)^T$ with $w_1^* = W_1 w_1$. Furthermore, to estimate $\psi = (\phi, \alpha, \gamma_1, \gamma_2)^T$ we still use the partial restricted likelihood $p_\tau(h_p)$.

9.4 Software and Examples Using R

For illustration, we present two practical examples. To fit the joint models, the `jmfit()` function in `frailtyHL` package is used.

9.4.1 Joint Analysis for Repeated Measures and a Single Event-Time Data: Renal Transplant Data

We consider the data from a clinical study to investigate the chronic renal allograft dysfunction in renal transplants. Since the time interval between the consecutive measurements differs from patient to patient, we focus on the mean creatinine levels over 6 months. In addition, a single terminating survival time (time to graft loss) in months is observed from each patient. During the study period, there were 13 graft losses due to the kidney dysfunction. For the other remaining patients, we assumed

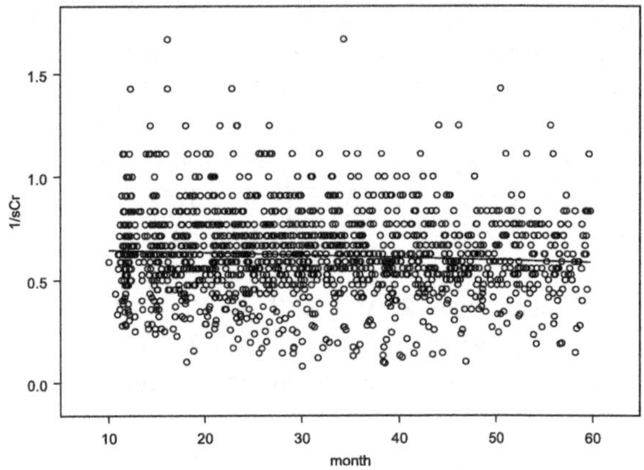

Fig. 9.1 Plot of 1/sCr against month; (———) fitted line, 1/sCr= 0.6566 − 0.0012×month

that the censoring occurred at the last follow-up time. The censoring rate is about 88%.

We are interested in investigating the effects of covariates on these two types of responses, i.e., serum creatinine (sCr) values and time to graft loss. Here, we consider month, gender, and age as covariates for sCr, and gender and age for the loss time; Gender is coded as 1 for male and as 0 for female. The reciprocal of sCr levels tends to decrease linearly over time, having possibly constant variance (Fig. 9.1). Thus, for the LMM, we use the values of 1/sCr as the response y_{ij}.

In order to fit the model of interest for the graft-loss time, Ha et al. (2003) assumed the Weibull frailty model for the graft-loss time t_i, which requires checking the distributional assumption for the baseline hazard. Using the procedures presented in this book, we can now fit a joint model allowing for an arbitrary baseline hazard. That is, with the response values of 1/sCr, we consider a LMM with three covariates,

$$y_{ij} = \beta_{10} + \beta_{11}\text{Month}_{ij} + \beta_{12}\text{Gender}_i + \beta_{13}\text{Age}_{ij} + v_i + \epsilon_{ij}, \tag{9.10}$$

where $v_i \sim N(0, \alpha)$ and $\epsilon_{ij} \sim N(0, \phi)$ are independent, and with the response of graft-loss time t_i, we consider a semiparametric frailty model with two covariates (Gender and Age).

$$\lambda_i(t|v_i) = \lambda_0(t)\exp(\beta_{21}\text{Gender}_i + \beta_{22}\text{Age}_i + \gamma v_i), \tag{9.11}$$

where $\lambda_0(\cdot)$ is an unknown baseline hazard function. The results from fitting the nonparametric joint model are summarized in the first portion of Table 9.2. In JM1, we allows for an arbitrary baseline hazard. Under JM1, the estimate $\widehat{\gamma} = -15.256$ gives a negative sign with a reference scale of 0, showing a negative correlation between 1/sCr and the hazard rate. That is, a patient with the larger 1/sCr would tend to have a lower hazard rate.

Based on the t tests, all three covariates (Month, gender, and age) have statistically significant effects on 1/sCr at a 5% significance level. In other words, the values of

Table 9.2 Results from fitting joint models (JM) and separate models (SM) with renal transplant data.

	JM1 (Nonparametric)		SM		JM2 (Weibull)	
Parameter	Estimate	SE	Estimate	SE	Estimate	SE
MM for 1/sCr						
Intercept	0.517	0.070	0.517	0.070	0.518	0.068
Month	−0.002	0.000	−0.002	0.000	−0.002	0.000
Gender (male)	−0.101	0.038	−0.100	0.038	−0.101	0.037
Age	0.007	0.002	0.007	0.002	0.007	0.002
ϕ	0.013	–	0.013	–	0.013	–
α	0.027	–	0.026	–	0.025	–
Frailty model for graft-loss time						
Intercept	–	–	–	–	−11.756	2.788
Gender (male)	1.509	0.949	−0.070	0.695	1.137	0.813
Age	−0.132	0.049	−0.050	0.034	−0.102	0.042
γ	−15.256	–	$\widehat{\text{var}}(v_i) = 0.676$	–	−11.704	–
					($\widehat{\tau} = 2.766$ SE = 0.617)	

JM2, joint model with Weibull baseline hazard (τ, shape parameter)

1/sCr decrease as time passes, males tend to have smaller values of 1/sCr than females do, and older patients tend to have larger values of 1/sCr.

With the response of graft-loss time, males tend to have a higher hazard rate than females, where the estimated relative risk is exp(1.509) = 4.52. However, the Gender effect is not significant at a 5% significance level. On the other hand, Age has a significantly negative effect on the hazard rate. It has been shown that the Age effect of the donor was positive, while that of the recipient was negative (Sung et al. 1998; Ha et al. 2003). The result is consistent here in that the Age effect in Table 9.2 is also negative.

We also separately fitted the LMM (9.10) for 1/sCr and the semiparametric frailty model (9.11) for graft-loss time. The results are summarized in the second portion of Table 9.2 under the heading of SM (separate model). Note that both JM1 and SM provide almost the same results for the LMM. However, both age and gender effects are nonsignificant in the separate analysis of the graft-loss time data, while age effect is significant in both joint analyses of JM1 and JM2. This means that information in the repeated measures from the same patient can be exploited for the analysis of the graft-loss time data.

For comparison, we included the results from the parametric joint model (Ha et al. 2003) with the Weibull baseline hazard, $\lambda_0(t) = \tau t^{\tau-1}\exp(\beta_{20})$, in the frailty model (9.11). In order to fit the model of interest for the graft-loss time, Ha et al. (2003) plotted $\log\{-\log \widehat{S}_0(t)\}$ versus $\log t$, which showed a linear trend confirming the Weibull assumption. The results are given in the third portion of Table 9.2 under the heading of JM2. With the response of 1/sCr, the results from both JM1 and JM2 are very similar, although JM1 yields larger SEs. With the response of graft-loss time, however, both results are somewhat different in that JM1 has larger absolute estimates and SEs, caused by the nonparametric estimation of the baseline hazard; in the Weibull frailty model, the results may be sensitive to different modelings of the baseline hazard. Between JM1 and JM2, there is not much difference in the fixed-effect estimates. In practice, however, JM1 might be preferred to JM2 because JM1 is robust against the baseline hazard assumption while JM2 is not.

- **R codes and output:**

Below are R codes and outputs from fitting the joint model (JM1) for the renal transplant data.

```
> data(renal,package="frailtyHL")
> data_conti <- renal
> data_conti <- subset(data_conti, month<=sur_time)
> data_surv <- subset(data_conti,first==1)
########################## HL #########################################
> jm1 <- jointmodeling(Model="mean",RespDist="gaussian",Link="identity",
+          LinPred=icr~month+sex+age+(1|id),RandDist="gaussian")
> jm2 <- jointmodeling(Model="mean",RespDist="FM",Link="log",
+          LinPred=Surv(sur_time,status)~sex+age+(1|id),RandDist="gaussian")
> jm <- list(jm1,jm2)
> data <- list(data_conti, data_surv)
> res <- jmfit(jm, data, Maxiter=200)
########################## Output ####################################
[1] "iterations :"
[1] 97
[1] "convergence :"
             [,1]
[1,] 9.923815e-05
[1] "Estimates for fixed effects"
             Estimate Std. Error  t_value p_value
(Intercept)   0.51745    0.07038  7.35240 0.00000
month        -0.00170    0.00025 -6.76273 0.00000
sex          -0.10088    0.03811 -2.64678 0.00813
age           0.00658    0.00174  3.77103 0.00016
sex           1.50883    0.94874  1.59035 0.11176
age          -0.13221    0.04869 -2.71539 0.00662
[1] "Estimates for dispersion parameters"
      phi_h alpha_h    gamma_h
[1,] 0.01323    0.027 -15.25595
```

9.4.2 *Joint Analysis of Repeated Measures and Competing-Risks Data: PBC Data*

Consider a data set from a clinical study on primary biliary cirrhosis (PBC) in the liver conducted by Mayo Clinic between 1974 and 1984 (Therneau and Grambsch 2000). A total of 424 PBC patients, referred to Mayo Clinic during that 10-year

Table 9.3 Results from fitting the joint competing-risks models (JM) and separate models (SM) for the PBC data

	JM		SM	
Parameter	Estimate	SE	Estimate	SE
LMM for the logarithm of serum bilirubin				
Intercept	0.592	0.094	0.586	0.092
Year	0.099	0.004	0.095	0.004
Drug	−0.125	0.128	−0.126	0.127
Gender	0.427	0.201	0.415	0.198
ϕ	0.241	–	0.242	–
α	1.219	–	1.181	–
Competing-risks frailty model for death				
Drug	−0.146	0.238	−0.017	0.175
Gender	0.794	0.338	0.671	0.229
γ_1	1.288	–	$\text{var}(v_i) = 0.108$	–
Competing-risks frailty model for transplantation				
Drug	−4.534	0.698	−0.386	0.384
Gender	0.116	0.593	0.146	0.623
γ_2	1.200	–	$\text{var}(v_i) = 0.306$	–

interval, met eligibility criteria for the randomized placebo-controlled trial of the drug D-penicillamine. Here, we consider 312 patients who participated in the randomized trial. We consider two event types:

Type 1 event is death (140 patients) and
Type 2 event is transplantation (29 patients).

The remaining 143 patients are censored at the last follow-up. The PBC data set is also available in the R package `JM` (Rizopoulos 2012).

Let y_{ij} be the logarithm of serum bilirubin (mg/dl) for the jth visit of the ith patient and let T_{ik} be the event time from cause k ($k = 1, 2$) of the ith patient. The event time is coded as `years` with event status, `status`. Here, the `status` is coded as "dead" for Type 1, "transplanted" for Type 2, and "alive" for censoring. For the joint model, we consider the LMM with three covariates (visiting year; drug (=1 for D-penicillamine,=0 for placebo); and gender (=1 for male,=0 for female)) and the frailty model (FM) with two covariates (gender and drug). Note here that "visiting year" is coded as `year`.

This model can be fitted by using the `jmfit()` function in the `frailtyHL` package; the R codes and outputs are provided below. The results are summarized in Table 9.3.

We first analyze the results from the joint models (JM) in Table 9.3. The estimates of the association parameters, $\widehat{\gamma_1} = 1.288$ and $\widehat{\gamma_2} = 1.200$, show all positive associations between y_{ij} and death and between y_{ij} and transplantation, respectively.

That is, $\widehat{\gamma_1} = 1.288$ means that a patient with a larger serum bilirubin (y_{ij}) shows a tendency to have a higher death rate, and $\widehat{\gamma_2} = 1.200$ indicates that a patient with a larger serum bilirubin (y_{ij}) tends to have a higher hazard rate of transplantation. From Table 9.3 and the R output, the estimated covariate effects are summarized as follows:

(i) For the serum bilirubin, the visiting year effect on the serum bilirubin is significantly positive (p value = 0.000). In other words, the value of serum bilirubin significantly increases as time passes. The gender effect is also significantly positive (p value = 0.033). That is, males have significantly larger serum bilirubin as compared to females. However, the drug effect is negative but not significant (p value = 0.328).
(ii) For the death rate, the gender effect is positively significant (p value = 0.019), implying that males have significantly higher death rate than females. The drug effect is still not significant (p value = 0.541).
(iii) For the transplant hazard rate, the gender effect is not significant (p value = 0.845), but the drug effect (i.e., D-penicillamine effect) is significantly negative (p value = 0.000), implying significant benefits for the PBC patients.

For a comparison, we fitted the three models separately (SM), i.e., LMM and two competing-risks frailty models. In Table 9.3, we observe that both LMM analyses from the joint model and separate model give almost the same results, while two competing-risks analyses are quite different as in Sect. 9.4.1. For example, the drug effect is not significant in the separate analysis for both types of events, whereas it is significant in joint analysis of time-to-transplant data. We again see that information in the repeated measures from the same patient can also be exploited for the analysis of competing-risks data.

- **R codes and output:**

Below are R codes and outputs for fitting the joint model for the PBC data.

```
# Joint model for repeated measures and competing-risks data
> data(pbc2,package="JM")     # repeated-measures data
> data(pbc2.id,package="JM") # competing-risks data
> pbc2$sex<-ifelse(pbc2$sex=="male",1,0)
> pbc2.id$sex<-ifelse(pbc2.id$sex=="male",1,0)
> pbc2$drug<-ifelse(pbc2$drug=="D-penicil",1,0)
> pbc2.id$drug<-ifelse(pbc2.id$drug=="D-penicil",1,0)

> jm1<-jointmodeling(Model="mean",RespDist="gaussian",Link="identity",
+   LinPred=log(serBilir)~year+drug+sex+(1|id),RandDist="gaussian")
> jm2<-jointmodeling(Model="mean",RespDist="FM",Link="log",
+   LinPred=Surv(years,status=="dead")~drug+sex+(1|id),
+   RandDist="gaussian")
> jm3<-jointmodeling(Model="mean",RespDist="FM",Link="log",
+   LinPred=Surv(years,status=="transplanted")~drug+sex+(1|id),
+   RandDist="gaussian")
> jm <- list(jm1,jm2,jm3)
> data_surv <- pbc2.id
> data <- list(pbc2, data_surv)
```

```
> res <- jmfit(jm, data, Maxiter=200)
[1]"iterations :"
[1] 25
[1]"convergence :"
             [,1]
[1,] 6.431976e-05


[1]"Estimates for fixed effects"
             Estimate Std. Error  t_value p_value
(Intercept)   0.59197    0.09363  6.32266 0.00000
year          0.09887    0.00431 22.95874 0.00000
drug         -0.12546    0.12835 -0.97753 0.32830
sex           0.42694    0.20081  2.12612 0.03349
drug         -0.14560    0.23816 -0.61136 0.54096
sex           0.79392    0.33794  2.34931 0.01881
drug         -4.53389    0.69789 -6.49662 0.00000
sex           0.11622    0.59252  0.19614 0.84450
[1]"Estimates for dispersion parameters"
       phi_h alpha_h gamma1_h gamma2_h
[1,] 0.24147 1.21878  1.28762  1.19992
```

9.5 Discussion

An advantage of the h-likelihood method is its easy extensibility to the joint models with multicomponent random effects, for which the integration to obtain the marginal likelihood is often intractable. The separate modeling does not consider the dependency between responses. Lee et al. (2017b) also provided multivariate analysis for the multiple outcomes from the HGLM. Thus, the joint models can easily extended to the multivariate analysis with more than two responses, including competing risks (Ha et al. 2017). The joint modeling enables information from all the responses to be exploited to improve inference on the regression parameter estimators, which is impossible from the separate analyses. Information gain from the other responses can be important in the analysis of scarce data.

The joint models considered in this chapter can be fitted via the marginal likelihood (ML) method, which is implemented as R packages using the GHQ (Rizopoulos 2012 and EM (Philipson et al. 2012). We have found (not shown here) that the h-likelihood method gives similar results to the ML method.

We assumed a shared frailty v_i in the joint models. Extension to correlated frailties is useful; it is easily implemented to allow for a bivariate normal distribution with a correlation between v_{i1} in the LMM and v_{i2} in the frailty model (Elashoff et al. 2008; Ha et al. 2017). Furthermore, the development of an extended joint model allowing for time-dependent covariates would also merit future research.

Chapter 10
Further Topics

We have previously presented the h-likelihood procedures for the analysis of survival data under competing risks. There are still many unresolved problems in this area. In this chapter, we present some further topics to highlight that the h-likelihood approach can be extended to more complex multistate survival data. We deal with competing-risks data with missing causes of failure and the semi-competing-risks data.

10.1 Competing-Risks Frailty Models with Missing Causes of Failure

In a clinical study, information on cause of death may not be observed for some subjects due to loss to follow-up or difficulties in determining the cause of death. When causes of failure are missing, the subjects with missing causes may be excluded from the analysis and the standard competing-risks analysis may be applied. However, such approaches lose information and may lead to biased results. In multicenter clinical trials, competing-risks problems with missing causes often occurs within a center.

In this section, under the missing at random (MAR) assumption, we present the h-likelihood approach to fitting the cause-specific competing-risks model with a univariate log-normal frailty in the presence of missing causes of failure. Here, we use the multiple imputation methods (Bakoyannis et al. 2010) to deal with missing causes of failure. Following Bakoyannis et al. (2010), we impute the missing causes of failure multiple times from the conditional distribution of failure type given the observed data, and then fit the cause-specific log-normal frailty model using the h-likelihood procedure presented in Chap. 6.

I.D. Ha et al., *Statistical Modelling of Survival Data with Random Effects*,
Statistics for Biology and Health, https://doi.org/10.1007/978-981-10-6557-6_10

10.1.1 Example: Bladder Cancer Data with Missing Causes of Failure

Consider the bladder cancer data again. For this analysis, we study the data set consisting of 396 patients with bladder cancer treated in 21 centers from the EORTC trial 30791, focusing on two competing endpoints, i.e., death from malignant disease (an event of interest) and death from other causes (competing events).

The descriptive statistics are given in Table 10.1. Here, patients with missing values of covariates of interest (age, gender, carcinoma in stiu (CIS) and grade) were excluded. Among 396 patients, 50 patients died from malignant disease, 94 died from other causes, and 211 patients were censored. The causes of death were not observed for 41 patients. Thus, 22.16% of 185 patients who died had missing causes of death. Recall that the numbers of patients per center varied from 3 to 78, with the mean of 18.9 and the median of 14.

Table 10.2 summarizes the results from the multiple imputation methods (MI) with $m = 10$ imputation and complete case analysis (CC) obtained by excluding patients with missing causes of death. In both MI and CC analyses, age at diagnosis, CIS, and grade were significant predictors of death from malignant disease, while

Table 10.1 Descriptive statistics for 396 patients in the bladder cancer data

Characteristic	Number of patients	Percentage of patients
Age at diagnosis		
≤65 years	183	46.2
>65 years	213	53.8
Gender		
Male	330	83.3
Female	66	16.7
Carcinoma in situ (CIS)		
No	372	93.9
Yes	24	6.1
Grade		
Grade1	191	48.2
Grade2	167	42.2
Grade3	38	9.6
Cause of death		
Malignant disease	50	12.6
Other	94	23.7
Alive	211	53.3
Missing	41	10.4

Table 10.2 The regression parameter estimates (standard errors) and p-values for death from malignant disease and from other causes in the cause-specific PH model with a univariate log-normal frailty; MI, multiple imputation; CC, complete case

22.16% missing	MI				CC			
	Death from malignant disease		Death from other causes		Death from malignant disease		Death from other causes	
Covariates	$\hat{\beta}_1$ (SE)	p-value	$\hat{\beta}_2$ (SE)	p-value	$\hat{\beta}_1$ (SE)	p-value	$\hat{\beta}_2$ (SE)	p-value
Age at diagnosis								
≤65 years	0 (–)	–	0 (–)	–	0 (–)	–	0 (–)	–
>65 years	0.954 (0.301)	0.002	1.299 (0.224)	<0.001	0.925 (0.307)	0.003	1.475 (0.246)	<0.001
Gender								
Male	0 (–)	–	0 (–)	–	0 (–)	–	0 (–)	–
Female	−0.672 (0.431)	0.120	−0.444 (0.275)	0.107	−0.940 (0.477)	0.049	−0.499 (0.303)	0.100
CIS								
No	0 (–)	–	0 (–)	–	0 (–)	–	0 (–)	–
Yes	1.044 (0.400)	0.009	0.239 (0.410)	0.560	1.214 (0.400)	0.002	0.057 (0.467)	0.903
Grade								
Grade1	0 (–)	–	0 (–)	–	0 (–)	–	0 (–)	–
Grade2	0.849 (0.314)	0.007	0.286 (0.200)	0.152	0.941 (0.330)	0.004	0.375 (0.219)	0.086
Grade3	1.318 (0.409)	0.001	−0.073 (0.421)	0.863	1.208 (0.443)	0.006	−0.057 (0.442)	0.897
Frailty parameter	$\hat{\sigma}^2 = 0.034$				$\hat{\sigma}^2 = 0.031$			

only age was a significant predictor of death from other causes. Note that gender was not a significant predictor of death from malignant disease in the MI method, while it was significant in the CC analysis. This confirms the simulation results by Lee et al. (2017a) that excluding patients with missing causes of death from the analyses might lead to biased results. They also showed that the h-likelihood procedure performs well, even if the imputation model is misspecified.

The variance estimate ($\hat{\sigma}^2$) of the random effect is 0.034 and 0.031 from the MI method and CC analysis, respectively, showing that there is little difference between the two analyses in the amount of variation in the baseline risk over centers. We tested the null hypothesis $H_0 : \sigma^2 = 0$ (i.e., $v_i = 0$ for all i with no center effect) using $p_\tau(h_p)$. Since the null hypothesis $H_0 : \sigma^2 = 0$ is on the boundary of the parameter space, the likelihood difference based on $-2p_\tau(h_p)$ between the cause-specific PH model without frailty and with frailty from the MI methods is $-2\{-953.855 - (-953.215)\} = 1.28$ (p-value=0.129). This indicates that the null hypothesis of no variation in the baseline risk across centers is not rejected.

Missing data are unobserved random variables, so that they can be treated as random effects. For application of the h-likelihood to missing data and imputation in general, see Lee et al. (2017b).

10.2 Frailty Models for Semi-competing-Risks Data

In this section, we consider the semi-competing-risks situation where a terminal event (e.g., death) censors a nonterminal event (e.g., disease recurrence), but not vice versa (Fig. 10.1). Here, a subject may experience both events that might be correlated. We show that the frailty models are useful for modelling such semi-competing-risks data. For inference, we propose the h-likelihood procedure, which is compared with the marginal likelihood approach.

We first review the classical semi-competing-risks model, which is also well known as the illness-death model.

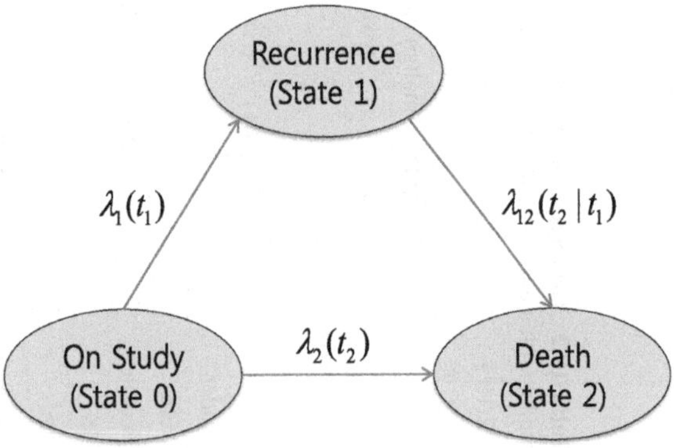

Fig. 10.1 A schematic diagram of semi-competing-risks data

10.2.1 Classical Semi-competing-Risks Model

Suppose that a subject may experience a terminal event (e.g., death) and/or a nonterminal event (e.g., disease recurrence). Let T_{i1} and T_{i2} be the nonterminal and terminal event times for the ith subject, respectively, and let C_i be the corresponding censoring time ($i = 1, \ldots, n$). If the subject fails before the nonterminal event occurs, we conventionally define $T_{i1} = \infty$. Then, we have the following observable data:

$$y_{i1} = T_{i1} \wedge y_{i2},\ y_{i2} = T_{i2} \wedge C_i,\ \delta_{i1} = I(T_{i1} \leq y_{i2}) \text{ and } \delta_{i2} = I(T_{i2} \leq C_i).$$

Note that $0 \leq y_{i1} \leq y_{i2}$. In particular, subjects can die from related or unrelated causes. Thus, there are four cases we can observe from each subject:

(i) if $(\delta_{i1}, \delta_{i2}) = (0, 0)$ (i.e., a subject still alive without recurrence),
$y_{i1} = y_{i2} = C_i$,
(ii) if $(\delta_{i1}, \delta_{i2}) = (0, 1)$ (i.e., a subject dies without recurrence),
$y_{i1} = y_{i2} = T_{i2}$,
(iii) if $(\delta_{i1}, \delta_{i2}) = (1, 0)$ (i.e., a subject still alive after recurrence),
$y_{i1} = T_{i1}$ and $y_{i2} = C_i$,
(iv) if $(\delta_{i1}, \delta_{i2}) = (1, 1)$ (i.e., a subject dies after recurrence),
$y_{i1} = T_{i1}$ and $y_{i2} = T_{i2}$.

Figure 10.1 again shows a schematic diagram of semi-competing-risks data, with three states (on study, recurrence, and death). The hazard functions in Fig. 10.1 are defined as follows:

$$\begin{aligned}
\lambda_1(t_1) &= \lim_{\Delta t \to 0} Pr\{t_1 \leq T_1 \leq t_1 + \Delta t | T_1 \geq t_1, T_2 \geq t_1\}/\Delta t,\ t_1 > 0,\\
\lambda_2(t_2) &= \lim_{\Delta t \to 0} Pr\{t_2 \leq T_2 \leq t_2 + \Delta t | T_1 \geq t_2, T_2 \geq t_2\}/\Delta t,\ t_2 > 0,\\
\lambda_{12}(t_2|t_1) &= \lim_{\Delta t \to 0} Pr\{t_2 \leq T_2 \leq t_2 + \Delta t | T_1 = t_1, T_2 \geq t_2\}/\Delta t,\ 0 < t_1 < t_2.
\end{aligned}$$

For simplicity, for $\lambda_{12}(t_2|t_1)$ we assume a Markov process where the transition probability from state 1 to state 2 does not depend on the duration in state 1 (Aalen et al. 2008). That is, we assume

$$\lambda_{12}(t_2|t_1) = \lambda_{12}(t_2), 0 < t_1 < t_2.$$

Note that for transition from state 1 to state 2, the left truncation time is t_1, the time at which the recurrence occurred. Let x_i be a p -dimensional vector of covariates for the ith subject. The semi-competing-risks regression model (Lawless 2003) is described as follows:

$$\lambda_{1i}(t_1; x_i) = \lambda_{01}(t_1)\exp(x_i^T\beta_1),\ t_1 > 0, \tag{10.1}$$
$$\lambda_{2i}(t_2; x_i) = \lambda_{02}(t_2)\exp(x_i^T\beta_2),\ t_2 > 0, \tag{10.2}$$
$$\lambda_{12i}(t_2; x_i) = \lambda_{03}(t_2)\exp(x_i^T\beta_3),\ 0 < t_1 < t_2, \tag{10.3}$$

where $\lambda_{01}(\cdot)$, $\lambda_{02}(\cdot)$ and $\lambda_{03}(\cdot)$ are the parametric or nonparametric baseline hazard functions. Let $\Lambda_{01}(t) = \int_0^t \lambda_{01}(s)ds$ and $\Lambda_{02}(t) = \int_0^t \lambda_{02}(s)ds$ be the baseline cumulative hazard functions corresponding to $\lambda_{01}(t)$ and $\lambda_{02}(t)$, respectively. Following Xu et al. (2010), the likelihood function based on the models (10.1)–(10.3) given the observed data $y_i^o = (y_{i1}, \delta_{i1}, y_{i2}, \delta_{i2})$ $(i = 1, \dots, n)$ is given by

$$\begin{aligned}
L &= \prod_{i=1}^n \lambda_{1i}(y_{i1})^{\delta_{i1}} \lambda_{2i}(y_{i2})^{\delta_{i2}(1-\delta_{i1})} \lambda_{12i}(y_{i2})^{\delta_{i1}\delta_{i2}} \\
&\quad \times \exp\left[-\int_0^{y_{i1}} \{\lambda_{1i}(t) + \lambda_{2i}(t)\}\, dt\right] \times \exp\left\{-\int_{y_{i1}}^{y_{i2}} \lambda_{12i}(t)\, dt\right\} \\
&= \prod_{i=1}^n \lambda_{01}(y_{i1})^{\delta_{i1}} \lambda_{02}(y_{i2})^{\delta_{i2}(1-\delta_{i1})} \lambda_{03}(y_{i2})^{\delta_{i1}\delta_{i2}} \\
&\quad \times \exp\{x_i^T\beta_1\delta_{i1} + x_i^T\beta_2\delta_{i2}(1-\delta_{i1}) + x_i^T\beta_3\delta_{i1}\delta_{i2}\} \\
&\quad \times \exp\left\{-\Lambda_{01}(y_{i1})e^{x_i^T\beta_1} - \Lambda_{02}(y_{i1})e^{x_i^T\beta_2} - \Lambda_{03}(y_{i1}, y_{i2})e^{x_i^T\beta_3}\right\},
\end{aligned} \tag{10.4}$$

where $\Lambda_{03}(s, t) = \Lambda_{03}(t) - \Lambda_{03}(s)$. If the forms of $\lambda_{01}(\cdot)$, $\lambda_{02}(\cdot)$ and $\lambda_{03}(\cdot)$ are parametric, then the MLEs for the parameters of interest $(\beta_1, \beta_2, \beta_3)$ are available by directly maximizing L in (10.4) via the numerical methods such as Newton–Raphson. If they are unspecified, the estimates can be obtained by maximizing L using the Breslow (1972) method for the baseline cumulative hazards (Andersen et al. 1997; Xu et al. 2010).

10.2.2 Fitting the Semi-competing-Risks Frailty Model

10.2.2.1 The Model

The classical model (10.1)–(10.3) can be extended to the frailty model which describes the dependency between nonterminal and terminal event times.

For simplicity, we consider the semi-competing-risks frailty model with a common frailty. Denote by u_i a shared unobserved frailty (random effect) for the ith subject. Following Xu et al. (2010), the semi-competing risk frailty model is described as follows. The conditional hazards (10.1)–(10.3) given u_i are expressed as

$$\lambda_{1i}(t_1|u_i; x_i) = \lambda_{01}(t_1)\exp(x_i^T\beta_1)u_i,\ t_1 > 0, \tag{10.5}$$
$$\lambda_{2i}(t_2|u_i; x_i) = \lambda_{02}(t_2)\exp(x_i^T\beta_2)u_i,\ t_2 > 0, \tag{10.6}$$
$$\lambda_{12i}(t_2|u_i; x_i) = \lambda_{03}(t_2)\exp(x_i^T\beta_3)u_i,\ 0 < t_1 < t_2, \tag{10.7}$$

where $\lambda_{01}(\cdot)$, $\lambda_{02}(\cdot)$ and $\lambda_{03}(\cdot)$ are the unspecified baseline hazard functions. Here, the frailties u_i are assumed to be unobserved realizations of an iid random variable with a density function having a frailty parameter α. The popular gamma and log-normal frailty models, respectively, assume gamma and log-normal distributions for u_i; for the gamma, we assume $E(u_i) = 1$ and $\text{var}(u_i) = \alpha$, and for the log-normal $v_i = \log u_i \sim N(0, \alpha)$.

10.2.2.2 Estimation Procedure Based on the H-likelihood

We now show how to derive the h-likelihood estimation procedure to fit the frailty model (10.5)–(10.7). The h-likelihood for the semi-competing-risks frailty model (10.5)–(10.7) is defined by

$$h = h(\beta, v, \lambda_0, \alpha) = \sum_i \ell_{1i} + \sum_i \ell_{2i}, \tag{10.8}$$

where $\ell_{1i} = \ell_{1i}(\beta, \lambda_0; y_i^o|u_i)$ is the logarithm of the conditional density function for $y_i^o = (y_{i1}, y_{i2}, \delta_{i1}, \delta_{i2})$ given u_i, i.e.,

$$\begin{aligned}\ell_{1i} = {}& \delta_{i1}\{\log\lambda_{01}(y_{i1}) + \eta_{i1}\} + \delta_{i2}(1 - \delta_{i1})\{\log\lambda_{02}(y_{i2}) + \eta_{i2}\}\\ &+ \delta_{i1}\delta_{i2}\{\log\lambda_{03}(y_{i2}) + \eta_{i3}\}\\ &-\{\Lambda_{01}(y_{i1})\exp(\eta_{i1}) + \Lambda_{02}(y_{i1})\exp(\eta_{i2}) + \Lambda_{03}(y_{i1}, y_{i2})\exp(\eta_{i3})\},\end{aligned}$$

and $\ell_{2i} = \ell_{2i}(\alpha; v_i)$ is the logarithm of the density function of $v_i = \log u_i$ with parameter α. Here, $\eta_{i1} = x_i^T\beta_1 + v_i$, $\eta_{i2} = x_i^T\beta_2 + v_i$ and $\eta_{i3} = x_i^T\beta_3 + v_i$, $\beta = (\beta_1^T, \beta_2^T, \beta_3^T)^T$ with $\beta_j = (\beta_{j1}, \ldots, \beta_{jp})^T$, $v = (v_1, \ldots, v_n)^T$, and $\Lambda_{03}(s, t) = \Lambda_{03}(t) - \Lambda_{03}(s)$. For the gamma frailty, we have $\ell_{2i} = \alpha^{-1}(v_i - u_i) - \log\Gamma(\alpha^{-1}) - \alpha^{-1}\log\alpha$, and for the log-normal frailty $\ell_{2i} = -\log(2\pi\alpha)/2 - v_i^2/2\alpha$.

Note that the functional forms of $\lambda_{0j}(\cdot)$ $(j = 1, 2, 3)$ are unknown. Let $\lambda_{0jk_j} = \lambda_{0j}(y_{j(k_j)}) > 0$ be a jump size at the observed event time $y_{j(k_j)}$, where $y_{j(k_j)}$ is the k_jth $(k_j = 1, \ldots, D_j)$ smallest distinct event time for each j. Let $y_{1(1)}, y_{1(2)}, \ldots, y_{1(D_1)}$ be ordered distinct recurrence times among data with $\delta_{i1} = 1$ (i.e., $(\delta_{i1}, \delta_{i2})$ =(1, 0) or (1, 1)), and $y_{2(1)}, y_{2(2)}, \ldots, y_{2(D_2)}$ be ordered distinct death times without recurrence among data with $(\delta_{i1}, \delta_{i2}) = (0, 1)$, and $y_{3(1)}, y_{3(2)}, \ldots, y_{3(D_3)}$ be ordered distinct death times following recurrence among data with $(\delta_{i1}, \delta_{i2}) = (1, 1)$. Again, we consider the baseline cumulative hazard function $\Lambda_{0j}(t)$ $(j = 1, 2, 3)$ as a step function with jumps at the observed event times

$$\Lambda_{0j}(t) = \sum_{k_j : y_{j(k_j)} \le t} \lambda_{0jk_j}, \; (j = 1, 2, 3). \tag{10.9}$$

Then $\sum_i \ell_{1i}$ in (10.8) can be rewritten as

$$\begin{aligned}\sum_i \ell_{1i} &= \sum_{k_1} d_{1(k_1)} \log \lambda_{01k_1} + \sum_i \delta_{i1}\eta_{i1} - \sum_{k_1} \lambda_{01k_1} \left\{ \sum_{i \in R_{(k_1)}} \exp(\eta_{i1}) \right\} \\ &\quad + \sum_{k_2} d_{2(k_2)} \log \lambda_{02k_2} + \sum_i \delta_{i2}(1 - \delta_{i1})\eta_{i2} \\ &\quad - \sum_{k_2} \lambda_{02k_2} \left\{ \sum_{i \in R_{(k_2)}} \exp(\eta_{i2}) \right\} + \sum_{k_3} d_{3(k_3)} \log \lambda_{03k_3} \\ &\quad + \sum_i \delta_{i1}\delta_{i2}\eta_{i3} - \sum_{k_3} \lambda_{03k_3} \left\{ \sum_{i \in R_{(k_3)}} \exp(\eta_{i3}) \right\},\end{aligned} \tag{10.10}$$

where $d_{j(k_j)}$ $(j = 1, 2, 3)$ is the number of events at $y_{j(k_j)}$, and

$$\begin{aligned}R_{(k_1)} &= R(y_{1(k_1)}) = \{i : y_{i1} \ge y_{1(k_1)}\}, \\ R_{(k_2)} &= R(y_{2(k_2)}) = \{i : y_{i1} \ge y_{2(k_2)}\}, \\ R_{(k_3)} &= R(y_{3(k_3)}) = \{i : y_{i1} < y_{3(k_3)} \le y_{i2}\},\end{aligned}$$

are the risk sets at $y_{1(k_1)}$, $y_{2(k_2)}$ and $y_{3(k_3)}$, respectively. We let $\lambda_{01} = (\lambda_{011}, \ldots, \lambda_{01D_1})^T$, $\lambda_{02} = (\lambda_{021}, \ldots, \lambda_{02D_2})^T$, and $\lambda_{03} = (\lambda_{031}, \ldots, \lambda_{03D_3})^T$. As the number of nuisance parameters λ_{0j}'s increases with the number of events, the function $\lambda_{0j}(t)$ is potentially of high dimension. Accordingly, for estimation of (β, v), the profiled h-likelihood h^* is used to eliminate λ_{0j} $(j = 1, 2, 3)$:

$$h^* = h|_{\lambda_{0j} = \widehat{\lambda}_{0j}} = \sum_i \ell^*_{1i} + \sum_i \ell_{2i}, \tag{10.11}$$

where

$$\widehat{\lambda}_{0jk_j}(\beta, v) = \frac{d_{j(k_j)}}{\sum_{i \in R_{(k_j)}} \exp(\eta_{ij})}, \; (j = 1, 2, 3)$$

are the solutions of the estimating equations, $\partial h / \partial \lambda_{0jk_j} = 0$, for $k_j = 1, \ldots, D_j$. Here, from (10.10) we have that

$$\begin{aligned}\sum_i \ell_{1i}^* &= \sum_i \ell_{1i}|_{\lambda_{0j}=\widehat{\lambda}_{0j}}\\ &= \sum_{k_1} d_{1(k_1)} \log \widehat{\lambda}_{01k_1} + \sum_i \delta_{i1}\eta_{i1} - \sum_{k_1} d_{1(k_1)}\\ &\quad + \sum_{k_2} d_{2(k_2)} \log \widehat{\lambda}_{02k_2} + \sum_i \delta_{i2}(1-\delta_{i1})\eta_{i2} - \sum_{k_2} d_{2(k_2)}\\ &\quad + \sum_{k_3} d_{3(k_3)} \log \widehat{\lambda}_{03k_3} + \sum_i \delta_{i1}\delta_{i2}\eta_{i3} - \sum_{k_3} d_{3(k_3)},\end{aligned}$$

which is proportional to the partial conditional likelihood ℓ_p, given by

$$\begin{aligned}\ell_p &= \sum_i \delta_{i1}\eta_{i1} - \sum_{k_1} d_{1(k_1)} \log\Big\{\sum_{i\in R_{(k_1)}} \exp(\eta_{i1})\Big\}\\ &\quad + \sum_i \delta_{i2}(1-\delta_{i1})\eta_{i2} - \sum_{k_2} d_{2(k_2)} \log\Big\{\sum_{i\in R_{(k_2)}} \exp(\eta_{i2})\Big\}\\ &\quad + \sum_i \delta_{i1}\delta_{i2}\eta_{i3} - \sum_{k_3} d_{3(k_3)} \log\Big\{\sum_{i\in R_{(k_3)}} \exp(\eta_{i3})\Big\}\end{aligned}$$

with the constant terms eliminated. This leads to the partial h-likelihood

$$h_p = \ell_p + \sum_i \ell_{2i}.$$

Thus, once we have h_p, the h-likelihood method presented in Chap. 4 can be directly extended to the semi-competing frailty model.

Remark 10.1 (i) For the semi-competing-risks model (10.5)–(10.7) with gamma frailty, in Appendix 10.4.1, we derive the marginal likelihood procedure which is equivalent to that of Xu et al. (2010). In Appendix 10.4.2, we also show that given α, the h-likelihood and marginal likelihood procedures give the same estimators as in the standard gamma frailty model (Ha et al. 2001; Ha and Lee 2003).

(ii) For estimation of the dispersion parameter α in the gamma frailty model, we practically use the second-order approximation $s_v(h_p) = p_v(h_p) - F(h)/24$, which is also an approximation of $p_{\log\lambda_0}(m)$ in (4.8) (Ha et al. 2010) because $p_v(h)$ and $p_{\beta,v}(h)$ are asymptotically equivalent (Noh and Lee 2007; Ha et al. 2007b). However, we have found that $s_v(h_p)$ sometimes gives a convergence problem in fitting the semi-competing-risks gamma frailty models. To overcome this problem, we further consider a higher order approximation using the h-likelihood, i.e., the fourth-order Laplace approximation in (10.23) (denoted by $m_v(h_p)$): see Appendix 10.4.3.

(iii) Furthermore, for the semi-competing-risks model (10.5)–(10.7) with log-normal frailty, the h-likelihood method can be easily implemented. However, the corresponding marginal likelihood may require an intractable integration over the frailty.

10.2.3 Example: Breast Cancer Data

For an illustration, we consider the breast cancer data (Sect. 1.2.6) including 2,572 eligible patients with follow-up and known pathological tumor size. The aim of this analysis is to investigate the effect of treatment on cancer recurrence and/or death, considering three event types: Type 1, cancer recurrence from study; Type 2, death without recurrence; Type 3, death after recurrence.

Table 10.3 gives the number of observed event types in this data set. Here 180 patients (7.00%) experienced Type 1, 535 patients (20.80%) did Type 2, 540 patients (21.00%) did Type 3, and the remaining 1317 patients (51.21%) had no events. Table 10.3 also shows the number of observed event types by two treatment arms.

Here, we consider three covariates of interest: treatment (x_{i1} is 1 for tamoxifen and 0 for placebo), tumor size (x_{i2}) and age (x_{i3}). For the analysis of data, we use the Markov model (10.1)–(10.3) without frailty and the model (10.5)–(10.7) with gamma frailty. For estimation of the gamma frailty model, we use the h-likelihood method, which is compared with the marginal likelihood method.

The fitted results are listed in Table 10.4. The results from the Markov and frailty models are very similar because the frailty parameter estimate ($\hat{\alpha} = 0.090$) is very small. Moreover, to test the absence of the frailty effect $H_0 : \alpha \equiv \text{var}(u_i) = 0$, the likelihood difference between the Markov model and the frailty model is $2\{s_v(h_p) - \ell_B\} = 0.7 < 2.71$, indicating that the frailty effect is not significant; using the marginal likelihood m, we also have $D = 2\{m - \ell_B\} = 0.4$. Here, ℓ_B is the Breslow's (partial) likelihood ($u_i = 1$ for the gamma frailty model and $v_i = 0$ for the log-normal frailty model for all i). That is, it is defined by

$$\ell_B = \lim_{\alpha \to 0} s_v(h_p),$$

Table 10.3 Observed event types by two treatment arms ($n = 2,572$ patients)

Types of event	Placebo	Tamoxifen	Total (%)
Type 1 (State 0 → State 1): Recurrence	108	72	180 (7.00)
Type 2 (State 0 → State 2): Death without recurrence	242	293	535 (20.80)
Type 3 (State 1 → State 2): Death after recurrence	331	209	540 (21.00)
No event (Censoring)	613	704	1,317 (51.21)

Table 10.4 Fitted results from the two semi-competing-risks models with the breast cancer data

Model	Time to recurrence	Time to death without recurrence	Time to death after recurrence
	Est. (SE)	Est. (SE)	Est. (SE)
Markov model			
Treatment	−0.543 (0.077)	0.058 (0.087)	0.329 (0.089)
Age	−0.015 (0.004)	0.090 (0.006)	0.007 (0.004)
Tumor size (x_3) $-2\ell_{pc} = 23886.1$	0.018 (0.002)	0.007 (0.003)	0.010 (0.003)
Frailty model (HL)			
Treatment	−0.552 (0.078)	0.046 (0.089)	0.332 (0.094)
Age	−0.015 (0.004)	0.090 (0.006)	0.008 (0.004)
Tumor size	0.018 (0.003)	0.008 (0.004)	0.011 (0.003)
Frailty α $-2s_v(h_p) = 23885.4$	0.090		
Frailty model (ML)			
Treatment	−0.549 (0.077)	0.050 (0.088)	0.332 (0.093)
Age	−0.015 (0.004)	0.090 (0.006)	0.007 (0.004)
Tumor size	0.018 (0.002)	0.008 (0.004)	0.011 (0.003)
Frailty α $-2m = 23885.7$	0.059		

Markov model, semi-competing-risks model without frailty
Frailty model, semi-competing-risks model with gamma frailty
HL, h-likelihood; ML, marginal likelihood
α, variance of the gamma frailty

which is the adjusted profile h-likelihood under the model (10.5)–(10.7) without the term u_i (Lee and Nelder 1996). In Table 10.4, the treatment effect (x_1) is significant on time to recurrence and time to death after recurrence, but not time to death without recurrence. For time to death without recurrence, the sign of treatment effect is positive; this coincides with the fact that more patients died without cancer recurrence in the tamoxifen group (293/535) than in the placebo group (242/535) in Table 10.3. We also see that the use of tamoxifen (tamoxifen=1) significantly reduces cancer recurrence (Type 1) but that it is not beneficial in terms of time to death after recurrence. In terms of other covariates, the age effect (x_2) is significant only on time to recurrence. The effect of tumor size (x_3) is positively significant on all three event types, implying that the event rate is significantly higher among patients whose tumor sizes were larger at surgery.

Now we restrict the data set only to older patients (i.e., $n = 1,776$ with age ≥ 50). The results are summarized in Table 10.5. Here, we present the results for $m_v(h_p)$ because the $s_v(h_p)$ method did not converge. We find that the frailty estimate is relatively large ($\hat{\alpha} = 1.315$). The LRT is $2\{m_v(h_p) - \ell_B\} = 9.8 > 2.71$, indicating that the frailty effect is significantly large, i.e., $\alpha > 0$. We also have $2(m - \ell_B) = 9.3$, selecting the frailty model. Treatment effects are overall similar to those in Table 10.4,

Table 10.5 Fitted results from the two semi-competing-risks models for old patients (age $\geq$50) in the breast cancer data

Model	Time to recurrence	Time to death without recurrence	Time to death after recurrence
	Est. (SE)	Est. (SE)	Est. (SE)
Markov model			
Treatment	−0.631 (0.097)	0.081 (0.092)	0.471 (0.112)
Tumor size $-2\ell_B = 16417.1$	0.023 (0.003)	0.005 (0.004)	0.006 (0.004)
Frailty model (HL)			
Treatment	−0.775 (0.116)	−0.101 (0.119)	0.472 (0.150)
Tumor size $-2m_v(h_p) = 16407.3$	0.032 (0.004)	0.016 (0.005)	0.015 (0.006)
α	1.315		
Frailty model (ML)			
Treatment	−0.787 (0.120)	−0.116 (0.126)	0.466 (0.160)
Tumor size	0.033 (0.005)	0.017 (0.006)	0.016 (0.006)
α $-2m = 16407.8$	1.444		

even though their signs in the frailty model have changed for time to death without recurrence. However, for time to death without recurrence and time to death after recurrence, the tumor size effect is not significant in the Markov model, whereas it is significant in the frailty model.

10.3 Discussion

Recently, the frailty modelling approaches to semi-competing-risks data have been widely studied (Xu et al. 2010; Zhang et al. 2013; Varadhan et al. 2014; Meira-Machado and Faria 2014; Jiang and Haneuse 2015; Lee et al. 2015b, 2016). In particular, Xu et al. (2010) proposed a marginal likelihood approach under the gamma frailty model. Zhang et al. (2013) and Lee et al. (2015b, 2016) have studied Bayesian approaches. However, the marginal likelihood and Bayesian approaches may involve evaluation of the intractable integrals over the random-effect distributions, which can be avoided by the h-likelihood approach.

To model semi-competing-risks data, we used only a univariate frailty based on three transitions in Fig. 10.1. Extension to models with correlated frailties would be an interesting future work. Furthermore, we have assumed a Markov process for such transition, but comparison with a semi-Markov assumption may also be interesting.

10.4 Appendix

10.4.1 Marginal Likelihood Estimation Procedure

The marginal likelihood, denoted by m, can be obtained by integrating out the frailties from the h-likelihood:

$$m = m(\beta, \lambda_0, \alpha) = \sum_i \log\left\{\int \exp(h_i)\, dv_i\right\}, \tag{10.12}$$

where $h_i = \ell_{1i} + \ell_{2i}$ is the contribution of the ith individual to h in (10.8). The marginal likelihood m often requires a numerical integration (e.g., for the log-normal frailty model).

However, for the gamma frailty model with $E(u_i) = 1$ and $\text{var}(u_i) = \alpha$, we have an explicit marginal likelihood as follows. Since the second term of the h-likelihood in (10.8) with gamma frailty is given by

$$\ell_{2i} = \ell_{2i}(\alpha; v_i) = \alpha^{-1}(v_i - u_i) + c(\alpha),$$

with $c(\alpha) = -\log\Gamma(\alpha^{-1}) - \alpha^{-1}\log\alpha$, from (10.8) and (10.12) we have

$$\begin{aligned}
m &= \sum_i [\delta_{i1}\{\log\lambda_{01}(y_{i1}) + x_i^T\beta_1\} + \delta_{i2}(1-\delta_{i1})\{\log\lambda_{02}(y_{i2}) + x_i^T\beta_2\} \\
&\quad + \delta_{i1}\delta_{i2}\{\log\lambda_{03}(y_{i2}) + x_i^T\beta_3\}] \\
&\quad - \sum_i [(\alpha^{-1} + \delta_{i+})\log(1+\alpha\mu_{i+}) - \log\{\alpha^{\delta_{i+}}\Gamma(\alpha^{-1}+\delta_{i+})/\Gamma(\alpha^{-1})\}] \\
&= \sum_{k_1} d_{1(k_1)}\log\lambda_{01k_1} + \sum_i \delta_{i1}(x_i^T\beta_1) + \sum_{k_2} d_{2(k_2)}\log\lambda_{02k_2} \\
&\quad + \sum_i \delta_{i2}(1-\delta_{i1})(x_i^T\beta_2) + \sum_{k_3} d_{3(k_3)}\log\lambda_{03k_3} + \sum_i \delta_{i1}\delta_{i2}(x_i^T\beta_3) \\
&\quad - \sum_i [(\alpha^{-1} + \delta_{i+})\log(1+\alpha\mu_{i+}) - \delta_{i1}\delta_{i2}\log(1+\alpha)],
\end{aligned} \tag{10.13}$$

where $\delta_{i+} = \delta_{i1} + \delta_{i2}$ and $\mu_{i+} = \sum_{j=1}^{3}\mu_{ij}$ with

$$\begin{aligned}
\mu_{i1} &= \Lambda_{01}(y_{i1})\exp(x_i^T\beta_1) = \sum_{k_1}\lambda_{01k_1} I(y_{1(k_1)} \le y_{i1})\exp(x_i^T\beta_1), \\
\mu_{i2} &= \Lambda_{02}(y_{i2})\exp(x_i^T\beta_2) = \sum_{k_2}\lambda_{02k_2} I(y_{2(k_2)} \le y_{i1})\exp(x_i^T\beta_2), \\
\mu_{i3} &= \Lambda_{03}(y_{i1}, y_{i2})\exp(x_i^T\beta_3) = \sum_{k_3}\lambda_{03k_3} I(y_{i1} < y_{3(k_3)} \le y_{i2})\exp(x_i^T\beta_3).
\end{aligned}$$

In fact, the marginal likelihood (10.13) is the same as that of Xu et al. (2010).

With gamma frailty, the score equations for β are given by

$$\frac{\partial m}{\partial \beta_1} = \sum_i \left\{ \delta_{i1} - \left(\frac{\alpha^{-1} + \delta_{i+}}{\alpha^{-1} + \mu_{i+}} \right) \mu_{i1} \right\} x_i, \tag{10.14}$$

$$\frac{\partial m}{\partial \beta_2} = \sum_i \left\{ \delta_{i2}(1 - \delta_{i1}) - \left(\frac{\alpha^{-1} + \delta_{i+}}{\alpha^{-1} + \mu_{i+}} \right) \mu_{i2} \right\} x_i, \tag{10.15}$$

$$\frac{\partial m}{\partial \beta_3} = \sum_i \left\{ \delta_{i1}\delta_{i2} - \left(\frac{\alpha^{-1} + \delta_{i+}}{\alpha^{-1} + \mu_{i+}} \right) \mu_{i3} \right\} x_i. \tag{10.16}$$

In particular, the solutions of $\partial m/\partial \lambda_{0jk_j} = 0$ $(j = 1, 2, 3)$ lead to the closed forms:

$$\widetilde{\lambda}_{0jk_j}(\beta, \alpha) = \frac{d_{j(k_j)}}{\sum_{i \in R_{(k_j)}} \exp(x_i^T \beta_j) \tilde{u}_i}, \tag{10.17}$$

where $\tilde{u}_i = (\alpha^{-1} + \delta_{i+})/(\alpha^{-1} + \mu_{i+})$. We see that the equations for (β, λ_{0j}) in (10.14)–(10.16) and (10.17) are extensions of those from the univariate gamma frailty models (Andersen et al. 1997). Finally, the score equation for the frailty parameter α is given by

$$\frac{\partial m}{\partial \alpha} = \sum_i \left\{ \delta_{i1}\delta_{i2}(1 + \alpha)^{-1} + \alpha^{-2} \log(1 + \alpha\mu_{i+}) - (\alpha^{-1} + \delta_{i+})\mu_{i+}(1 + \alpha\mu_{i+})^{-1} \right\}.$$

Then the fixed parameters (β, α) can be estimated using the Newton–Raphson method, with the second derivatives $-\partial^2 m/\partial \alpha^2$.

10.4.2 Comparison of H-Likelihood with Marginal Likelihood

We assume that α is known. Recall that given (β, v), the score equations $\partial h/\partial \lambda_{0jk_j} = 0$ $(j = 1, 2, 3)$ provide the nonparametric MHLEs, i.e.,

$$\widehat{\lambda}_{0jk_j}(\beta, v) = \frac{d_{j(k_j)}}{\sum_{i \in R_{(k_j)}} \exp(x_i^T \beta_j) u_i}.$$

From the MHL estimating equations $\partial h_p/\beta_j = 0$, the score equations for β, under gamma frailty, become

$$\frac{\partial h}{\partial \beta_1}\Big|_{\lambda_{01}=\widehat{\lambda}_{01}} = \sum_i \left\{ \delta_{i1} - \mu_{i1}u_i \right\} x_i \Big|_{\lambda_{01}=\widehat{\lambda}_{01}}, \tag{10.18}$$

$$\frac{\partial h}{\partial \beta_2}\Big|_{\lambda_{02}=\widehat{\lambda}_{02}} = \sum_i \left\{ \delta_{i2}(1-\delta_{i1}) - \mu_{i2}u_i \right\} x_i \Big|_{\lambda_{02}=\widehat{\lambda}_{02}}, \tag{10.19}$$

$$\frac{\partial h}{\partial \beta_3}\Big|_{\lambda_{03}=\widehat{\lambda}_{03}} = \sum_i \left\{ \delta_{i1}\delta_{i2} - \mu_{i3}u_i \right\} x_i \Big|_{\lambda_{03}=\widehat{\lambda}_{03}}. \tag{10.20}$$

From

$$\frac{\partial h}{\partial v_i} = (\delta_{i+} - \mu_{i+}u_i) + \alpha^{-1} - \alpha^{-1}u_i = 0,$$

we have

$$\hat{u}_i = \frac{\alpha^{-1} + \delta_{i+}}{\alpha^{-1} + \mu_{i+}}, \tag{10.21}$$

which also becomes $E(u_i|y_i^o)$ because the conditional distribution of u_i given the observed data $y_i^o = (y_{i1}, y_{i2}, \delta_{i1}, \delta_{i2})$ is again gamma. Here $\delta_{i+} = \delta_{i1} + \delta_{i2}$ and $\mu_{i+} = \mu_{i1} + \mu_{i2} + \mu_{i3}$. It can be easily seen that the score Eqs. (10.18)–(10.20) with (10.21) are equivalent to the Eqs. (10.14)–(10.16) with (10.17), which are given by

$$\frac{\partial m}{\partial \beta_1}\Big|_{\lambda_{01}=\widetilde{\lambda}_{01}} = \sum_i \left\{ \delta_{i1} - \mu_{i1}\left(\frac{\alpha^{-1} + \delta_{i+}}{\alpha^{-1} + \mu_{i+}}\right) \right\} x_i \Big|_{\lambda_{01}=\widetilde{\lambda}_{01}},$$

$$\frac{\partial m}{\partial \beta_2}\Big|_{\lambda_{02}=\widetilde{\lambda}_{02}} = \sum_i \left\{ \delta_{i2}(1-\delta_{i1}) - \mu_{i2}\left(\frac{\alpha^{-1} + \delta_{i+}}{\alpha^{-1} + \mu_{i+}}\right) \right\} x_i \Big|_{\lambda_{02}=\widetilde{\lambda}_{02}},$$

$$\frac{\partial m}{\partial \beta_3}\Big|_{\lambda_{03}=\widetilde{\lambda}_{03}} = \sum_i \left\{ \delta_{i1}\delta_{i2} - \mu_{i3}\left(\frac{\alpha^{-1} + \delta_{i+}}{\alpha^{-1} + \mu_{i+}}\right) \right\} x_i \Big|_{\lambda_{03}=\widetilde{\lambda}_{03}}.$$

Accordingly, given α the MHLEs for β are the same as the MLEs.

10.4.3 Fourth-order Laplace approximation

Following Tierney and Kadane (1986) and Lee et al. (2017b), we can show that with the gamma frailty, the fourth-order Laplace approximation (denoted by $m_v(h)$) to the marginal likelihood $m = \int \exp(h)dv$ is given by

$$m_v(h) = s_v(h) - F^*(h), \tag{10.22}$$

where $s_v(h) = p_v(h) - F(h)/24$ is the second-order Laplace approximation to m, and $F(h) = -2\sum_i(\alpha^{-1} + \delta_{i+})^{-1}$ and $F^*(h) = (1/360)\sum_i(\alpha^{-1} + \delta_{i+})^{-3}$. Note here that

$$p_v(h) = \left[h - \frac{1}{2} \log \det\{H(h; v)/(2\pi)\} \right] \Bigg|_{v=\widehat{v}},$$

where $H(h; v) = -\partial^2 h/\partial v^2$ and $\widehat{v}$ solves $\partial h/\partial v = 0$. Then it becomes

$$m_v(h) = p_v(h) + (1/12) \sum_i (\alpha^{-1} + \delta_{i+})^{-1} - (1/360) \sum_i (\alpha^{-1} + \delta_{i+})^{-3}.$$

It can be seen that $m_v(h)$ in (10.22) is equivalent to approximating m by the fourth-order Stirling approximation

$$\log \Gamma(x) \doteq (x - 1/2) \log(x) + \log(2\pi)/2 - x + 1/(12x) - 1/(360x^3).$$

Accordingly, we suggest a modified h-likelihood based on h_p, defined by

$$m_v(h_p) = s_v(h_p) - F^*(h), \tag{10.23}$$

which becomes a higher order approximation to m_p. Here $s_v(h_p) = p_v(h_p) - F(h)/24$. Note that $s_v(h)$ and $s_v(h_p)$ are the second-order Laplace approximations to m in (4.7) and m_p in (4.8), respectively and that $m_v(h)$ and $m_v(h_p)$ are the fourth-order Laplace approximations to m and m_p, respectively.

Appendix
Formula for Fitting Fixed and Random Effects

We outline a unified formula for estimating the fixed and random effects $\tau = (\beta^T, v^T)^T$ in the models used in each chapter. This consists of two procedures, the IWLS and ILS equations, for the HGLMs and semiparametric frailty models, respectively.

A.1 IWLS Procedures

$$\boxed{(\mathbf{P}^T\mathbf{V}\mathbf{P})\hat{\tau} = \mathbf{P}^T\mathbf{V}\mathbf{y_0}}$$

(1) HGLM (Chap. 3)

$$\mathbf{P} = \begin{pmatrix} X & Z \\ \mathbf{0} & I_q \end{pmatrix}, \quad \mathbf{V} = \begin{pmatrix} W & \mathbf{0} \\ \mathbf{0} & Q \end{pmatrix} \text{ and } \mathbf{y_0} = (w^T, R^T Q^{-1})^T$$

(Note) $R = \mathbf{0}$ under normal random effects.

(2) AFT model with normal random effects (Chap. 8)

$$\mathbf{P} = \begin{pmatrix} X & Z \\ \mathbf{0} & I_q \end{pmatrix}, \quad \mathbf{V} = \begin{pmatrix} W^* & \mathbf{0} \\ \mathbf{0} & Q \end{pmatrix} \text{ and } \mathbf{y_0} = (w^{*T}, \mathbf{0}^T)^T$$

A.2 ILS Procedures

$$\boxed{(\mathbf{P}^T\mathbf{V}\mathbf{P})\hat{\tau} = \mathbf{P}^T\mathbf{y_0^*}}$$

I.D. Ha et al., *Statistical Modelling of Survival Data with Random Effects*,
Statistics for Biology and Health, https://doi.org/10.1007/978-981-10-6557-6

(1) Cox-PH model (Chap. 2)

$$\mathbf{P} = X,\ \mathbf{V} = W^* \text{ and } \mathbf{y}_0^* = w^*$$

(2) Simple frailty model (Chap. 4)

$$\mathbf{P} = \begin{pmatrix} X & Z \\ \mathbf{0} & I_q \end{pmatrix},\quad \mathbf{V} = \begin{pmatrix} W^* & \mathbf{0} \\ \mathbf{0} & Q \end{pmatrix} \text{ and } \mathbf{y}_0^* = (w^{*T}, R^T Q^{-1})^T.$$

(Note) $R = \mathbf{0}$ under normal random effects (or log-frailties).
(3) Multicomponent frailty model (Chap. 5)

$$\mathbf{P} = \begin{pmatrix} X & \mathbf{Z} \\ \mathbf{0} & \mathbf{I_q} \end{pmatrix},\quad \mathbf{V} = \begin{pmatrix} W^* & \mathbf{0} \\ \mathbf{0} & \mathbf{Q} \end{pmatrix} \text{ and } \mathbf{y}_0^* = (w^{*T}, \mathbf{0}^T)^T.$$

Here, $\mathbf{Z} = (Z_1, \cdots, Z_k)$, $\mathbf{I_q} = \text{BD}(I_{q_1}, \ldots, I_{q_k})$, and $\mathbf{Q} = \text{BD}(Q_1, \ldots, Q_k)$ with $Q_r = -\partial^2 \ell_2 / \partial v^{(r)2} = \Sigma_r^{-1}$ $(r = 1, \ldots, k)$.
(4) Competing-risks frailty model (Chap. 6).
(4-1) Cause-specific hazards frailty model

$$\mathbf{P} = \begin{pmatrix} \mathbf{X} & \mathbf{Z} \\ \mathbf{0} & I_{k^*} \end{pmatrix},\quad \mathbf{V} = \begin{pmatrix} \mathbf{W}^* & \mathbf{0} \\ \mathbf{0} & Q \end{pmatrix} \text{ and } \mathbf{y}_0^* = (w^{*T}, \mathbf{0}^T)^T.$$

Here, $k^* = K \times q$ with the number of events $K = 2$, $\mathbf{X} = \text{BD}(X, X)$, $\mathbf{Z} = \text{BD}(Z, Z)$, and $\mathbf{W}^* = \text{BD}(W_1^*, W_2^*)$.
(4-2) Subhazards frailty model

$$\mathbf{P} = \begin{pmatrix} X & Z \\ \mathbf{0} & I_{q^*} \end{pmatrix},\quad \mathbf{V} = \begin{pmatrix} W^* & \mathbf{0} \\ \mathbf{0} & Q \end{pmatrix} \text{ and } \mathbf{y}_0^* = (w^{*T}, \mathbf{0}^T)^T.$$

Here $q^* = q \times m$ with the number of frailty terms m, and W^* and w^* depend on the IPCW weights w_{ij}.
(5) Joint survival model (Chap. 9).
(5-1) Repeated measures and a univariate event time

$$\mathbf{P} = \begin{pmatrix} \mathbf{X} & \mathbf{Z} \\ \mathbf{0} & I_q \end{pmatrix},\quad \mathbf{V} = \begin{pmatrix} \mathbf{W} & \mathbf{0} \\ \mathbf{0} & Q \end{pmatrix} \text{ and } \mathbf{y}_0^* = (w_1^T, w_2^T, \mathbf{0}^T)^T.$$

Here, $\mathbf{X} = \text{BD}(X_1, X_2)$, $\mathbf{Z} = (Z_1^T, \gamma Z_2^T)^T$, and $\mathbf{W} = \text{BD}(W_1, W_2)$.
(5-2) Repeated measures and competing-risks event times

$$\mathbf{P} = \begin{pmatrix} \mathbf{X} & \mathbf{Z} \\ \mathbf{0} & I_q \end{pmatrix},\quad \mathbf{V} = \begin{pmatrix} \mathbf{W} & \mathbf{0} \\ \mathbf{0} & Q \end{pmatrix} \text{ and } \mathbf{y}_0^* = (w_1^T, w_2^T, w_3^T, \mathbf{0}^T)^T.$$

Here, $\mathbf{X} = \text{BD}(X_1, X_2, X_2)$, $\mathbf{Z} = (Z_1^T, \gamma_1 Z_2^T, \gamma_2 Z_2^T)^T$ and $\mathbf{W} = \text{BD}(W_1, W_2, W_3)$.

(6) Semi-competing-risks model (Chap. 10)

$$\mathbf{P} = \begin{pmatrix} \mathbf{X} & \mathbf{Z} \\ \mathbf{0} & I_q \end{pmatrix}, \quad \mathbf{V} = \begin{pmatrix} \mathbf{W}^* & \mathbf{0} \\ \mathbf{0} & Q \end{pmatrix} \text{ and } \mathbf{y}_\mathbf{0}^* = (w_1^{*T}, w_2^{*T}, w_3^{*T}, R^T Q^{-1})^T.$$

Here, $\mathbf{X} = \mathrm{BD}(X, X, X)$, $\mathbf{Z} = (Z^T, Z^T, Z^T)^T$ and $\mathbf{W}^* = \mathrm{BD}(W_1^*, W_2^*, W_3^*)$, and $w_j^* = W_j^* \eta_j + (\delta_j + \mu_j)$ for $j = 1, 2, 3$.

(7) Variable selection in the frailty model with competing risks (Chap. 7)

$$(\mathbf{P}^T \mathbf{V} \mathbf{P} + n \Sigma_{\gamma,w})\hat{\tau} = \mathbf{P}^T \mathbf{y}_\mathbf{0}^*,$$

where $\Sigma_{\gamma,w} = \mathrm{diag}\{J'_{\gamma,w}(|\beta_j|)/|\beta_j|\}$.

References

Aalen OO (1978) Nonparametric inference for a family of counting process. Ann Stat 6:534–545
Aalen OO (1980) A model for non-parametric regression analysis of counting processes. Lecture notes in statistics, vol 2. Springer, New-York, pp 1–25
Aalen OO (1988) Heterogeneity in survival analysis. Stat Med 7:1121–1137
Aalen OO (1992) Modelling heterogeneity in survival analysis by the compound Poisson distribution. Ann Appl Probab 2:951–972
Aalen OO, Borgan O, Gjessing HK (2008) Survival and event history analysis. Springer, New York
Abbring JH, van den Berg GJ (2003) The identifiability of the mixed proportional hazards competing risks model. J R Stat Soc B 65:701–710
Agresti A, Caffo B, Ohman-Strickland P (2004) Example in which misspecification of a random effects distribution reduces efficiency, and possible remedies. Comput Stat Data Anal 47:639–653
Aitkin MA (1981) A note on the regression analysis of censored data. Technometrics 23:161–163
Aitkin M, Clayton D (1980) The fitting of exponential, Weibull and extreme value distributions to complex censored survival data using GLIM. Appl Stat 29:156–163
Akaike H (1973) Information theory and an extension of maximum likelihood principle. In: Petrov BN, Csáki F (eds) Proceedings 2nd international symposium on information theory. Akadémiai Kiadó, Budapest, pp 267–281
Andersen EB (1970) Asymptotic properties of conditional maximum-likelihood estimators. J R Stat Soc B 32:283–301
Andersen PK, Gill RD (1982) Cox's regression model for counting processes: a large sample study. Ann Stat 10:1100–1120
Andersen PK, Borgan Ø, Gill RD, Keiding N (1993) Statistical models based on counting processes. Springer, New York
Andersen J, Goetghebeur E, Ryan L (1996) Missing cause of death information in the analysis of survival data. Stat Med 15:2191–2201
Andersen PK, Klein JP, Knudsen K, Palacios RT (1997) Estimation of variance in Cox's regression model with shared gamma frailties. Biometrics 53:1475–1484
Androulakis E, Koukouvinos C, Vonta F (2012) Estimation and variable selection via frailty models with penalized likelihood. Stat Med 31:2223–2239
Bakoyannis G, Siannis F, Touloumi G (2010) Modelling competing risks data with missing cause of failure. Stat Med 29:3172–3185
Barker P, Henderson R (2005) Small sample bias in the gamma frailty model for univariate survival. Lifetime Data Anal 11:265–284
Barlow WE, Prentice RL (1988) Residuals for relative risk regression. Biometrika 75:65–74

I.D. Ha et al., *Statistical Modelling of Survival Data with Random Effects*,
Statistics for Biology and Health, https://doi.org/10.1007/978-981-10-6557-6

Batchelor JR, Hackett M (1970) HLA matching in treatment of burned patients with skin allografts. Lancet 2:581–583

Bayarri MJ, DeGroot MH, Kadane JB (1988) What is the likelihood function? (with discussion). Ststistical decision theory and related topics IV, Springer, New York

Bennett S (1983) Analysis of survival data by the proportional odds model. Stat Med 2:273–277

Beyersmann J, Dettenkofer M, Bertz H, Schumacher M (2007) A competing risks analysis of bloodstream infection after stem-cell transplantation using subdistribution hazardsa and cause-specific hazards. Stat Med 26:5360–5369

Beyersmann J, Latouche A, Buchholz A, Schumacher M (2009) Simulating competing risks data in survival analysis. Stat Med 28:956–971

Birnbaum A (1962) On the foundations of statistical inference (with discussion). J Am Stat Assoc 57:269–306

Bjørnstad JF (1990) Predictive likelihood: a review (with discussion). Stat Sci 5:242–265

Bjørnstad JF (1996) On the generalization of the likelihood function and likelihood principle. J Am Stat Assoc 91:791–806

Blachman NM, Christensen R, Utts JM (1996) Letter with corrections to theoriginal article by Christensen R, Utts J (1992). Am Stat 50:98–99

Booth JG, Hobert JP (1998) Standard errors of prediction in generalized linear mixed models. J Am Stat Assoc 93:262–272

Breiman L (1996) Heuristics of instability and stabilization in model selection. Ann Stat 24:2350–2383

Breslow NE (1972) Discussion of Professor Cox's paper. J R Stat Soc B 34:216–217

Breslow NE (1974) Covariance analysis of censored survival data. Biometrics 30:89–99

Breslow NE, Clayton DG (1993) Approximate inference in generalized linear mixed model. J Am Stat Assoc 88:9–25

Breslow NE, Crowley J (1974) A large sample study of the life table and product-limit estimates under random censorship. Ann Stat 2:437–453

Burke K, MacKenzie G (2017) Multi-parameter regression survival modeling: an alternative to proportional hazards. Biometrics 73:678–686

Buckley J, James I (1979) Linear regression with censored data. Biometrika 66:429–436

Burnham KP, Anderson DR (2002) Model selection and multimodel inference: a practical information-theoretic approach, 2nd edn. Springer, New York

Burnham KP, Anderson DR (2004) Multimodel inference. Sociol Methods Res 33:261–304

Burr IW (1942) Cumulative frequency functions. Ann Math Stat 13:215–232

Butler SM, Louis TA (1992) Random effects models with non-parametric priors. Stat Med 11:1981–2000

Cai J, Fan J, Li R, Zhou H (2005) Variable selection for multivariate failure time data. Biometrika 92:303–316

Cao C, Shi JQ, Lee Y (2017) Robust functional regression model for marginal mean and subject-specific inferences. Stat Methods Med Res (in press)

Carlin BP, Gelfand AE (1990) Approaches for empirical Bayes confidence intervals. J Am Stat Assoc 85:105–114

Carlin BP, Louis TA (2000) Bayes and empirical bayes methods for data analysis, 2nd edn. Chapman and Hall, London

Chen Y-H (2010) Semiparametric marginal regression analysis for dependent competing risks under an assumed copula. J R Stat Soc B 72:235–251

Cheng SC, Wei LJ, Ying Z (1995) Analysis of transformation models with censored data. Biometrika 82:835–845

Cheng S, Fine J, Wei L (1998) Prediction of cumulative incidence function under the proportional hazards model. Biometrics 54:219–228

Cheng Y, Fine J, Kosorok M (2009) Nonparametric association analysis of exchangeable clustered competing risks data. Biometrics 65:385–393

Chernoff H (1954) On the distribution of the likelihood ratio. Ann Math Stat 25:573–578

Chiou SH, Kang S, Yan J (2014) Fitting accelerated failure time models in routine survival analysis with R package aftgee. J Stat Softw 61:1–23
Christensen R, Utts J (1992) Bayesian resolution of the 'exchange paradox'. Am Stat 46:274–276
Christian NJ (2011) Hierarchical likelihood inference on clustered competing risk data. PhD thesis, Department of Biostatistics, University of Pittsburgh
Christian NJ, Ha ID, Jeong J-H (2016) Hierarchical likelihood inference on clustered competing risks data. Stat Med 35:251–267
Clayton DG (1978) A model for association in bivariate life tables and its application in epidemiological studies of familial tendency in chronic disease incidence. Biometrika 65:141–151
Clayton DG, Cuzick J (1985) The EM algorithm for Cox's regression model using GLIM. Appl Stat 34:148–156
Collet D (2015) Modelling survival data in medical research, 3rd edn. Chapman and Hall, London
Cox DR (1972) Regression models and life tables (with Discussion). J R Stat Soc B 74:187–220
Cox DR (1975) Partial likelihood. Biometrika 62:269–276
Cox DR, Hinkley DV (1974) Theoretical Statistics. Chapman and Hall, London
Cox DR, Oakes D (1984) Analysis of survival data. Chapman and Hall, London
Cox DR, Reid N (1987) Parameter orthogonality and approximate conditional inference (with discussion). J R Stat Soc B 49:1–39
Cox DR, Snell EJ (1968) A general definition of residuals. J R Stat Soc B 30:248–275
Crowder MJ, Kimber AC, Smith RL, Sweeting TJ (1991) Statistical analysis of reliability data. Chapman and Hall, London
Cullis BR, Smith AB, Thompson R (2004) Perspectives of ANOVA, REML and a general linear mixed model. In honour of Professor John Nelder, FRS, 53–94
Donohue M, Xu R (2012) phmm: proportional hazards mixed-effects model. http://CRAN.R-project.org/package=phmm R package version 0.7-4
Donohue M, Overholser D, Xu R, Vaida F (2011) Conditional Akaike information under generalized linear and proportional hazards mixed models. Biometrika 98:685–700
Duchateau L, Janssen P (2008) The frailty model. Springer, New York
Elashoff RM, Li G, Li N (2008) A joint model for longitudinal measurements and survival data in the presence of multiple failure types. Biometrics 64:762–771
Elbers C, Ridder G (1982) True and spurious duration dependence: the identifiability of the proportional hazard model. Rev Econ Stud 49:403–409
Efron B (1977) The efficiency of Cox's likelihood function for censored data. J Am Stat Assoc 72:557–565
Efron B (2013) Bayes' theorem in the 21st century. Science 340:1177–1178
Ettinger DS, Finkelstein DM, Abeloff MD et al (1990) A randomized comparison of standard chemotherapy versus alternating chemotherapy and maintenance versus no maintenance therapy for extensive-stage small-cell lung cancer: a phase III study of the Eastern Cooperative Oncology Group. J Clin Oncol 8:230–240
Fan J (1997) Comments on "Wavelets in statistics: a review" by A. Antoniadis. J Ital Stat Soc 6:131–138
Fan J, Li R (2001) Variable selection via nonconcave penalized likelihood and its oracle properties. J Am Stat Assoc 96:1348–1360
Fan J, Li R (2002) Variable selection for Cox's proportional hazards model and frailty model. Ann Stat 30:74–99
Fan J, Lv J (2010) A selective overview of variable selection in high dimensional feature space. Statistica Sinica 20:101–148
Farewell VT (1982) The use of mixture models for the analysis of survival data with long-term survivors. Biometrics 38:1041–1046
Fine JP, Gray RJ (1999) A proportional hazards model for the subdistribution of a competing risk. J Am Stat Assoc 94:496–509
Fine JP, Jiang H, Chappell R (2001) On semi-competing risks data. Biometrika 88:907–919

Fisher B, Costantino J, Redmond C, Poisson R, Bowman D, Couture J, Dimitrov NV, Wolmark N, Wickerham DL, Fisher ER, Margolese R, Robidoux A, Shibata H, Terz J, Paterson AHG, Feldman MI, Farrar W, Evans J, Lickley HL, Ketner M (1989) A randomized clinical trial evaluating tamoxifen in the treatment of patients with node-negative breast cancer who have estrogen receptor-positive tumors. N Engl J Med 320:479–484

Fisher B, Dignam J, Bryant J, DeCillis A, Wickerham DL, Wolmark N, Costantino J, Redmond C, Fisher ER, Bowman DM, Deschnes L, Dimitrov NV, Margolese RG, Robidoux A, Shibata H, Terz J, Paterson AH, Feldman MI, Farrar W, Evans J, Lickley HL (1996) Five versus more than five years of tamoxifen therapy for breast cancer patients with negative lymph nodes and estrogen receptor- positive tumors. J Natl Cancer Inst 88:1529–1542

Fisher B, Costantino J, Wicerham D, Redmond C, Kavanah M, Cronin W, Vogel V, Robidoux A, Dimitrov N, Atkins J, Daly M, Wieand S, Tan-Chiu E, Ford L, Wolmark N (1998) Tamoxifen for prevention of breast cancer: Report of the national surgical adjuvant breast and bowel project p-1 study. J Natl Cancer Inst 90:1371–1388

Fisher RA (1922) On the mathematical foundations of theoretical statistics. Philos Trans R Soc A 222:309–368

Fleming TR, Harrington DP (1991) Counting processes and survival analysis. Wiley, New York

Fong Daniel YT, Lam KF, Lawless JF, Lee YW (2001) Dynamic random effects models for times between repeated events. Lifetime Data Anal 7:345–362

Gail MH, Santner TJ, Brown CC (1980) An analysis of comparative carcinogenesis experiments based on multiple times to tumor. Biometrics 36:255–266

Gamst A, Donohue M, Xu R (2009) Asymptotic properties and empirical evaluation of the NPMLE in the proportional hazards mixed-effects models. Statistica Sinica 19:997–1011

Gardner M (1982) Aha! Gotcha: paradoxes to puzzle and delight. W. H. Freeman & Co, San Francisco

Gao G, Tsiatis AA (2005) Semiparametric estimators for the regression coefficients in the linear transformation competing risks model with missing cause of failure. Biometrika 92:875–891

Gastrointestinal Tumor Study Group (1982) A comparison of combination chemotherapy and combined modality therapy for locally advanced gastric carcinoma. Cancer 49:1771–1777

Gehan EA (1965) A generalized Wilcoxon test for comparing arbitrarily singly censored samples. Biometrika 52:203–223

Goetghebeur E, Ryan L (1995) Analysis of competing risks survival data when some failure types are missing. Biometrika 82:821–834

Goldstein H (1995) Multilevel statistical models. Arnold, London

Gonzalez JR, Rondeau V, Mazroui Y, Mauguen A, Diakité A (2012) frailtypack: frailty models using a semi-parametrical penalized likelihood estimation or a parametrical estimation. R package version 2.2–2:3

Gooley TA, Leisenring W, Crowley J, Storer BE (1999) Estimation of failure probabilities in the presence of competing risks: new representations of old estimators. Stat Med 18:695–706

Gorfine M, Hsu L (2011) Frailty-based competing risks model for multivariate survival data. Biometrics 67:415–426

Grambsch P, Therneau T (1994) Proportional hazards tests and diagnostics based on weighted residuals. Biometrika 81:515–526

Gray RJ (1988) A class of K-sample tests for comparing the cumulative incidence of a competing risk. Ann Stat 16:1141–1154

Gray RJ (1992) Flexible methods for analyzing survival data using splines, with applications to breast cancer prognosis. J Am Stat Assoc 87:942–951

Gray RJ (1994) A Bayesian analysis of institutional effects in multicenter cancer clinical trial. Biometrics 50:244–253

Greven S, Kneib T (2010) On the behaviour of marginal and conditional AIC in linear mixed models. Biometrika 97:773–789

Gu MG, Sun L, Huang C (2004) A universal procedure for parametric frailty models. J Stat Comput Simul 74:1–13

Guo X, Carlin BP (2004) Separate and joint modeling of longitudinal and event time data using standard computer packages. Am Stat 58:16–24

Ha ID (2007) Discussion of Zeng and Lin's paper. J Roy Stat Soc B 69:549–550

Ha ID (2008) A general mixed linear model with left-censoring data. Commun Korean Stat Soc 15:969–976

Ha ID (2012) Comparison of estimation methods for semi-parametric frailty models. Quant Bio-Sci 31:39–45

Ha ID, Lee Y (2003) Estimating frailty models via Poisson hierarchical generalized linear models. J Comput Graph Stat 12:663–681

Ha ID, Lee Y (2005a) Multilevel mixed linear models for survival data. Lifetime Data Anal 11:131–142

Ha ID, Lee Y (2005b) Comparison of hierarchical likelihood versus orthodox best linear unbiased predictor approaches for frailty models. Biometrika 92:717–723

Ha ID, MacKenzie G (2010) Robust frailty modelling using non-proportional hazards models. Stat Model 10:315–332

Ha ID, Lee Y, Song J-K (2001) Hierarchical likelihood approach for frailty models. Biometrika 88:233–243

Ha ID, Lee Y, Song J-K (2002) Hierarchical likelihood approach for mixed linear models with censored data. Lifetime Data Anal 8:163–176

Ha ID, Park T, Lee Y (2003) Joint modelling of repeated measures and survival time data. Biomed J 45:647–658

Ha ID, Lee Y, MacKenzie G (2007a) Model selection for multi-component frailty models. Stat Med 26:4790–4807

Ha ID, Lee Y, Pawitan Y (2007b) Genetic mixed linear models for twin survival data. Behav Genet 37:621–630

Ha ID, Noh M, Lee Y (2010) Bias reduction of likelihood estimators in semi-parametric frailty models. Scand J Stat 37:307–320

Ha ID, Sylvester R, Legrand C, MacKenzie G (2011) Frailty modelling for survival data from multi-centre clinical trials. Stat Med 30:28–37

Ha ID, Noh M, Lee Y (2012) frailtyHL: a package for fitting frailty models with h-likelihood. R J 4:307–320

Ha ID, Noh M, Kim J, Lee Y (2018) frailtyHL: frailty models using h-likelihood. http://CRAN.R-project.org/package=frailtyHL. R package version 2.1

Ha ID, Lee M, Oh S, Jeong J-H, Sylvester R, Lee Y (2014a) Variable selection in subdistribution hazard frailty models with competing risks data. Stat Med 33:4590–4604

Ha ID, Pan J, Oh S, Lee Y (2014b) Variable selection in general frailty models using penalized h-likelihood. J Comput Graph Stat 23:1044–1060

Ha ID, Christian NJ, Jeong J-H, Park J, Lee Y (2016a) Analysis of clustered competing risks data using subdistribution hazard models with multivariate frailties. Stat Methods Med Res 25:2488–2505

Ha ID, Vaida F, Lee Y (2016b) Interval estimation of random effects in proportional hazards models with frailties. Stat Methods Med Res 25:936–953

Ha ID, Noh M, Lee Y (2017) H-likelihood approach for joint modelling of longitudinal outcomes and time-to-event data. Biom J 59:1122–1143

Harrington DP, Fleming TR (1982) A class of rank test procedures for censored survival data. Biometrika 69:553–566

Harville DA (1977) Maximum likelihood approaches to variance component estimation and to related problems. J Am Stat Assoc 72:320–338

Henderson CR (1975) Best linear unbiased estimation and prediction under a selection model. Biometrics 31:423–447

Henderson R, Oman P (1999) Effect of frailty on marginal regression estimates in survival analysis. J Roy Stat Soc B 61:367–379

Henderson R, Shimakura S, Gorst D (2002) Modeling spatial variation in leukemia survival data. J Am Stat Assoc 97:965–972
Herring AH, Ibrahim JG, Lipsitz SR (2002) Frailty models with missing covariates. Biometrics 58:98–109
Hodges JS, Sargent DJ (2001) Counting degrees of freedom in hierarchical and other richly-parameterised models. Biometrika 88:367–379
Hougaard P (1991) Modelling heterogeneity in survival data. J Appl Probab 28:695–701
Hougaard P (1999) Fundamentals of survival data. Biometrics 55:13–22
Hougaard P (2000) Analysis of multivariate survival data. Springer, New York
Hsu L, Gorfine M, Malone K (2007) On robustness of marginal regression coefficient estimates and hazard functions in multivariate survival analysis of family data when the frailty distribution is mis-specified. Stat Med 26:4657–4678
Huang X, Wolfe R (2002) A frailty model for informative censoring. Biometrics 58:510–520
Huang X, Zhang N (2008) Regression survival analysis with an assumed copula for dependent censoring: a sensitivity analysis approach. Biometrics 64:1090–1099
Huang J, Breheny P, Ma S (2012) A selective review of group selection in high-dimensional models. Stat Sci 27:481–499
Huffer FW, McKeague IW (1991) Weighted least squares estimation for Aalen's additive risk model. J Am Stat Assoc 86:114–129
Hughes JP (1999) Mixed effects models with censored data with application to HIV RNA levels. Biometrics 55:625–629
Hunter D, Li R (2005) Variable selection using MM algorithms. Ann Stat 33:1617–1642
Hutton JL, Monaghan PF (2002) Choice of parametric accelerated life and proportional hazards models for survival data: asymptotic results. Lifetime Data Anal 8:375–393
Jeong JH (2014) Statistical inference on residual life. Springer, New York
Jiang J (1997) Wald consistency and the method of sieves in REML estimation. Ann Stat 25:1781–1803
Jiang F, Haneuse S (2015) Simulation of semicompeting risk survival data and estimation based on multistate frailty model. Harvard University Biostatistics Working Paper Series188
Jin Z (2016) Semiparametric accelerated failure time model for the analysis of right censored data. Commun Stat Appl Methods 23:467–478
Johansen S (1983) An extension of Cox's regression model. Int Stat Rev 51:165–74
Johnson BA, Lin DY, Zeng D (2008) Penalized estimating functions and variable selection in semiparametric regression models. J Am Stat Assoc 103:672–680
Kalbfleisch JD, Prentice RL (1980) The statistical analysis of failure time data, 2nd edn. Wiley, New York
Kalbfleisch JD, Prentice RL (2002) The statistical analysis of failure time data, 2nd edn. Wiley, New York
Kaplan EL, Meier P (1958) Nonparametric estimation from incomplete observations. J Am Stat Assoc 53:457–481
Kass RE, Steffey D (1989) Approximate Bayesian inference in conditionally independent hierarchical models (parametric empirical Bayes models). J Am Stat Assoc 84:717–726
Katsahian S, Boudreau C (2011) Estimating and testing for center effects in competing risks. Stat Med 30:1608–1617
Katsahian S, Resche-Rigon M, Chevret S, Porcher R (2006) Analysing multicentre competing risks data with a mixed proportional hazards model for the subdistribution. Stat Med 25:4267–4278
Keiding N (1992) Independent delayed entry. Survival analysis state of the art. Kluwer Academic Publishers, Boston
Klein JP (1992) Semiparametric estimation of random effects using Cox model based on the EM algorithm. Biometrics 48:795–806
Klein JP, Moeschberger ML (2003) Survival analysis: techniques for censored and truncated data, 2nd edn. Springer, New York

Klein JP, Pelz C, Zhang M (1999) Modelling random effects for censored data by a multivariate normal regression model. Biometrics 55:497–506
Komarek A, Lesaffre E, Legrand C (2007) Baseline and treatment effect heterogeneity for survival times between centers using a random effects accelerated failure time model with flexible error distribution. Stat Med 26:5457–5472
Kosorok MR, Lee BL, Fine JP (2004) Robust inference for univariate proportional hazards frailty regression models. Ann Stat 32:1448–1491
Kuha J (2004) AIC and BIC. Sociol Methods Res 33:188–229
Kuk AYC, Chen CH (1992) A mixture model combining logistic regression with proportional hazards regression. Biometrika 79:531–541
Kuk D, Varadhan R (2013) Model selection in competing risks regression. Stat Med 32:3077–3088
Kullback S, Leibler RA (1951) On information and sufficiency. Ann Math Stat 22:79–86
Kwon S, Oh S, Lee Y (2016) The use of random-effect models for high-dimensional variable selection models. Comput Stat Data Anal 103:401–412
Lachin JM, Foulkes MA (1986) Evaluation of sample size and power for analyses of survival with allowance for nouniform patient entry, losses to follow-up, noncompliance, and stratification. Biometrics 42:507–519
Lai TZ, Ying Z (1994) A missing information principle and M-estimators in regression analysis with censored and truncated data. Ann Stat 22:1222–1255
Laird NM, Louis TA (1987) Empirical Bayes confidence intervals based on bootstrap samples. J Am Stat Assoc 82:739–750
Lambert P, Collett D, Kimber A, Johnson R (2004) Parametric accelerated failure time models with random effects and an application to kidney transplant survival. Stat Med 23:3177–3192
Lancaster T (1990) The econometric analysis of transition data. Cambridge University Press, Cambridge
Latouche A, Porcher R, Chevret S (2004) Sample size formula for proportional hazards modelling of competing risks. Stat Med 23:3263–3274
Lawless JF (2003) Statistical models and methods for lifetime data, 2nd edn. Wiley, New York
Lee Y (2002) Fixed-effect versus random-effect models for evaluating therapeutic preferences. Stat Med 21:2325–2330
Lee Y, Ha ID (2010) Orthodox BLUP versus h-likelihood methods for inferences about random effects in Tweedie mixed models. Stat Comput 20:295–303
Lee Y, Kim G (2016) H-likelihood predictive intervals for unobservables. Int Stat Rev 84:487–505
Lee Y, Nelder JA (1996) Hierarchical generalized linear models (with discussion). J R Stat Soc B 58:619–678
Lee Y, Nelder JA (2001a) Hierarchical generalised linear models: a synthesis of generalised linear models, random-effect models and structured dispersions. Biometrika 88:987–1006
Lee Y, Nelder JA (2001b) Modelling and analysing correlated non-normal data. Stat Model 1:3–16
Lee Y, Nelder JA (2003) Extended-REML estimators. J Appl Stat 30:845–856
Lee Y, Nelder JA (2005) Likelihood for random-effects (with discussion). Stat Oper Res Trans 29:141–182
Lee Y, Nelder JA (2006) Double hierarchical generalized linear models (with discussion). J R Stat Soc C 55:139–185
Lee Y, Nelder JA (2009) Likelihood inference for models with unobservables: another view (with discussion). Stat Sci 24:255–293
Lee Y, Oh HS (2014) A new sparse variable selection via random-effect model. J Multivar Anal 125:89–99
Lee K, Thompson SG (2008) Flexible parametric models for random-effects distributions. Stat Med 27:418–434
Lee ET, Wang WW (2003) Statistical methods for survival data analysis, 3rd edn. Wiley, New York
Lee Y, Nelder JA, Pawitan Y (2006) Generalised linear models with random effects: unified analysis via h-likelihood. CRC Press, Boca Raton

Lee D, Lee W, Lee Y, Pawitan Y (2010) Super sparse principal component analysis for high-throughput genomic data. BMC Bioinform 11:296
Lee D, Lee W, Lee Y, Pawitan Y (2011a) Sparse partial least-squares regression and its applications to high-throughput data analysis. Chemometr Intell Lab Syst 109:1–8
Lee W, Lee D, Lee Y, Pawitan Y (2011b) Sparse canonical covariance analysis for high-throughput data. Stat Appl Genet Mol Biol 10:1–24
Lee S, Pawitan Y, Lee Y (2015a) A random-effect model approach for group variable selection. Comput Stat Data Anal 89:147–157
Lee KH, Haneuse S, Schrag D, Dominici F (2015b) Bayesian semiparametric analysis of semicompeting risks data: investigating hospital readmission after a pancreatic cancer diagnosis. J R Stat Soc C 64:253–273
Lee KH, Dominici F, Schrag D, Haneuse S (2016) Hierarchical models for semi-competing risks data with application to quality of end-of-life care for pancreatic cancer. J Am Stat Assoc 111:1075–1095
Lee M, Ha ID, Lee Y (2017a) Frailty modeling for clustered competing risks data with missing cause of failure. Stat Methods Med Res 26:356–373
Lee Y, Nelder JA, Pawitan Y (2017b) Generalised linear models with random effects: unified analysis via h-likelihood, 2nd edn. Chapman and Hall, Boca Raton
Legrand C, Ducrocq V, Janssen P, Sylvester R, Duchateau L (2005) A Bayesian approach to jointly estimate centre and treatment by centre heterogeneity in a proportional hazards model. Stat Med 24:3789–3804
Leng C, Ma S (2007) Accelerated failure time models with nonlinear covariates effects. Aust N Z J Stat 49:155–172
Leppik IE et al (1985) A double-blind crossover evaluation of progabide in partial seizures. Neurology 35:285
Li H, Lahiri P (2010) An adjusted maximum likelihood method for solving small area estimation problems. J Multivar Anal 101:882–892
Liang H, Wu H, Zou G (2008) A note on conditional AIC for linear mixed-effects models. Biometrika 95:773–778
Lim J, Jeong J (2014) Cause-specific quantile residual life regression. Submitted
Lin DY, Ying Z (1994) Semiparametric analysis of the additive risk model. Biometrika 81:61–71
Litiére S, Alonso A, Molenberghs G (2008) The impact of a misspecified random-effects distribution on the estimation and the performance of inferential procedures in generalized linear mixed models. Stat Med 27:3125–3144
Liu L, Wolfe RA, Huang X (2004) Shared frailty models for recurrent events and a terminal event. Biometrics 60:747–756
Longford NT (1993) Random coefficient models. Oxford University Press, New York
Lu W, Liang Y (2008) Analysis of competing risks data with missing cause of failure under additive hazards model. Statistica Sinica 18:219–234
Lu K, Tsiatis AA (2001) Multiple imputation methods for estimating regression coefficients in the competing risks model with missing cause of failure. Biometrics 57:1191–1197
Ma R, Jørgensen B (2007) Nested generalized linear mixed models: orthodox best linear unbiased predictor approach. J R Stat Soc B 69:625–641
Ma R, Krewski D, Burnett RT (2003) Random effects Cox models: a Poisson modelling approach. Biometrika 90:157–169
MacKenzie G (1996) Regression models for survival data: the generalised time dependent logistic family. J R Stat Soc D 45:21–34
MacKenzie G (1997) On a non-proportional hazards regression model for repeated medical random counts. Stat Med 16:1831–1843
Manola JB, Gray RJ (2011) When bad things happen to good studies. J Clin Oncol 29:3497–3502
Mantel N, Bohidar NR, Ciminera JL (1977) Mantel-Haenszel analysis of litter-matched time-to-response data, with modifications for recovery of interlitter information. Cancer Res 37:3863–3868

Mathiasen PE (1979) Prediction functions. Scand J Stat 6:1–21
McCullagh P, Nelder JA (1989) Generalized linear models, 2nd edn. Chapman and Hall, London
McGilchrist CA (1993) REML estimation for survival models with frailty. Biometrics 49:221–225
McGilchrist CA (1994) Estimation in generalized mixed models. J R Stat Soc B 56:61–69
McGilchrist CA, Aisbett CW (1991) Regression with frailty in survival analysis. Biometrics 47:461–466
Meira-Machado L, Faria S (2014) A simulation study comparing modeling approaches in an illness-death multi-state model. Commun Stat Simul Comput 43:929–946
Mell LK, Jeong J (2010) Pitfalls of using composite primary end points in the presence of competing risks. J Clin Oncol 28:4297–4299
Mell LK, Lau SK, Rose BS, Jeong J (2012) Reporting of cause-specific treatment effects in cancer clinical trials with competing risks: a systematic review. Contemp Clin Trials 33:920–924
Meng XL (1994) Multiple imputation inferences with uncongenial sources of input (with discussion). Stat Sci 9:538–573
Meng X-L (2009) Decoding the H-likelihood. Stat Sci 24:280–293
Meng X-L (2011) What's the H in H-likelihood: a holy grail or an achilles' heel? (with discussion). Bayesian Stat 9:473–500
Miller RG (1981) Survival analysis. Wiley, New York
Morris CN (1983) Parametric empirical Bayes inference: theory and application. J Am Stat Assoc 78:47–55
Morris CN (2006) Mixed model prediction and small area estimation. Test 15:72–76
Murphy SA, van der Vaart AW (2000) On profile likelihood. J Am Stat Assoc 95:449–465
Neale MC, Cardon LR (1992) Methodology for genetic studies of twin and families. Kluwer Academic, Dordrecht
Nelder JA, Wedderburn RWM (1972) Generalized linear models. J R Stat Soc A 135:370–384
Nelson W (1969) Hazard plotting for incomplete failure data. J Qual Technol 1:27–52
Nelson W (1972) Theory and applications of hazard plotting for censored failure data. Technometrics 14:945–966
Ng ETM, Cook RJ (2000) A comparison of some random effect models for parameter estimation in recurrent events. Math Comput Modell 32:11–26
Ng T, Lee W, Lee Y (2016) Change-point estimators with true identification property. Bernoulli (in press)
Nielsen GG, Gill RD, Andersen PK, Sørensen TIA (1992) A counting process approach to maximum likelihood estimation in frailty models. Scand J Stat 19:25–44
Noh M, Lee Y (2007) REML estimation for binary data in GLMMs. J Multivar Anal 98:896–915
Noh M, Ha ID, Lee Y (2006) Dispersion frailty models and HGLMs. Stat Med 25:1341–1354
Oakes D (1989) Bivariate survival models induced by frailties. J Am Stat Assoc 84:487–493
Oakes D, Dasu T (1990) A note on residual life. Biometrika 77:409–410
Orbe J, Ferreira E, Nez-Antn V (2002) Comparing proportional hazards and accelerated failure time models for survival analysis. Stat Med 21:3493–3510
Othus M, Li Y (2009) Marginalized frailty models for multivariate survival data. Harvard University Biostatistics Working Paper Series
Paik MC, Lee Y, Ha ID (2015) Frequentist inference on random effects based on summarizability. Statistica Sinica 25:1107–1132
Pan W, Kooperberg C (1999) Linear regression for bivariate censored data via multiple imputation. Stat Med 18:3111–3121
Pan W, Louis TA (2000) A linear mixed-effects model for multivariate censored data. Biometrics 56:160–166
Parner E (1998) Asymptotic theory for the correlated Gamma-frailty model. Ann Stat 26:183–214
Patel K, Kay R, Rowell L (2006) Comparing proportional hazards and accelerated failure time models: an application in influenza. Pharm Stat 5:213–224
Patterson HD, Thompson R (1971) Recovery of interblock information when block sizes are unequal. Biometrika 58:545–554

Pawitan Y (2001) In all likelihood: statistical modelling and inference using likelihood. Clarendon and Press, Oxford

Pawitan Y, Lee Y (2017) Wallet game: probability, likelihood and extended likelihood. Am Stat 71:120–122

Pawitan Y, Reilly M, Nilsson E, Cnattingius S, Lichtenstein P (2004) Estimation of genetic and environmental factors for binary traits using family data. Stat Med 23:449–465

Pearson K (1920) The fundamential problems of practical statistics. Biometrika 13:1–16

Peng L, Fine JP (2007) Nonparametric quantile inference with competingrisks data. Biometrika 94:735–744

Pepe MS, Mori M (1993) Kaplan-Meier, marginal or conditional probability curves in summarizing competing risks failure time data? Stat Med 12:737–751

Peto R, Peto J (1972) Asymptotically efficient rank invariant procedures. J R Stat Soc A 135:185–207

Pettitt AN (1986) Censored observations, repeated measures and mixed effects models: an approach using the EM algorithm and normal errors. Biometrika 73:635–643

Philipson P, Sousa I, Diggle P, Williamson P, Kolamunnage-Dona R, Henderson R (2012) joineR: joint modelling of repeated measurements and time to event data. https://CRAN.R-project.org/package=joineR

Pickles A, Crouchley R (1995) A comparison of frailty models for multivariate survival data. Stat Med 14:1447–1461

Pintilie M (2006) Competing risks: a practical perspective. Wiley, Chichester

Pintilie M (2007) Analysing and interpreting competing risk data. Stat Med 26:1360–1367

Prenen L, Braekers R, Duchateau L (2017) Extending the Archimedean copula methodology to model multivariate survival data grouped in clusters of variable size. J R Stat Soc B 79:483–505

Prentice R, Kalbfleisch JD, Peterson AV, Flournoy N, Farewell VT, Breslow NE (1978) The analysis of failure times in the presence of competing risks. Biometrics 34:541–554

Putter H, Fiocco M, Geskus RB (2007) Tutorial in biostatistics: competing risks and multi-state models. Stat Med 26:2389–2430

Rabe-Hesketh S, Pickles A, Skrondal A (2001) Gllamm: a general class of multilevel models and a Stata program. Multilevel Model Newsl 13:17–23

Rabe-Hesketh S, Skrondal A, Pickles A (2002) Reliable estimation of generalized linear mixed models using adaptive quadrature. Stata J 2:1–21

Radchenko P, James GM (2008) Variable inclusion and shrinkage algorithms. J Am Stat Assoc 103:1304–1315

Reid N (1994) A conversation with Sir David Cox. Stat Sci 9:439–455

Ripatti S, Palmgren J (2000) Estimation of multivariate frailty models using penalized partial likelihood. Biometrics 56:1016–1022

Rizopoulos D (2012) Joint models for longitudinal and time-to-event data, with applications in R. Chapman and Hall, Boca Raton

Robins JM, Wang N (2000) Inference for imputation estimators. Biometrika 87:113–124

Robinson GK (1991) That BLUP is a good thing: the estimation of random effects (with discussion). Stat Sci 6:15–51

Rondeau V, Commenges D, Joly P (2003) Maximum penalized likelihood estimation in a gamma-frailty model. Lifetime Data Anal 9:139–153

Rondeau V, Mathoulin-Pelissier S, Jacqmin-Gadda H, Brouste V, Soubeyran P (2007) Joint frailty models for recurring events and death using maximum penalized likelihood estimation: application on cancer events. Biostatistics 8:708–721

Rondeau V, Michiels S, Liquet B, Pignon JP (2008) Investigating trial and treatment heterogeneity in an individual patient data meta-analysis of survival data by means of the penalized maximum likelihood approach. Stat Med 27:1894–1910

Ruan PK, Gray RJ (2008) Analyses of cumulative incidence functions via non-parametric multiple imputation. Stat Med 27:5709–5724

Rubin DB (1976) Inference and missing data. Biometrika 63:581–592

Rubin DB (1987) Multiple imputation for nonresponse in surveys. Wiley, New York
Rubin DB, Schenker N (1991) Multiple imputation in health-care databases: an overview and some applications. Stat Med 15:585–598
Sakamoto Y, Ishiguro M, Kitagawa G (1986) Akaike information criterion statistics. KTK Scientific Publisher, Tokyo
Sastry NA (1997) Nested frailty model for survival data, with application to study of child survival in Northeast Brazil. J Am Stat Assoc 92:426–435
Schall R (1991) Estimation in generalized linear models with random effects. Biometrika 78:719–727
Schoenfeld D (1982) Partial residuals for the proportional hazards regression model. Biometrika 69:239–241
Searle SR, Casella G, McCulloch CE (1992) Variance components. Wiley, New York
Self SG, Liang KY (1987) Asymptotic properties of maximum likelihood estimators and likelihood ratio tests under nonstandard conditions. J Am Stat Assoc 82:605–610
Sham PC (1998) Statistics in human genetics. Arnold, London
Shih JH, Louis TA (1995) Inferences on the association parameter in copula models for bivariate survival data. Biometrics 51:1384–1399
Sklar A (1959) Fonctions de répartition à n dimensions et leurs marges. Publ Inst Stat Univ Paris 8:229–231
Smyth GK (2002) An efficient algorithm for REML in heteroscedastic regression. J Comput Graph Stat 11:1–12
Spiegelhalter DJ, Best NG, Carlin BP, van der Linde A (2002) Bayesian measures of model complexity and fit (with discussion). J R Stat Soc B 64:583–639
Stablein DM, Koutrouvelis IA (1985) A two-sample test sensitive to crossing hazards in uncensored and singly censored data. Biometrics 41:643–652
Stram DO, Lee JW (1994) Variance components testing in the longitudinal mixed effects model. Biometrics 50:1171–1177
Sylvester R, van der Meijden APM, Oosterlinck W, Witjes J, Bouffioux C, Denis L, Newling DWW, Kurth K (2006) Predicting recurrence and progression in individual patients with stage Ta T1 bladder cancer using EORTC risk tables: a combined analysis of 2596 patients from seven EORTC trials. Eur Urol 49:466–477
Sung K-H, Kahng KW, Kang CM, Kwak JY, Park TS, Lee SY (1998) Study on the factors affecting the chronic renal allograft dysfunction. Korean J Nephrol 17:483–493
Tanner MA (1993) Tools for statistical inference, 2nd edn. Springer, New York
Thall PF, Vail SC (1990) Some covariance models for longitudinal count data with overdispersion. Biometrics 46:657–671
Therneau TM (2010) Survival: survival analysis, including penalised likelihood. http://CRAN.R-project.org/package=survival. R package version 2.36-2
Therneau TM (2011) coxme: mixed effects Cox models. http://CRAN.R-project.org/package=coxme. R package version 2.2-1
Therneau TM, Grambsch PM (2000) Modeling survival data: extending the Cox model. Springer, New York
Therneau T, Grambsch P, Fleming T (1990) Martingale based residuals for survival models. Biometrika 77:147–160
Therneau TM, Grambsch PM, Pankratz VS (2003) Penalized survival models and frailty. J Comput Graph Stat 12:156–175
Therneau TM, Crowson C, Atkinson E (2016) Using time dependent covariates and time dependent coefficients in the Cox model. https://cran.r-project.org/web/packages/survival/vignettes/timedep.pdf
Tibshirani R (1996) Regression shrinkage and selection via the Lasso. J R Stat Soc B 58:267–288
Tibshirani R (1997) The LASSO method for variable selection in the Cox model. Stat Med 16:385–395

Tierney L, Kadane JB (1986) Accurate approximations for posterior moments and marginal densities. J Am Stat Assoc 81:82–86
Tobin J (1958) Estimation of relationship for limited dependent variables. Econometrica 26:24–36
Tsiatis A (1975) A nonidentifiability aspect of the problem of competing risks. Proc Natl Acad Sci 72:20–22
Vaida F, Blanchard S (2005) Conditional Akaike information for mixed-effects models. Biometrika 92:351–370
Vaida F, Xu R (2000) Proportional hazards model with random effects. Stat Med 19:3309–3324
van Houwelingen HC, Putter H (2012) Dynamic prediction in clinical survival analysis. Chapma and Hall, London
Vaupel JW, Manton KG, Stallard E (1979) The impact of heterogeneity in individual frailty on the dynamics of mortality. Demography 16:439–454
Varadhan R, Xue Q-L, Bandeen-Roche K (2014) Semicompeting risks in aging research: methods, issues and needs. Liftime Data Anal 20:538–562
Verbeke G, Lesaffre E (1997) The effects of misspecifying the random-effects distribution in linear mixed models for longitudinal data. Comput Stat Data Anal 23:541–556
Verbeke G, Molenberghs G (2003) The use of score test for inference on variance components. Biometrics 59:254–262
Verbeke G, Molenberghs G (2009) Linear mixed models for longitudinal data. Springer, New York
Verweij JM, Van Houwelingen HC (1994) Penalized likelihood in Cox regression. Stat Med 13:2427–2436
Vu HTV, Knuiman MW (2002) A hybrid ML-EM algorithm for calculation of maximum likelihood estimates in semiparametric shared frailty models. Comput Stat Data Anal 40:173–187
Vu HTV, Segal MR, Knuiman MW, James IR (2001) Asymptotic and small sample statistical properties of random frailty variance estimates for shared gamma frailty models. Commun Stat-Simul Comput 30:581–595
Wang H, Li R, Tsai CL (2007) Tuning parameter selectors for the smoothly clipped absolute deviation method. Biometrika 94:553–568
Whitehead J (1980) Fitting Cox's regression model to survival data using GLIM. Appl Stat 29:268–275
Wienke A (2011) Frailty models in survival analysis. Chapman and Hall, London
Wienke A, Holm NV, Christensen K, Skytthe A, Vaupel JW, Yashin AI (2003) The heritability of cause-specific mortality: a correlated gamma-frailty model applied to mortality due to respiratory disease in Danish twins born 1870–1930. Stat Med 22:3873–3887
Wolfinger RD (1993) Covariance structure selection in general mixed models. Commun Stat-Simul Comput 22:1079–1106
Wolfinger RD (1999) Fitting nonlinear mixed models with the new NLMIXED procedure. Proc Joint Stat Meet 287
Xiang L, Ma X, Yau KKW (2011) Mixture cure model with random effects for clustered and interval-censored survival data. Stat Med 30:995–1006
Xu R, Vaida F, Harrington DP (2009) Using profile likelihood for semiparametric model selection with application to proportional hazards mixed models. Statistica Sinica 19:819–842
Xu J, Kalbfleisch JD, Tai B (2010) Statistical analysis of illness-death processes and semicompeting riskd data. Biometrics 66:716–725
Xue X (2001) Analysis of childhood brain tumour data in New York city using frailty. Stat Med 20:3459–3473
Xue X, Brookmeyer R (1996) Bivariate frailty model for the analysis of multivariate survival time. Lifetime Data Anal 2:277–289
Yamaguchi T, Ohashi Y (1999) Investigating centre effects in a multi-centre clinical trial of superficial bladder cancer. Stat Med 18:1961–1971
Yashin AI, Iachine IA (1995) How long can humans live? Lower bound for biological limit of human longevity calculated from Danish twin data using correlated frailty model. Mech Ageing Dev 80:147–169

Yashin AI, Iachine IA, Harris JR (1999) Half of the variation in susceptibility to mortality is genetic: findings from Swedish twin survival data. Behav Genet 29:11–19

Yau KKW (2001) Multilevel models for survival analysis with random effects. Biometrics 57:96–102

Yau KKW, Kuk AYC (2002) Robust estimation in generalized linear mixed models. J R Stat Soc B 64:101–117

Yau KKW, McGilchrist CA (1998) ML and REML estimation in survival analysis with time dependent correlated frailty. Stat Med 17:1201–1213

Yau KKW, Ng ASK (2001) Long-term survivor mixture model with random effects: application to a multi-centre clinical trial of carcinoma. Stat Med 20:1591–1607

Yu D, Yau KK (2012) Conditional Akaike information criterion for generalized linear mixed models. Comput Stat Data Anal 56:629–644

Yu D, Zhang X, Yau KK (2013) Information based model selection criteria for generalized linear mixed models with unknown variance component parameters. J Multivar Anal 116:245–262

Yuan M, Lin Y (2006) Model selection and estimation in regression with grouped variables. J R Stat Soc B 68:49–67

Yun S, Lee Y (2004) Comparison of hierarchical and marginal likelihood estimators for binary outcomes. Comput Stat Data Anal 45:639–650

Zeng D, Lin DY (2007) Maximum likelihood estimation in semiparametric regression models with censored data. J R Stat Soc B 69:507–564

Zhang C-H (2010) Nearly unbiased variable selection under minimax concave penalty. Ann Stat 38:894–942

Zhang HH, Lu W (2007) Adaptive Lasso for Cox's proportional hazards model. Biometrika 94:691–703

Zhang Y, Li R, Tsai CL (2010) Regularization parameter selections via generalized information criterion. J Am Stat Assoc 105:312–323

Zhang Y, Chen M-H, Ibrahim JG, Zeng D, Chen Q, Pan Z, Xue X (2013) Bayesian gamma frailty models for survival data with semi-competing risks and treatment switching. Liftime Data Anal 20:76–105

Zhou B, Latouche A (2015) crrSC: competing risks regression for stratified and clustered data. R package version 1:1

Zhou B, Fine J, Latouche A, Labopin M (2012) Competing risks regression for clustered data. Biostatistics 13:371–383

Zou H (2006) The adaptive Lasso and its oracle properties. J Am Stat Assoc 101:1418–1429

Zou H, Hastie T (2005) Regularization and variable selection via the elastic net. J R Stat Soc B 67:301–320

Index

I.D. Ha et al., *Statistical Modelling of Survival Data with Random Effects*,
Statistics for Biology and Health, https://doi.org/10.1007/978-981-10-6557-6